Emerging Topics in Statistics and Biostatistics

Carlos A. Coelho • Ding-Geng Chen
Editors

Statistical Modeling and Applications

Multivariate, Heavy-Tailed, Skewed Distributions and Mixture Modeling, Volume 2

Springer

Editors
Carlos A. Coelho
Department of Mathematics
NOVA University of Lisbon
Caparica, Portugal

Ding-Geng Chen
College of Health Solutions
Arizona State University
Phoenix, AZ, USA

ISSN 2524-7735 ISSN 2524-7743 (electronic)
Emerging Topics in Statistics and Biostatistics
ISBN 978-3-031-69621-3 ISBN 978-3-031-69622-0 (eBook)
https://doi.org/10.1007/978-3-031-69622-0

This Springer imprint is published by the registered company Springer Nature Switzerland AG
The registered company address is: Gewerbestrasse 11, 6330 Cham, Switzerland

Preface

In an era defined by the seamless integration of data and sophisticated analytical and modeling techniques, the quest for advanced statistical modeling and methodologies has never been more pertinent. This book entitled "Statistical Modeling and Applications: Multivariate, Heavy-Tailed, Skewed Distributions and Mixture Modeling, Volume 2" represents a concerted effort to bridge the gap between theoretical advancements and practical applications in the realm of Statistical Science, namely in the area of Statistical Modeling. It also aims to present a wide range of emerging topics in mathematical and statistical modeling written by a group of distinguished researchers from top-tier universities and research institutes in order to offer broader opportunities to foster and stimulate further collaborations in the areas of mathematics and statistics.

The book has 11 chapters, divided in two parts, with Part I comprising five chapters dealing with the application of Multivariate Analysis techniques and multivariate distributions to a set of different situations, and Part II consisting of six chapters which address the modeling of a number of interesting phenomena through the use of Heavy-Tailed, Skewed, Circular-Linear and Mixture Distributions and also Neural Networks.

Part I—"Multivariate Statistical Modeling and Applications" includes five chapters (Chaps. 1–5) which deal with the application of several Multivariate Analysis modeling techniques and multivariate distributions to a number of different practical problems.

Chapter 1, by Sílvia Guerra, Fernanda Cipriano, and Isabel Natário, invites the reader to deep dive into the world of "Random Gaussian fields and systems of stochastic partial differential equations" where the authors address the use of certain systems of stochastic partial differential equations to generate multivariate Gaussian random fields. The study emphasizes the Bayesian framework to obtain posterior distributions of Gaussian Markov random fields using the Finite Element Method. The goal is to compile and detail the necessary calculations to find the posterior distributions for the multivariate case, facilitating future applications which may lead to a better comprehensive understanding of complex spatial dependencies and interrelations.

Chapter 2, "A Poly-cylindrical Bayesian network for clustering oceanography data" is authored by Thasmika Mohan and Najmeh Rad, and it ventures into the dynamic field of oceanography, presenting the use of a pioneering multivariate cylindrical or poly-cylindrical distribution to cluster oceanographic data. They apply a Bayesian network combined with the Expectation-Maximization algorithm to teach a finite mixture model to efficiently handle the clustering of multivariate cylindrical observations, that is, observations that involve directional-linear variables. The authors refer this methodology as critical to study ocean dynamics and predicting oceanographic phenomena.

The importance of solar energy in our sustainable future cannot be overstated, and Chap. 3, by Priyanka Nagar, Francois von Holtzhausen, Andriette Bekker, Ding-Geng Chen, and Najmeh Rad, offers "A Copulas-Based Approach to Statistical Modelling of Solar Irradiance". The authors propose a statistical model to estimate and predict solar irradiance, in particular its variability, using a multivariate probability distribution constructed via a copula. This distribution accounts for the dependency between Direct Normal Irradiance and Direct Horizontal Irradiance in modeling the clear sky irradiance indexes. The model is applied to data from two South African Universities radiometric network stations. By accounting for the dependencies between various irradiance indexes, the authors enhanced the ability to predict and optimize solar energy utilization, a critical endeavor in the face of global energy challenges.

Chapter 4, entitled "Two-Sample Intraclass Correlation Coefficient Tests for Matrix-Valued Data" by Yuli Liang, Chengcheng Hao, and Deliang Dai, deals with hypothesis testing for the equality of correlation parameters under a model with a Kronecker product covariance structure. It introduces new tests using the ratio of independent F distributions and compares them with existing tests, providing practical recommendations based on error probabilities and power. These tests are applied to a real data set and they are tailored to improve accuracy and reliability in practical applications, thereby expanding our toolkit for advanced statistical inference.

Chapter 5, by Ricardo Moura, Carlos A. Coelho, and Bimal Sinha, entitled "Evolution of the Generation and Analysis of Single Imputation Synthetic Datasets in Statistical Disclosure Control" presents an overview of the evolution of techniques used to generate and analyze single imputation synthetic datasets in Statistical Disclosure Control. The chapter's focus is on procedures to analyze singly imputed synthetic datasets generated from a multivariate normal distribution, using Plug-in Sampling, a model-based approach. The chapter aims to contribute to a better understanding of these methods and assist researchers in taking informed decisions regarding the generation and analysis of singly imputed synthetic datasets for statistical purposes, while maintaining privacy. This chapter is a testimony to the ongoing efforts to safeguard individual privacy, while maintaining the integrity and usefulness of statistical data.

Part II—"Modeling with Spherically Symmetric, Heavy-Tailed, Skewed, Circular-Linear, Mixture Distributions and Neural Networks" includes six chapters (Chaps. 6–11) which deal with the application of a number of different distributions and techniques to a number of different modeling applications.

Chapter 6, entitled "Some Empirical Findings on Neural Network-Based Forecasting When Subjected to Autoregressive Resampling" authored by Johannes T. Ferreira, Dominic Wrbka, and Ane van der Merwe, explores neural network-based forecasting for stationary autoregressive data resampled using methods like the moving block bootstrap and stationary bootstrap. This study illustrates the benefits and relevance of using a neural network-based approach in the forecasting of autoregressive models, and highlights the drawbacks when relying on the usual bootstrap approach.

Income data analysis takes center stage in Chap. 7, "Enriched Lognormal Models for Income Data" by Seite Makgai, Johannes T. Ferreira, Jason Pillay, Andriette Bekker, and Filipe Marques, where the authors introduce enriched lognormal models to capture the complexities of income distributions. The authors use an enriched Lognormal model which incorporates regression functions that explore dependencies between income and relevant covariates and also consider a finite mixture model for bimodal income distributions, introducing a novel extension using the modal curve. The approach is validated through simulations and illustrated by practical applications, offering new perspectives on socioeconomic and health data analysis, with implications for social research and policy-making.

In Chap. 8, entitled "A New Approach to Estimate Semi-parametric Gaussian Mixtures of Regressions with Varying Mixing Proportions" the authors Sphiwe B. Skhosana, Salomon M. Millard, and Frans H. J. Kanfer address the estimation problems in a semi-parametric model for Gaussian mixtures of regressions with varying mixing proportions (SPGMRVPs) and non-parametric mixing probabilities. This is a very flexible model and framework, but this flexibility does not come without its challenges. While traditional methods of estimation are not guaranteed to produce reliable estimates of the model, a local-likelihood approach for estimating the non-parametric functions requires the maximization of a set of local-likelihood functions. But then, using the Expectation-Maximization (EM) algorithm to separately maximize each local-likelihood function may lead to label switching. The authors propose a unified model-based approach to address the label-switching problem by reformulating the SPGMRVPs model and develop one-step backfitting estimates of the parametric and non-parametric terms. The effectiveness of the proposed approach is shown by using Monte Carlo simulations and its practical utility through an environmental data analysis.

Chapter 9, entitled "Computational Comparisons of Two-Component Mixtures Using Lindley-Type Models" by O. van Heerden, S. Makgai, A. Bekker, and J.T. Ferreira explores finite mixtures of Lindley-type models for lifetime data analysis that include, as components, well-known cases from literature as well as a proposed alternative parsimonious mixture model. Simulation studies investigate the effect of the additional parameter(s) of the finite mixtures of Lindley-type models, when compared to the finite mixture of the one parameter Lindley model. The EM algorithm is considered as a suitable method for the parameter estimation. The practical implementation of these mixture models is illustrated through real data examples from the survival analysis domain.

In Chap. 10, entitled "Baranchik-Type Estimators Under Modified Balanced Loss Functions" the authors Lahoucine Hobbad, Idir Ouassou, and Mohamed Alahiane focus on the estimation of the mean vector of a spherically symmetric distribution, and establish a condition under which the Baranchik-type estimator dominates the maximum likelihood estimator and achieves minimaxity, with these results extending the results of Hobbad et al. (2021) to situations where the random vector is not continuous with respect to the Lebesgue measure, thus not having a probability density function. This chapter extends classical results to broader contexts, enriching the theoretical underpinnings of statistical estimation.

Chapter 11, "Modelling the Movement of a South African Cheetah Using a Hidden Markov Model and Circular-Linear Regression" by Bianca Gascoyne, Najmeh Nakhaei Rad, and Andriette Bekker presents an interesting study on cheetahs movement using hidden Markov models and circular-linear regression, using GPS data. The authors apply a hidden Markov model to decode probable states of the cheetah's movement and circular-linear regression to predict movement directions based on linear predictors. The analysis aids in understanding how human activity affects large predators' behavior, and the importance of this study builds on the fact that Cheetah numbers are decreasing drastically and analyzing these types of datasets can aid in the efforts to conserve this species, while studying GPS data gives insight into the habitats and migratory patterns of animals. This approach may contribute not only to wildlife conservation efforts but also to show the versatility of statistical modeling in ecological research.

As we present this compilation of advanced statistical methodologies and their diverse applications, our hope is that this book will serve both as a valuable reference and a source of inspiration for researchers, practitioners, and students alike. The chapters within reflect a commitment to pushing the boundaries of statistical science, fostering innovation, and addressing the complex challenges of our data-driven world.

We extend our deepest gratitude to all contributors, whose expertise and dedication have made this book possible. The successful publication of this book would have been impossible without the invaluable support and contributions of many people. We would like to express our heartfelt gratitude to them. Firstly and most important, we extend our sincere appreciation to all the chapter authors for their high-quality contributions to this book. Furthermore, and most sincerely, we want to thank the reviewers for their professional service and contribution to this book and also Anne Rubio for her professional assistance in communicating with the contributing authors throughout the revision and production processes. Moreover, we also owe a special debt of gratitude to Faith Su, the Editor for Statistics Books at Springer, for her exceptional guidance and support throughout the entire publication process. Their expertise and dedication were instrumental in bringing this book to fruition.

This book was initiated as a professional product, when Professor Ding-Geng Chen was invited by "Centro de Matemática e Aplicações" of NOVA University of Lisbon (NOVA Math) for his academic visit during the period from April 30, 2023, to May 12, 2023. With this publication, Professor Chen would like to thank all the researchers and staff members of NOVA Math for their hospitality and professional collaborations.

To strive for continuous improvement, we are open to receive any suggestions from our readers for further enhancements. Please feel free to contact the editors. Your input is always invaluable and we look forward to receiving your comments on this book.

May this work ignite further inquiry and collaboration in the ever-evolving field of statistical modeling.

Caparica, Portugal — Carlos A. Coelho
Phoenix, AZ, USA — Ding-Geng Chen

Contents

About the Editors

Dr. Carlos A. Coelho is a Full Professor of Statistics at the Mathematics Department of NOVA School of Science and Technology of NOVA University of Lisbon. He holds a Ph.D. in Biostatistics from The University of Michigan, Ann Arbor, MI, USA, where he was a Fulbrighter. His main area of research is Multivariate Analysis, namely the development of likelihood ratio tests for elaborate covariance structures and for MANOVA models, also with elaborate covariance structures, together with the study of the exact distribution and the development of near-exact distributions for the associated test statistics. Related with this area, other areas of interest are Mathematical Statistics and Distribution Theory, as well as Estimation, Univariate and Multivariate Linear, Generalized Linear and Mixed Models. More recently, he also got interested in tests for high-dimensionality and the application of Multivariate Analysis techniques to Statistical Disclosure Control problems. Carlos A. Coelho has served as Associate Editor in the Editorial Boards of *Discussiones Mathematicae—Probability and Statistics*, *REVSTAT-Statistical Journal*, the *Journal of Interdisciplinary Mathematics*, and the *Journal of Applied Statistics* and currently serves in the Editorial Boards of the *Journal of Statistical Theory and Practice* and the *American Journal of Mathematical and Management Sciences*. He is also Associate Editor of the Springer Book series "Emerging Topics in Statistics and Biostatistics" and a member of the International Council of the "Business World" Library of the Tsenov Academy of Economics (Svishtov, Bulgaria). Currently, Carlos A. Coelho also serves as President of Fulbrighters Portugal, the Portuguese Fulbright Alumni Association.

Dr. Ding-Geng Chen is a fellow of the American Statistical Association and is currently the executive director and a professor of biostatistics at the College of Health Solutions, Arizona State University. He is also an extraordinary professor and the SARChI research chair in biostatistics at the University of Pretoria, an honorary professor at the University of KwaZulu-Natal, South Africa. Dr. Chen was a professor in Biostatistics at the University of North Carolina at Chapel Hill, a professor in biostatistics at the University of Rochester Medical School, and the Karl E. Peace Endowed Eminent Scholar Chair in Biostatistics at Georgia Southern University. He is a senior biostatistics consultant for biopharmaceuticals and government agencies with extensive expertise in biostatistics, clinical trials, and public health statistics and data sciences. Dr. Chen has more than 200 referred professional publications, co-authored/co-edited 40 books on clinical trial methodology, meta-analysis, data science, causal inference, and public health research.

List of Contributors

Mohamed Alahiane Cadi Ayyad University, National School of Applied Sciences, Marrakesh, Morocco

Andriette Bekker Department of Statistics, University of Pretoria, Pretoria, South Africa
DSI-NRF Centre of Excellence in Mathematical and Statistical Sciences (CoE-MaSS), and Centre for Environmental Studies, Department of Geography, Geoinformatics and Meteorology, University of Pretoria, Pretoria, South Africa

Fernanda Cipriano NOVA MATH & Department of Mathematics, NOVA School of Science and Technology, NOVA University Lisbon, Lisbon, Portugal

Carlos A. Coelho Mathematics Department, Nova School of Science and Technology, Caparica, Portugal
NOVA Math – Center for Mathematics and Applications, Caparica, Portugal

Deliang Dai Department of Economics and Statistics, Linnaeus University, Växjö, Sweden

Johannes T. Ferreira Department of Statistics, University of Pretoria, Pretoria, South Africa
National Institute for Theoretical and Computational Sciences (NITheCS), South Africa

Bianca Gascoyne Department of Statistics, University of Pretoria, Pretoria, South Africa

Sílvia Guerra NOVA MATH, NOVA School of Science and Technology & NOVA School of Business and Economics, NOVA University Lisbon, Lisbon, Portugal
ISEG – Lisbon School of Economics and Management, Universidade de Lisboa, Lisboa, Portugal

Chengcheng Hao Department of Data Science, School of Statistics and Information, Shanghai University of International Business and Economics, Shanghai, China

Ockert van Heerden Department of Statistics, University of Pretoria, Pretoria, South Africa

Lahoucine Hobbad Cadi Ayyad University, National School of Applied Sciences, Marrakesh, Morocco

Francois von Holtzhausen Department of Statistics, University of Pretoria, Pretoria, South Africa

Frans H. J. Kanfer Department of Statistics, University of Pretoria, Pretoria, South Africa

Yuli Liang Department of Economics and Statistics, Linnaeus University, Växjö, Sweden

Seite Makgai Department of Statistics, University of Pretoria, Pretoria, South Africa

Filipe J. Marques Departamento de Matemática, Faculdade de Ciências e Tecnologia da Universidade Nova de Lisboa, Caparica, Portugal

Ane van der Merwe Department of Statistics, University of Pretoria, Pretoria, South Africa

Salomon M. Millard Department of Statistics, University of Pretoria, Pretoria, South Africa

Thasmika Mohan Department of Statistics, University of Pretoria, Pretoria, South Africa

Ricardo Moura Mathematics Department, Lisbon School of Economics & Management (ISEG), Lisbon, Portugal
NOVA Math – Center for Mathematics and Applications, Caparica, Portugal
Portuguese Navy Research Center (CINAV), Alfeite, Portugal

Priyanka Nagar Department of Statistics and Actuarial Science, Stellenbosch University, Stellenbosch, South Africa

Isabel Natário NOVA MATH & Department of Mathematics, NOVA School of Science and Technology, NOVA University Lisbon, Lisbon, Portugal

Idir Ouassou Cadi Ayyad University, National School of Applied Sciences, Marrakesh, Morocco

Jason Pillay Department of Statistics, University of Pretoria, Pretoria, South Africa

Najmeh Nakhaei Rad Department of Statistics, University of Pretoria, Pretoria, South Africa

Bimal Sinha Department of Mathematics and Statistics, University of Maryland, Baltimore County, MD, USA

Center for Statistical Research and Methodology, U.S. Census Bureau, Washington, DC, USA

Sphiwe B. Skhosana Department of Statistics, University of Pretoria, Pretoria, South Africa

Dominic Wrbka Department of Statistics, University of Pretoria, Pretoria, South Africa

List of Reviewers

Vahideh Ahrari Department of Statistics, Shahrekord University, Shahrekord, Iran

Mahboubeh Akbari Department of Statistics, University of Pretoria, Pretoria, South Africa

Masoumeh Akbari Department of Statistics, University of Mazandaran, Babolsar, Mazandaran, Iran

Mohammad Arashi Department of Statistics, Faculty of Mathematical Sciences, Ferdowsi University of Mashhad, Mashhad, Iran

Tanita Botha Faculty of Health, Deakin University, Melbourne, Australia

Yakup Murat Bulut Department of Statistics, Eskişehir Osmangazi University, Eskişehir, Turkey

Gauri Datta Department of Statistics, University of Georgia, Athens, GA, USA

Vahid Fakoor Department of Statistics, Faculty of Mathematical Sciences, Ferdowsi University of Mashhad, Mashhad, Iran

Martin Klein Division of Biometrics VIII, Office of Biostatistics, Office of Translational Sciences, Center for Drug Evaluation and Research, U.S. Food and Drug Administration, Silver Spring, MD, USA

Augustyn Markiewicz Department of Mathematical and Statistical Methods, Poznań University of Life Sciences, Poznań, Poland

Thomas Mathew Mathematics and statistics Department, University of Maryland, Baltimore County, MD, USA

Rachdi Mustapha Laboratoire AGEIS EA 7407, University Grenoble-Alpes, Grenoble, France

Priyanka Nagar Department of Statistics and Actuarial Science, Stellenbosch University, Stellenbosch, South Africa

Tapan Nayak Department of Statistics, George Washington University, Washington, DC, USA

Salomi Du Plessis Department of Statistics, Faculty of Natural and Agricultural Science, University of Pretoria, Pretoria, South Africa

Milan Stehlik Linz Institute of Technology, Johannes Kepler University, Linz, Austria

Anna Szczepańska-Álvarez Department of Mathematical and Statistical Methods, Poznań University of Life Sciences, Poznań, Poland

Albert Whata Department of Mathematical and Computer Sciences, Sol Plaatje University, Kimberley, South Africa

Jeffrey Wilson Department of Economics, W. P. Carey School of Business, Arizona State University, Tempe, AZ, USA

Maria Antónia Turkman Department of Statistics and Operations Research, Faculty of Sciences, University of Lisbon, Lisboa, Portugal

Part I
Multivariate Statistical Modeling and Applications

Chapter 1
Random Gaussian Fields and Systems of Stochastic Partial Differential Equations

Sílvia Guerra, Fernanda Cipriano, and Isabel Natário

1.1 Introduction

The Matérn covariance function plays a central role in modelling continuous spatial variation, contributing for a class of models that describes successfully the observed spatial variation in data from applications such as atmospheric and environmental sciences, forestry, etc. (see Gelfand et al. 2010). It is assumed that the spatial phenomena is driven by a spatial process, namely a Gaussian field (GF). Recall that $\{x(s), s \in \mathcal{D} \subseteq \mathbb{R}^d\}$ is said to be a continuously indexed (scalar) Gaussian field if all finite collections $\{x(s_1), \ldots, x(s_n)\}$ are jointly Gaussian distributed. We are interested in stationary processes, that is, the mean and covariance function do not depend on the location s, and isotropic, that is, the covariance function depends only on the Euclidean distance between two locations. Regarding this setting, Matérn (1960) developed a class of models for the covariance structure for these processes, which takes the form

$$\mathrm{Cov}(x(s), x(s')) = \frac{\sigma^2}{2^{\nu-1}\Gamma(\nu)}(k\|s - s'\|)^{\nu} K_{\nu}(k\|s - s'\|)$$

S. Guerra (✉)
NOVA MATH, NOVA School of Science and Technology & NOVA School of Business and Economics, NOVA University Lisbon, Lisbon, Portugal
e-mail: silvia.guerra@novasbe.pt

ISEG – Lisbon School of Economics and Management, Universidade de Lisboa, Lisbon, Portugal

F. Cipriano · I. Natário
NOVA MATH & Department of Mathematics, NOVA School of Science and Technology, NOVA University Lisbon, Lisbon, Portugal
e-mail: mfsm@fct.unl.pt; icn@fct.unl.pt

C. A. Coelho, D.-G. Chen (eds.), *Statistical Modeling and Applications*, Emerging Topics in Statistics and Biostatistics, https://doi.org/10.1007/978-3-031-69622-0_1

where K_ν is the modified Bessel function of second kind and order ν, and $\Gamma(\cdot)$ is the *Gamma* function.

Whittle (1954) and Whittle (1963) proved that the stationary solution of the stochastic partial differential equation (SPDE)

$$b(k^2 - \Delta)^{\alpha/2} x(s) = \mathcal{W}(s) \,, \tag{1.1}$$

with $s \in \mathbb{R}^2$, $\alpha = \nu + d/2$, and $k, \nu, b > 0$, is a GF with zero mean and Matérn covariance, with marginal variance given by

$$\sigma^2 = \frac{\Gamma(\nu)}{\Gamma\left(\nu + \frac{d}{2}\right)(4\pi)^{d/2} k^{2\nu} b^2} \,.$$

Moreover, $\mathcal{W}(s)$ is a Gaussian white noise with unit variance, Δ is the Laplacian in $\mathbb{R}^2$,

$$\Delta = \frac{\partial^2}{\partial u_1^2} + \frac{\partial^2}{\partial u_1^2} \,,$$

and $(k^2 - \Delta)^{\alpha/2}$ is a fractional differential operator, defined through the wave number spectrum of the stationary solution $x(s)$. In order to do that, first we notice that

$$\mathcal{F}\{b(k^2 - \Delta)x\}(\lambda) = b(k^2 + \|\lambda\|^2)\mathcal{F}\{x\}(\lambda)$$

so the Fourier transform of a fractional differential operator is defined as

$$\mathcal{F}\{b(k^2 - \Delta)^{\alpha/2} x\}(\lambda) = b(k^2 + \|\lambda\|^2)^{\alpha/2}\mathcal{F}\{x\}(\lambda) \,.$$

This way, it is possible to calculate the wave number spectrum of the stationary solution $x(s)$, which is given by

$$f(\lambda) = \frac{1}{(2\pi)^2 b^2 (k^2 + \|\lambda\|^2)^\alpha}$$

(see Lindgren et al. 2011 and Guerra 2022).

The parameter ν is related to the smoothness of the solution $x(s)$, higher values mean higher smoothness of the solution (see Krainski et al. 2018). This parameter is poorly identified by data, so it is kept fixed (see Lindgren et al. 2011). In our work, we have $d = 2$, and we consider $\alpha = 2$ as in Lindgren et al. (2011), so we get $\nu = 1$. Moreover, the parameter k is a scale parameter that shapes the spatial correlation range, through the empirical derived formula

$$\rho = \frac{\sqrt{8\nu}}{k}$$

called *practical range*. We have that ρ is the distance at which the correlation between $x(s)$ and $x(s')$ is around 0.1 (see Krainski et al. 2018). Finally, the parameter b controls the variance of the process.

Lindgren et al. (2011) apply the Finite Element Method (FEM) and solve SPDE (1.1) in the weak sense, therefore they are able to approximate the solution of the SPDE, $x(s)$, by a Gaussian Markov random field (GMRF), $\mathbf{x}$, with sparse precision matrix. We can find the detailed calculations in Guerra (2022). A spatial model for continuous spatial variation or geostatistical data assumes that the phenomenon of interest, described by the Matérn Gaussian field, is observed with error in some locations, resulting in observations that can be used to fit the model. If considered under a Bayesian framework, the estimation can be more efficiently done under an Integrated Nested Laplace Approach (INLA), following Rue et al. (2009).

Hu et al. (2013) and Hu and Steinsland (2016) extend Lindgren et al. (2011) results to random vector fields, $x(s) = (x_1(s), x_2(s))$. They consider systems of SPDEs,

$$\begin{cases} b_{11}(k_{11}^2 - \Delta)^{\alpha_{11}/2} x_1(s) + b_{12}(k_{12}^2 - \Delta)^{\alpha_{21}/2} x_2(s) = f_1(s) \\ b_{21}(k_{21}^2 - \Delta)^{\alpha_{21}/2} x_1(s) + b_{22}(k_{22}^2 - \Delta)^{\alpha_{22}/2} x_2(s) = f_2(s) \end{cases} \tag{1.2}$$

where f_1, f_2 are noise processes, to obtain GMRF approximations of multivariate Gaussian random fields (GRF).

The main goal of this work is to explicitly present the calculations of the referred approximations for the multivariate case, as they are not gathered together, or not fully detailed elsewhere in the literature, for instance, how do we obtain the matricial formulas in (1.14), (1.20), or (1.24), that end up in result (1.49), and also an explanation of the simplified Laplace approximation, that simplifies the calculation of (1.33). This can prove to be very useful for future applications of this methodology.

1.2 GMRF Approximation of a Gaussian Random Field

Hu et al. (2013) establishes that the solution of the system of SPDE (1.2) for $b_{12} = 0$ is a Gaussian random field (GRF),

$$x(s) = (x_1(s), x_2(s)) \sim MVN(\mathbf{0}, \mathbf{\Sigma}(s))$$

where $\mathbf{\Sigma}(s)$ is a non-negative definite matrix given by

$$\mathbf{\Sigma}(s) = \begin{bmatrix} \text{Cov}(x_1(s), x_1(s)) & \text{Cov}(x_1(s), x_2(s)) \\ \text{Cov}(x_2(s), x_1(s)) & \text{Cov}(x_2(s), x_2(s)) \end{bmatrix}.$$

They also point out that the parameter b_{21} controls the correlation for the two components of the vector field $x(s)$, for instance, the smaller b_{21}, the quicker correlation between the two components decreases. Moreover, if the noise processes are white noise processes, then the correlation is positive if $b_{21} < 0$ and negative otherwise.

Following Hu et al. (2013), we consider the following system of SPDEs,

$$\begin{cases} b_{11}(k_{11}^2 - \Delta)^{\alpha_{11}/2} x_1(s) = \mathcal{W}_1(s) \\ b_{21}(k_{21}^2 - \Delta)^{\alpha_{21}/2} x_1(s) + b_{22}(k_{22}^2 - \Delta)^{\alpha_{22}/2} x_2(s) = \mathcal{W}_2(s) \end{cases}. \tag{1.3}$$

Let us now define $\mathcal{L}(\boldsymbol{\theta})$ as the matrix operator given by

$$\mathcal{L}(\boldsymbol{\theta}) = \begin{bmatrix} \mathcal{L}_{11}(\boldsymbol{\theta}) & \mathbf{0} \\ \mathcal{L}_{21}(\boldsymbol{\theta}) & \mathcal{L}_{22}(\boldsymbol{\theta}) \end{bmatrix}$$

where $\mathcal{L}_{ij}(\boldsymbol{\theta}) = b_{ij}(k_{ij}^2 - \Delta)^{\alpha_{ij}/2}$, $\alpha_{ij} = 2$, and $\boldsymbol{\theta} = (b_{11}, b_{21}, b_{22}, k_{11}, k_{21}, k_{22})$ is the vector of parameters. System (1.3) can be written in matricial form as

$$\begin{aligned} \mathcal{L}(\boldsymbol{\theta})x(s) &= f(s) \\ \Leftrightarrow \begin{bmatrix} \mathcal{L}_{11}\tilde{x}_1 \\ \mathcal{L}_{21}\tilde{x}_1 + \mathcal{L}_{22}\tilde{x}_2 \end{bmatrix} &= \begin{bmatrix} \mathcal{W}_1 \\ \mathcal{W}_2 \end{bmatrix} \end{aligned} \tag{1.4}$$

where $s \in \mathbb{R}^2$, and $f(s) = (\mathcal{W}_1(s), \mathcal{W}_2(s))$ is a vector of standard independent white noise processes.

1.2.1 The Finite Element Method

We will consider the matricial form (1.4) in order to solve the system of SPDEs in the weak sense, following the FEM methodology. In order to do that, we consider a domain $D \subset \mathbb{R}^2$ with smooth boundary ∂D, chosen accordingly to the region where we want to approximate the solution $x(s)$, and define a triangulation of that domain, also called *mesh*. Let $\{n_i : 1 = 1, \ldots, m\}$ be the mesh nodes. We define a basis of finite element functions $\{\varphi_i : i = 1, \ldots, m\}$ in the following way: for each node n_i, we define a piece-wise linear function φ_i such that $\varphi(n_i) = 1$, linearly decreasing to zero in each triangle with one vertex in n_i. Notice that these base element functions are scalar, but our GRF is a vector field, so Hu et al. (2013) consider the base finite elements $\{\varphi_i : i = 1, \ldots, m\}$ and define a basis for the solution space for $x(s) = (x_1(s), x_2(s))$ in the following way

$$\{\psi_1, \cdots, \psi_{2m}\} = \left\{ \begin{bmatrix} \varphi_1 \\ 0 \end{bmatrix}, \ldots, \begin{bmatrix} \varphi_m \\ 0 \end{bmatrix}, \begin{bmatrix} 0 \\ \varphi_1 \end{bmatrix}, \ldots, \begin{bmatrix} 0 \\ \varphi_m \end{bmatrix} \right\}. \tag{1.5}$$

Consider $f, g \in L^2(D)$. We define $\langle f, g \rangle$ as

$$\langle f, g \rangle = \int_D f(s)g(s)\,ds$$

and we extend the definition to the space of all generalized functions defined on D. We are interested in a finite approximation $\tilde{x}(s)$ of the solution of the SPDE system (1.2) given by

$$\tilde{x}(s) = (\tilde{x}_1(s), \tilde{x}_2(s)) = \sum_{j=1}^{2m} x_j \psi_j(s) \tag{1.6}$$

where $\mathbf{x} = (x_1, \ldots, x_{2m})$ is a Gaussian vector of weights, to be determined. To solve the SPDE system (1.3) in the weak sense, we consider the matricial form (1.4), apply both sides to the test functions $\{\psi_1, \cdots, \psi_{2m}\}$, and consider a Neumann boundary condition,

$$[\langle \mathcal{L}\tilde{x}, \psi_i \rangle]_{i=1}^{2m} = [\langle f, \psi_i \rangle]_{i=1}^{2m} \quad , \quad \frac{\partial \tilde{x}(s)}{\partial \mathbf{n}}|_{\partial D} = 0\,, \tag{1.7}$$

where $s \in D$ and $\mathbf{n}$ is the normal vector on the boundary ∂D.

Let $i = 1, \ldots, m$. From (1.5) and (1.7), we obtain the following,

$$\langle \mathcal{L}\tilde{x}, \psi_i \rangle = \langle f, \psi_i \rangle \Leftrightarrow \begin{bmatrix} \langle \mathcal{L}_{11}\tilde{x}_1, \varphi_i \rangle \\ \langle \mathcal{L}_{21}\tilde{x}_1 + \mathcal{L}_{22}\tilde{x}_2, 0 \rangle \end{bmatrix} = \begin{bmatrix} \langle \mathcal{W}_1, \varphi_i \rangle \\ \langle \mathcal{W}_2, 0 \rangle \end{bmatrix}$$

$$\Rightarrow \langle \mathcal{L}_{11}\tilde{x}_1, \varphi_i \rangle = \langle \mathcal{W}_1, \varphi_i \rangle \tag{1.8}$$

and

$$\langle \mathcal{L}\tilde{x}, \psi_{m+i} \rangle = \langle f, \psi_{m+i} \rangle \Leftrightarrow \begin{bmatrix} \langle \mathcal{L}_{11}\tilde{x}_1, 0 \rangle \\ \langle \mathcal{L}_{21}\tilde{x}_1 + \mathcal{L}_{22}\tilde{x}_2, \varphi_i \rangle \end{bmatrix} = \begin{bmatrix} \langle \mathcal{W}_1, 0 \rangle \\ \langle \mathcal{W}_2, \varphi_i \rangle \end{bmatrix}$$

$$\Rightarrow \langle \mathcal{L}_{21}\tilde{x}_1 + \mathcal{L}_{22}\tilde{x}_2, \varphi_i \rangle = \langle \mathcal{W}_2, \varphi_i \rangle\,. \tag{1.9}$$

From the approximated solution formula in (1.6) we have

$$\tilde{x}_1(s) = \sum_{j=1}^{m} x_j \varphi_j(s) \tag{1.10}$$

therefore, applying formula (1.10) in the left side of (1.8), Green's first identity, and the Neumann condition in (1.7), we obtain the following

$$
\begin{aligned}
\langle \mathcal{L}_{11}\tilde{x}_1(s), \varphi_i\rangle &= \langle b_{11}(k_{11} - \Delta)\tilde{x}_1(s), \varphi_i(s)\rangle \\
&= b_{11}k_{11}^2 \int_D \tilde{x}_1\varphi_i(s)ds + b_{11}\int_D \nabla\tilde{x}_1\nabla\varphi_i(s)ds \\
&= \sum_{j=1}^{m} b_{11}k_{11}^2 x_j \int_D \varphi_i(s)\varphi_j(s)ds + \sum_{j=1}^{m} b_{11}x_j \int_D \nabla\varphi_i(s)\nabla\varphi_j(s)ds \\
&= b_{11}k_{11}^2 \sum_{j=1}^{m}\langle\varphi_i, \varphi_j\rangle x_j + b_{11}\sum_{j=1}^{m}\langle\nabla\varphi_i, \nabla\varphi_j\rangle x_j
\end{aligned}
\tag{1.11}
$$

for $i \in \{1, ...m\}$. See the Appendix for a detailed explanation. From (1.8) and (1.11), we obtain the following,

$$
b_{11}k_{11}^2 \sum_{j=1}^{m}\langle\varphi_i, \varphi_j\rangle x_j + b_{11}\sum_{j=1}^{m}\langle\nabla\varphi_i, \nabla\varphi_j\rangle x_j = \langle\mathcal{W}_1, \varphi_i\rangle \tag{1.12}
$$

for $\{i \in \{1, \dots, m\}\}$. Gathering the result in (1.12) in matricial form, we obtain

$$
\begin{aligned}
&b_{11}k_{11}\begin{bmatrix}\langle\varphi_1, \varphi_1\rangle & \cdots & \langle\varphi_1, \varphi_m\rangle \\ \vdots & \ddots & \vdots \\ \langle\varphi_m, \varphi_1\rangle & \cdots & \langle\varphi_m, \varphi_m\rangle\end{bmatrix} \cdot \begin{bmatrix}x_1 \\ \vdots \\ x_m\end{bmatrix} + \\
&+ b_{11}\begin{bmatrix}\langle\nabla\varphi_1, \nabla\varphi_1\rangle & \cdots & \langle\nabla\varphi_1, \nabla\varphi_m\rangle \\ \vdots & \ddots & \vdots \\ \langle\nabla\varphi_m, \nabla\varphi_1\rangle & \cdots & \langle\nabla\varphi_m, \nabla\varphi_m\rangle\end{bmatrix} \cdot \begin{bmatrix}x_1 \\ \vdots \\ x_m\end{bmatrix} = \begin{bmatrix}\langle\mathcal{W}_1, \varphi_1\rangle \\ \vdots \\ \langle\mathcal{W}_1, \varphi_m\rangle\end{bmatrix}.
\end{aligned}
\tag{1.13}
$$

Consider the vector $\mathbf{b}_1 = (\langle\mathcal{W}_1, \varphi_1\rangle, \cdots, \langle\mathcal{W}_1, \varphi_m\rangle)$ of Gaussian random variables, and define matrices $\mathbf{C}$, $\mathbf{G}$ as

$$
\mathbf{C} = \begin{bmatrix}\langle\varphi_1, \varphi_1\rangle & \cdots & \langle\varphi_1, \varphi_m\rangle \\ \vdots & \ddots & \vdots \\ \langle\varphi_m, \varphi_1\rangle & \cdots & \langle\varphi_m, \varphi_m\rangle\end{bmatrix}, \quad \mathbf{G} = \begin{bmatrix}\langle\nabla\varphi_1, \nabla\varphi_1\rangle & \cdots & \langle\nabla\varphi_1, \nabla\varphi_m\rangle \\ \vdots & \ddots & \vdots \\ \langle\nabla\varphi_m, \nabla\varphi_1\rangle & \cdots & \langle\nabla\varphi_m, \nabla\varphi_m\rangle\end{bmatrix},
$$

then the matricial equation in (1.13) can be re-written as

$$
b_{11}(k_{11}^2\mathbf{C} + \mathbf{G})\mathbf{x}_1 = \mathbf{b}_1 \tag{1.14}
$$

where $\mathbf{x}_1 = (x_1, \cdots, x_m)$ is the vector with the first m components of the vector of weights $\mathbf{x}$. So by now, we have a matricial equation that allow us to determine some of the components of the vector of weights, $\mathbf{x}$. Next, we determine a matricial formula for the remaining components of $\mathbf{x}$, namely, $\mathbf{x}_2 = (x_{m+1}, \ldots, x_{2m})$.

Let us consider now the second component of our vector field $x(s)$. From (1.6) we have

$$\tilde{x}_2(s) = \sum_{j=1}^{m} x_{m+j}\varphi_j(s) \,. \tag{1.15}$$

Notice that

$$\begin{aligned}\langle \mathcal{L}_{21}\tilde{x}_1(s) + \mathcal{L}_{22}\tilde{x}_2(s), \varphi_i\rangle &= \langle \mathcal{L}_{21}\tilde{x}_1(s), \varphi_i\rangle + \langle \mathcal{L}_{22}\tilde{x}_2(s), \varphi_i\rangle = \\ &= \langle b_{21}(k_{21} - \Delta)\tilde{x}_1(s), \varphi_i(s)\rangle + \langle b_{22}(k_{22} - \Delta)\tilde{x}_2(s), \varphi_i(s)\rangle \,. \end{aligned} \tag{1.16}$$

Again, applying (1.10) and (1.15), Green's first identity, and the Neumann condition in (1.7), we obtain, as before,

$$\begin{aligned}\langle \mathcal{L}_{21}\tilde{x}_1(s), \varphi_i\rangle &= \langle b_{21}(k_{21} - \Delta)\tilde{x}_1(s), \varphi_i(s)\rangle \\ &= b_{21}k_{21}^2 \sum_{j=1}^{m} \langle \varphi_i, \varphi_j\rangle x_j + b_{21} \sum_{j=1}^{m} \langle \nabla\varphi_i, \nabla\varphi_j\rangle x_j \end{aligned} \tag{1.17}$$

and

$$\begin{aligned}\langle \mathcal{L}_{22}\tilde{x}_2(s), \varphi_i\rangle &= \langle b_{22}(k_{22} - \Delta)\tilde{x}_2(s), \varphi_i(s)\rangle \\ &= b_{22}k_{22}^2 \sum_{j=1}^{m} \langle \varphi_i, \varphi_j\rangle x_{m+j} + b_{22} \sum_{j=1}^{m} \langle \nabla\varphi_i, \nabla\varphi_j\rangle x_{m+j} \end{aligned} \tag{1.18}$$

for $i \in \{1, \ldots m\}$. Therefore, from (1.9), (1.17) and (1.18), we obtain

$$\begin{aligned} & b_{21}k_{21}^2 \sum_{j=1}^{m} \langle \varphi_i, \varphi_j\rangle x_j + b_{21} \sum_{j=1}^{m} \langle \nabla\varphi_i, \nabla\varphi_j\rangle x_j + \\ & \quad + b_{22}k_{22}^2 \sum_{j=1}^{m} \langle \varphi_i, \varphi_j\rangle x_{m+j} + b_{22} \sum_{j=1}^{m} \langle \nabla\varphi_i, \nabla\varphi_j\rangle x_{m+j} = \langle \mathcal{W}_2, \varphi_i\rangle \end{aligned} \tag{1.19}$$

for $\{i \in \{1, \ldots, m\}\}$. As before, gathering the result in (1.19) in matricial form, we obtain

$$b_{21}(k_{21}^2\mathbf{C} + \mathbf{G})\mathbf{x}_1 + b_{22}(k_{22}^2\mathbf{C} + \mathbf{G})\mathbf{x}_2 = \mathbf{b}_2 \tag{1.20}$$

where $\mathbf{b}_2 = (\langle \mathcal{W}_2, \varphi_1\rangle, \cdots, \langle \mathcal{W}_2, \varphi_m\rangle)$ is a vector of Gaussian random variables. Following Hu et al. (2013), define $\mathbf{K}_{ij} = b_{ij}(k_{ij}^2\mathbf{C}+\mathbf{G})$, for $i, j \in \{1, 2\}$ with $i \geq j$. This way, the matricial equations in (1.14) and (1.20) can be rewritten with block matrices as

$$\begin{bmatrix} \mathbf{K}_{11} & \mathbf{0} \\ \mathbf{K}_{21} & \mathbf{K}_{22} \end{bmatrix} \cdot \begin{bmatrix} \mathbf{x}_1 \\ \mathbf{x}_2 \end{bmatrix} = \begin{bmatrix} \mathbf{b}_1 \\ \mathbf{b}_2 \end{bmatrix} . \tag{1.21}$$

Let $\mathbf{K}$ be the block matrix in (1.21), and consider $\mathbf{b} = (\mathbf{b}_1, \mathbf{b}_2)$. Notice that, by definition of white noise, $\langle \mathcal{W}_l, \varphi_i\rangle$ is a Gaussian random variable with zero mean and covariance given by

$$\text{Cov}(\langle \mathcal{W}_l, \varphi_i\rangle, \langle \mathcal{W}_l, \varphi_j\rangle) = \langle \varphi_i, \varphi_j\rangle , \tag{1.22}$$

for all $i, j \in \{1, \ldots, m\}$, and $l \in \{1, 2\}$ (see Rozanov 1998). Moreover, since $\mathcal{W}_1$ and $\mathcal{W}_2$ are independent distributed white noise processes, we have the following,

$$\text{Cov}(\langle \mathcal{W}_1, \varphi_i\rangle, \langle \mathcal{W}_2, \varphi_j\rangle) = 0 \tag{1.23}$$

for all $i, j \in \{1, \ldots, m\}$. Therefore,

$$\text{Cov}(\mathbf{b}) = \begin{bmatrix} \mathbf{C} & \mathbf{0} \\ \mathbf{0} & \mathbf{C} \end{bmatrix} .$$

Following Lindgren et al. (2011), as the matrix $\mathbf{C}$ is dense, for computational purposes, namely, to get a sparse precision matrix, it is possible to substitute $\mathbf{C}$ by the diagonal matrix $\tilde{\mathbf{C}}$, where $\tilde{\mathbf{C}}_{ii} = \langle \varphi, 1\rangle$. So we consider

$$\text{Cov}(\mathbf{b}) = \mathbf{D} , \quad \text{where} \quad \mathbf{D} := \begin{bmatrix} \tilde{\mathbf{C}} & \mathbf{0} \\ \mathbf{0} & \tilde{\mathbf{C}} \end{bmatrix} .$$

Then, from (1.21),

$$\begin{aligned} &\text{Cov}(\mathbf{Kx}) = \text{Cov}(\mathbf{b}) \\ \Leftrightarrow\ &\mathbf{K}^T\text{Cov}(\mathbf{x})\mathbf{K} = \mathbf{D} \\ \Leftrightarrow\ &\mathbf{K}^{-1}\mathbf{Q}(\boldsymbol{\theta})(\mathbf{K}^T)^{-1} = \mathbf{D}^{-1} \\ \Leftrightarrow\ &\mathbf{Q}(\boldsymbol{\theta}) = \mathbf{KD}^{-1}\mathbf{K}^T \end{aligned} \tag{1.24}$$

where $\mathbf{Q}(\boldsymbol{\theta})$ denotes the precision matrix for the vector of weights $\mathbf{x}$. Moreover, from (1.21), since $\mathbf{b}$ is a vector of Gaussian variables with zero mean, then $\mathbf{x}$ is also a vector of Gaussian variable with zero mean. We conclude that the conditional distribution of $\mathbf{x}$ is given by

$$\pi(\mathbf{x}|\boldsymbol{\theta}) = (2\pi)^{-m}|\mathbf{Q}(\boldsymbol{\theta})|^{1/2}\exp\left\{-\frac{1}{2}\mathbf{x}^T\mathbf{Q}(\boldsymbol{\theta})\mathbf{x}\right\}. \tag{1.25}$$

Therefore, following Rue and Held (2005) for the definition of GMRF, we conclude that the vector $\mathbf{x}$ is a GMRF.

1.3 Latent Gaussian Model with One Vector Spatial Effect

The main goal of this section is to follow Rue et al. (2009) and detail all calculations that are not explicit in the literature, that supports the INLA methodology, considering a vector field. In their work, the authors discuss how to perform Bayesian inference in a subclass of hierarchical models, namely, *latent Gaussian models*. They use an integrated nested Laplace approximation to compute accurate approximations to the posterior marginal distributions, making it possible to perform Bayesian inference in a fast and automatic way.

Consider for instance a vector field $v(s) = (v_1(s), v_2(s))$. Consider the vector of observations $\mathbf{y} = (y_{1,1}, \ldots, y_{1,n}, y_{2,1}, \ldots, y_{2,n}) = (\mathbf{y}_1, \mathbf{y}_2)$, where $\mathbf{y}_1 = (y_{1,1}, \ldots, y_{1,n})$ are observations of the first component v_1, and $\mathbf{y}_2 = (y_{2,1}, \ldots, y_{2,n})$ are observations of the second component v_2, both taken at the same locations $\{s_i \in D \subset \mathbb{R}^2 : i = 1, \ldots, n\}$. Hierarchical models combine different levels of information, for instance the observations $y_i = (y_{1,i}, y_{2,i})$ are assumed to follow a certain distribution $\pi(y_i|\mu_i, \boldsymbol{\theta}_1)$, that depends on a certain set of parameters $(\mu_i, \boldsymbol{\theta}_1)$, which are not known. Bayesian inference considers all unknown quantities as random variables, so, in one hand, the set of parameters $\boldsymbol{\theta}_1$ is assumed to follow a certain prior distribution, and, in the other hand, the parameter $\mu_i = (\mu_{1,i}, \mu_{2,i})$, usually taken as the expected value $E[y_i]$, reflects the spatial dependency, as it is assumed to depend on a GRF $x(s)$,

$$\mu_i = \alpha + x(s_i) = (\alpha_1, \alpha_2) + (x_1(s_i), x_2(s_i))$$

where α_1 and α_2 stand for the intercept of each component. The GRF $x(s)$ is approximated by a GMRF $\mathbf{x}$, provided by the FEM, that will depend on a set of parameters $\boldsymbol{\theta}_2$. Moreover, the set of parameters $\boldsymbol{\theta} = (\boldsymbol{\theta}_1, \boldsymbol{\theta}_2)$ is assumed to follow a certain prior distribution $\pi(\boldsymbol{\theta})$, that will depend, itself on a certain set of hyperparameters.

In this work, we consider a latent Gaussian model with one vector spatial effect, as explain above, that is

$$\begin{aligned} y_i &\sim \pi(y_i|\mu_i, \boldsymbol{\theta}) \\ \mu_i &= \alpha + x(s_i) \\ \mathbf{x}|\boldsymbol{\theta} &\sim MNV(\mathbf{0}, \mathbf{Q}^{-1}(\boldsymbol{\theta})) \\ \boldsymbol{\theta} &\sim \pi(\boldsymbol{\theta}) \end{aligned} \tag{1.26}$$

where the GMRF $\mathbf{x}$ is assumed to follow a Gaussian distribution as in (1.25), with $\mathbf{0}$ mean and precision matrix $\mathbf{Q}(\boldsymbol{\theta})$ given by (1.24).

1.3.1 Link Between the GMRF x, and the GRF $x(s)$

It is crucial to understand how do we relate the GMRF that arises from a triangulation of a certain domain D, with the GRF at the location of the observations, $x(s_i)$.

The domain $D \subset \mathbb{R}^2$ should be chosen accordingly with the set of the locations, $\{s_i \in D \subset \mathbb{R}^2 : i = 1, \ldots, n\}$. Then we define a mesh on the domain D, so we have a set of m nodes $\{n_i \in D \subset \mathbb{R}^2 : i = 1, \ldots, m\}$ that define the basis element functions $\{\varphi_1, \ldots, \varphi_m\}$, which are used to define the two-dimensional basis element functions $\{\psi_1, \ldots, \psi_{2m}\}$, as shown in (1.5). The GMRF $\mathbf{x}$ is the vector of weights that approximate the GRF $x(s)$, as we saw in (1.6). Notice that for each node n_i, with $i \in \{1, \ldots, m\}$, we have the following

$$\tilde{x}(n_i) = (x_i, x_{m+i})$$

meaning that the first component of the GRF $x_1(s)$, evaluated at the nodes, $(x_1(n_1), \ldots, x_1(n_m))$, is approximated by $(x_1, \ldots, x_m)$, and the second component $x_2(s)$, evaluated at the nodes, $(x_2(n_1), \ldots, x_2(n_m))$, is approximated by $(x_{m+1}, \ldots, x_{2m})$.

Let $\mathbf{x} = (\mathbf{x}_1, \mathbf{x}_2)$, where $\mathbf{x}_1 = (x_1, \ldots, x_m)$ and $\mathbf{x}_2 = (x_{m+1}, \ldots, x_{2m})$. If the locations of the observations coincide with the nodes of the mesh, then the fitted values for $\mathbf{y}$ are given, according to (1.26), by

$$\hat{\mathbf{y}}_1 = \alpha_1^* \mathbf{1} + \mathbf{x}_1^* \quad \text{and} \quad \hat{\mathbf{y}}_2 = \alpha_2^* \mathbf{1} + \mathbf{x}_2^*$$

where $\mathbf{1}$ is an n dimensional vector of 1's, and α_1^* and α_2^* are estimated values for the intercepts. Moreover, since the distribution of $\mathbf{x} = (\mathbf{x}_1, \mathbf{x}_2)$ depends also on a certain set of parameters $\boldsymbol{\theta}$ to be determined, $\mathbf{x}^* = (\mathbf{x}_1^*, \mathbf{x}_2^*)$ is an estimation for $\mathbf{x}$, later explained. If the locations s_i are different from the mesh nodes, we need a projector matrix A_{proj} that projects the GMRF $\mathbf{x}$ from the mesh nodes to the set of the locations. This matrix is defined as follows,

$$[A_{\text{proj}}]_{ij} = \varphi_j(s_i) \tag{1.27}$$

that is, each entry $\{i, j\}$ is the base element function φ_j evaluated at the location s_i. Then the fitted values for the observations will be given by

$$\hat{\mathbf{y}}_1 = \alpha_1^* \mathbf{1} + A_{\text{proj}}\mathbf{x}_1^* \quad \text{and} \quad \hat{\mathbf{y}}_2 = \alpha_2^* \mathbf{1} + A_{\text{proj}}\mathbf{x}_2^* \,. \tag{1.28}$$

1.3.2 Bayesian Inference and Posterior Marginals

In this section we follow Blangiardo and Cameletti (2015) and Rue et al. (2009) and present all calculations needed to perform Bayesian inference under INLA approach and considering a vector field, and obtain the posterior conditional marginals $\pi(\mathbf{x}|\boldsymbol{\theta}, \mathbf{y})$ and $\pi(\boldsymbol{\theta}|\mathbf{y})$.

Again, let $\mathbf{y} = (\mathbf{y}_1, \mathbf{y}_2)$ be the vector of observations $(y_{1,1}, \ldots, y_{1,n}, y_{2,1}, \ldots, y_{2,n})$ as before, and consider the latent Gaussian model (1.26). The observations are assumed to be conditionally independent given $\mathbf{x}$ and the parameters $\boldsymbol{\theta}$ (let us assume that the $\boldsymbol{\theta}$ includes the intercepts α_1 and α_2, for simplicity of the text), and belonging to the same distribution family,

$$\pi(\mathbf{y}|\mathbf{x}, \boldsymbol{\theta}) = \prod_{i=1}^{n} \pi(y_{1,i}|\mathbf{x}, \boldsymbol{\theta}) \prod_{i=1}^{n} \pi(y_{2,i}|\mathbf{x}, \boldsymbol{\theta}) \tag{1.29}$$

Recall that the conditional distribution for the GMRF is given by

$$\pi(\mathbf{x}|\boldsymbol{\theta}) = (2\pi)^{-m} |\mathbf{Q}(\boldsymbol{\theta})|^{1/2} \exp\left\{-\frac{1}{2}\mathbf{x}^T \mathbf{Q}(\boldsymbol{\theta})\mathbf{x}\right\}. \tag{1.30}$$

where the precision matrix is given by (1.24),

$$\mathbf{Q}(\boldsymbol{\theta}) = \mathbf{K}\mathbf{D}^{-1}\mathbf{K}^T \,.$$

Applying Bayes's Theorem,

$$\begin{aligned}\pi(\mathbf{x}, \boldsymbol{\theta}|\mathbf{y}) &= \frac{\pi(\mathbf{x}, \boldsymbol{\theta}, \mathbf{y})}{\pi(\mathbf{y})} \\ &= \frac{\pi(\mathbf{y}|\mathbf{x}, \boldsymbol{\theta})\pi(\mathbf{x}, \boldsymbol{\theta})}{\pi(\mathbf{y})}\end{aligned}$$

$$= \frac{\pi(\mathbf{y}|\mathbf{x}, \boldsymbol{\theta})\pi(\mathbf{x}|\boldsymbol{\theta})\pi(\boldsymbol{\theta})}{\pi(\mathbf{y})}$$
$$\propto \pi(\mathbf{y}|\mathbf{x}, \boldsymbol{\theta})\pi(\mathbf{x}|\boldsymbol{\theta})\pi(\boldsymbol{\theta}) .$$

Considering (1.29) and (1.30), we obtain

$$\pi(\mathbf{x}, \boldsymbol{\theta}|\mathbf{y}) \propto \pi(\boldsymbol{\theta})\pi(\mathbf{x}|\boldsymbol{\theta}) \prod_{i=1}^{n} \pi(y_{1,i}|\mathbf{x}, \boldsymbol{\theta}) \prod_{i=1}^{n} \pi(y_{2,i}|\mathbf{x}, \boldsymbol{\theta})$$

therefore

$$\pi(\mathbf{x}, \boldsymbol{\theta}|\mathbf{y}) \propto \pi(\boldsymbol{\theta})|\mathbf{Q}(\boldsymbol{\theta})|^{1/2} \exp\left\{-\frac{1}{2}\mathbf{x}^T \mathbf{Q}(\boldsymbol{\theta})\mathbf{x} + \right.$$
$$\left. + \sum_{i=1}^{n} \log(\pi(y_{1,i}|\mathbf{x}, \boldsymbol{\theta})) + \sum_{i=1}^{n} \log(\pi(y_{2,i}|\mathbf{x}, \boldsymbol{\theta}))\right\} . \quad (1.31)$$

Given the prior distributions for $\pi(\boldsymbol{\theta})$, $\pi(\mathbf{x}|\boldsymbol{\theta})$ and the distribution $\pi(\mathbf{y}|\mathbf{x}, \boldsymbol{\theta})$, we use Bayesian inference and the Laplace approximation to obtain the posterior distributions $\pi(\boldsymbol{\theta}|\mathbf{y})$ and $\pi(x_i|\mathbf{y})$.

1.3.3 Laplace Approximation and the INLA Methodology

Te INLA methodology calculates numerical approximations to the posterior distributions $\pi(\boldsymbol{\theta}|\mathbf{y})$ and $\pi(x_i|\mathbf{y})$, based on the Gaussian approximation given by the Laplace method. Notice the following,

$$\pi(\boldsymbol{\theta}|\mathbf{y}) = \frac{\pi(\mathbf{x}, \boldsymbol{\theta}|\mathbf{y})}{\pi(\mathbf{x}|\boldsymbol{\theta}, \mathbf{y})}$$
$$= \frac{\pi(\mathbf{x}, \boldsymbol{\theta}, \mathbf{y})}{\pi(\mathbf{x}|\boldsymbol{\theta}, \mathbf{y})\pi(\mathbf{y})}$$
$$= \frac{\pi(\mathbf{y}|\mathbf{x}, \boldsymbol{\theta})\pi(\mathbf{x}|\boldsymbol{\theta})\pi(\boldsymbol{\theta})}{\pi(\mathbf{x}|\boldsymbol{\theta}, \mathbf{y})\pi(\mathbf{y})}$$

therefore

$$\pi(\boldsymbol{\theta}|\mathbf{y}) \propto \frac{\pi(\mathbf{y}|\mathbf{x}, \boldsymbol{\theta})\pi(\mathbf{x}|\boldsymbol{\theta})\pi(\boldsymbol{\theta})}{\pi(\mathbf{x}|\boldsymbol{\theta}, \mathbf{y})} \quad (1.32)$$

In the case where the observations are non–Gaussian, the conditional distribution $\pi(\mathbf{x}|\boldsymbol{\theta}, \mathbf{y})$ is substituted by the Gaussian approximation given by the Laplace method, denoted by $\pi_G(\mathbf{x}|\boldsymbol{\theta}, \mathbf{y})$, and (1.32) is evaluated considering $\mathbf{x} = \mathbf{x}^*(\boldsymbol{\theta})$, meaning that the approximation for $\pi(\boldsymbol{\theta}|\mathbf{y})$ is given by

$$\tilde{\pi}(\boldsymbol{\theta}|\mathbf{y}) \propto \left.\frac{\pi(\mathbf{y}|\mathbf{x}, \boldsymbol{\theta})\pi(\mathbf{x}|\boldsymbol{\theta})\pi(\boldsymbol{\theta})}{\pi_G(\mathbf{x}|\boldsymbol{\theta}, \mathbf{y})}\right|_{\mathbf{x}=\mathbf{x}^*(\boldsymbol{\theta})}$$

where $\mathbf{x}^*(\boldsymbol{\theta})$ is the mode of $\pi_G(\mathbf{x}|\boldsymbol{\theta}, \mathbf{y})$. Moreover, to approximate the conditional distributions $\pi(x_i|\boldsymbol{\theta}, \mathbf{y})$, consider the following

$$\begin{aligned}\pi(\mathbf{x}_{-i}|x_i, \boldsymbol{\theta}, \mathbf{y}) &= \frac{\pi(\mathbf{x}, \boldsymbol{\theta}, \mathbf{y})}{\pi(x_i, \boldsymbol{\theta}, \mathbf{y})} \\ &= \frac{\pi(\mathbf{x}, \boldsymbol{\theta}, \mathbf{y})}{\pi(x_i|\boldsymbol{\theta}, \mathbf{y})\pi(\boldsymbol{\theta}, \mathbf{y})}\end{aligned}$$

where $\mathbf{x}_{-i}$ is the vector $\mathbf{x}$ without the component x_i. Therefore,

$$\begin{aligned}\pi(x_i|\boldsymbol{\theta}, \mathbf{y}) &= \frac{\pi(\mathbf{x}, \boldsymbol{\theta}, \mathbf{y})}{\pi(\mathbf{x}_{-i}|x_i, \boldsymbol{\theta}, \mathbf{y})\pi(\boldsymbol{\theta}, \mathbf{y})} \\ &\propto \frac{\pi(\mathbf{y}|\mathbf{x}, \boldsymbol{\theta})\pi(\mathbf{x}|\boldsymbol{\theta})\pi(\boldsymbol{\theta})}{\pi(\mathbf{x}_{-i}|x_i, \boldsymbol{\theta}, \mathbf{y})} \,.\end{aligned}$$

The conditional distribution $\pi(\mathbf{x}_{-i}|x_i, \boldsymbol{\theta}, \mathbf{y})$ is substituted by the Gaussian approximation $\tilde{\pi}_G(\mathbf{x}_{-i}|x_i, \boldsymbol{\theta}, \mathbf{y})$, in order to obtain the Laplace approximation

$$\tilde{\pi}(x_i|\boldsymbol{\theta}, \mathbf{y}) \propto \left.\frac{\pi(\mathbf{x}, \boldsymbol{\theta}, \mathbf{y})}{\tilde{\pi}_G(\mathbf{x}_{-i}|x_i, \boldsymbol{\theta}, \mathbf{y})}\right|_{\mathbf{x}_{-i}=\mathbf{x}^*_{-i}(\boldsymbol{\theta})} \tag{1.33}$$

where $\mathbf{x}^*_{-i}(\boldsymbol{\theta})$ denotes is the mode of $\tilde{\pi}_G(\mathbf{x}_{-i}|x_i, \boldsymbol{\theta}, \mathbf{y})$. In order to reduce the computational cost, Rue et al. (2009) use a simplified Laplace approximation in (1.33), which is based in a Taylor's series expansion of $\tilde{\pi}(x_i|\boldsymbol{\theta}, \mathbf{y})$.

To obtain the marginal distributions $\pi(\theta_j|\mathbf{y})$, and $\pi(x_i|\mathbf{y})$, Rue et al. (2009) apply numerical integration to

$$\tilde{\pi}(\theta_j|\mathbf{y}) = \int \tilde{\pi}(\boldsymbol{\theta}|\mathbf{y})\, d\boldsymbol{\theta} \,, \quad \text{and} \quad \tilde{\pi}(x_i|\mathbf{y}) = \int \tilde{\pi}(x_i|\boldsymbol{\theta}, \mathbf{y})\tilde{\pi}(\boldsymbol{\theta}|\mathbf{y})\, d\boldsymbol{\theta} \,.$$

1.3.3.1 Simplified Laplace Approximation

The simplified Laplace method uses second order Taylor expansion to provide a Gaussian approximation to $\tilde{\pi}(\mathbf{x}|\boldsymbol{\theta}, \mathbf{y})$. For simplicity of calculations, we consider the case where $m = n$, with the mesh nodes given by the locations of the observations. In this case, recall that the Gaussian vector of weights is $\mathbf{x} = (x_1, \ldots, x_{2n})$. Also, in order to simplify notation, let $(y_1, \ldots, y_{2n})$ denote the vector of observations $\mathbf{y} = (y_{1,1}, \ldots, y_{1,n}, y_{2,1}, \ldots, y_{2,n})$, that is, for $i \in \{1, \ldots, n\}$ we have $y_i = y_{1,i}$ and $y_{n+i} = y_{2,i}$.

We have the following,

$$\pi(\mathbf{x}|\boldsymbol{\theta}, \mathbf{y}) = \frac{\pi(\mathbf{x}, \boldsymbol{\theta}|\mathbf{y})}{\pi(\boldsymbol{\theta}|\mathbf{y})} \propto \pi(\mathbf{x}, \boldsymbol{\theta}|\mathbf{y})$$

so, considering (1.31), we obtain

$$\begin{aligned}\pi(\mathbf{x}|\boldsymbol{\theta}, \mathbf{y}) &\propto \pi(\boldsymbol{\theta})|\mathbf{Q}(\boldsymbol{\theta})|^{1/2} \exp\left\{-\frac{1}{2}\mathbf{x}^T\mathbf{Q}(\boldsymbol{\theta})\mathbf{x} + \sum_{i=1}^{2n} \log(\pi(y_i|\mathbf{x}, \boldsymbol{\theta}))\right\} \\ &\propto \exp\left\{-\frac{1}{2}\mathbf{x}^T\mathbf{Q}(\boldsymbol{\theta})\mathbf{x} + g(\mathbf{x})\right\}, \end{aligned} \tag{1.34}$$

where $g(\mathbf{x}) := \sum_{i=1}^{2n} g_i(x_i)$, with $g_i(x_i) = \log\{\pi(y_i|x_i, \boldsymbol{\theta})\}$ is a function that does not depend of x_j for $j \neq i$. Therefore,

$$\nabla g(\mathbf{x}) = \left[\left.\frac{\partial \log\{\pi(y_1|x_1, \boldsymbol{\theta})\}}{\partial x_1}\right|_{x_1} \quad \cdots \quad \left.\frac{\partial \log\{\pi(y_{2n}|x_{2n}, \boldsymbol{\theta})\}}{\partial x_{2n}}\right|_{x_{2n}}\right]$$

and the Hessian matrix $\mathbf{H}$ of $g(\mathbf{x})$ is diagonal,

$$\mathbf{H}(\mathbf{x}) = \operatorname{diag}\left(\left.\frac{\partial^2 \log\{\pi(y_1|x_1, \boldsymbol{\theta})\}}{\partial x_1^2}\right|_{x_1}, \quad \cdots \quad, \left.\frac{\partial^2 \log\{\pi(y_{2n}|x_{2n}, \boldsymbol{\theta})\}}{\partial x_{2n}^2}\right|_{x_{2n}}\right).$$

Let $\mathbf{x}^* = \mathbf{x}^*(\boldsymbol{\theta})$ be the mode of $\pi(\mathbf{x}|\boldsymbol{\theta}, \mathbf{y})$. The second order Taylor approximation for $g(\mathbf{x})$ around the mode is

$$g(\mathbf{x}) \approx g(\mathbf{x}^*) + \nabla g(\mathbf{x}^*)(\mathbf{x} - \mathbf{x}^*) + \frac{1}{2}(\mathbf{x} - \mathbf{x}^*)^T\mathbf{H}(\mathbf{x}^*)(\mathbf{x} - \mathbf{x}^*). \tag{1.35}$$

Considering (1.34) and (1.35),

$$\begin{aligned}\pi(\mathbf{x}|\boldsymbol{\theta}, \mathbf{y}) \approx \exp\Bigg\{&-\frac{1}{2}\mathbf{x}^T\mathbf{Q}(\boldsymbol{\theta})\mathbf{x} + g(\mathbf{x}^*) + \nabla g(\mathbf{x}^*)(\mathbf{x} - \mathbf{x}^*) \\ &+\frac{1}{2}(\mathbf{x} - \mathbf{x}^*)^T\mathbf{H}(\mathbf{x}^*)(\mathbf{x} - \mathbf{x}^*)\Bigg\}\end{aligned}$$

$$
\begin{aligned}
&= \exp\{g(\mathbf{x}^*)\} \exp\left\{-\frac{1}{2}\mathbf{x}^T \mathbf{Q}(\boldsymbol{\theta})\mathbf{x} + \nabla g(\mathbf{x}^*)(\mathbf{x}-\mathbf{x}^*)\right. \\
&\quad \left. + \frac{1}{2}(\mathbf{x}-\mathbf{x}^*)^T \mathbf{H}(\mathbf{x}^*)(\mathbf{x}-\mathbf{x}^*)\right\} \\
&\propto \exp\left\{-\frac{1}{2}\mathbf{x}^T \mathbf{Q}(\boldsymbol{\theta})\mathbf{x} + \nabla g(\mathbf{x}^*)\mathbf{x} + \frac{1}{2}(\mathbf{x}-\mathbf{x}^*)^T \mathbf{H}(\mathbf{x}^*)(\mathbf{x}-\mathbf{x}^*)\right\}.
\end{aligned} \tag{1.36}
$$

In one hand, we have,

$$
\begin{aligned}
(\mathbf{x}-\mathbf{x}^*)^T \mathbf{Q}(\mathbf{x}-\mathbf{x}^*) &= \mathbf{x}^T \mathbf{Q}(\mathbf{x}-\mathbf{x}^*) - (\mathbf{x}^*)^T \mathbf{Q}(\mathbf{x}-\mathbf{x}^*) \\
&= \mathbf{x}^T \mathbf{Q}\mathbf{x} - 2\,(\mathbf{x}^*)^T \mathbf{Q}\mathbf{x} + (\mathbf{x}^*)^T \mathbf{Q}\mathbf{x}^*
\end{aligned} \tag{1.37}
$$

where $\mathbf{Q} = \mathbf{Q}(\boldsymbol{\theta})$ for simplicity of notation. In the other hand, the mode is the maximum of the density probability function (for a differentiable density function), which implies[1]

$$
\begin{aligned}
\nabla_{\mathbf{x}}\left(-\frac{1}{2}\mathbf{x}^T \mathbf{Q}\mathbf{x} + \nabla g(\mathbf{x}^*)\mathbf{x} + \frac{1}{2}(\mathbf{x}-\mathbf{x}^*)^T \mathbf{H}(\mathbf{x}^*)(\mathbf{x}-\mathbf{x}^*)\right)\Bigg|_{\mathbf{x}=\mathbf{x}^*} &= \mathbf{0} \\
\Leftrightarrow \left(-\mathbf{x}^T \mathbf{Q} + \nabla g(\mathbf{x}^*) + (\mathbf{x}-\mathbf{x}^*)^T \mathbf{H}(\mathbf{x}^*)\right)\Big|_{\mathbf{x}=\mathbf{x}^*} &= \mathbf{0} \\
\Leftrightarrow (\mathbf{x}^*)^T \mathbf{Q} = \nabla g(\mathbf{x}^*)
\end{aligned} \tag{1.38}
$$

From (1.37) and (1.38) we obtain

$$
(\mathbf{x}-\mathbf{x}^*)^T \mathbf{Q}(\mathbf{x}-\mathbf{x}^*) = \mathbf{x}^T \mathbf{Q}\mathbf{x} - 2\,\nabla g(\mathbf{x}^*)\mathbf{x} + \nabla g(\mathbf{x}^*)\mathbf{x}^*\,, \tag{1.39}
$$

therefore, from (1.36) and (1.39), and taking $\mathbf{D} = -\mathbf{H}$, we conclude that

$$
\pi(\mathbf{x}|\boldsymbol{\theta}, \mathbf{y}) \propto \exp\left\{-\frac{1}{2}(\mathbf{x}-\mathbf{x}^*)^T (\mathbf{Q}+\mathbf{D})(\mathbf{x}-\mathbf{x}^*)\right\}, \tag{1.40}
$$

which is the kernel of a multivariate Normal distribution with mode $\mathbf{x}^* = \mathbf{x}^*(\boldsymbol{\theta})$ and covariance matrix $(\mathbf{Q}(\boldsymbol{\theta}) + \mathbf{D})^{-1}$.

The iterative process to approximate the mode $\mathbf{x}^*(\boldsymbol{\theta})$ and the covariance $(\mathbf{Q}(\boldsymbol{\theta}) + \mathbf{D})^{-1}$ is the following: consider an initial guess for the mode, $\mathbf{x}^{(0)} = (x_1^{(0)}, \ldots, x_{2n}^{(0)})$, and for $i \in \{1, \ldots, 2n\}$, approximate each $g_i(x_i)$ around $x_i^{(0)}$ using the second order

[1] Alternatively, we could have calculated the gradient of (1.36).

Taylor approximation for each $g_i(x_i)$,

$$g_i(x_i) \approx g(x_i^{(0)}) + (x_i - x_i^{(0)}) \left.\frac{\partial g_i}{\partial x_i}\right|_{x_i=x_i^{(0)}} - \frac{1}{2}(x_i - x_i^{(0)})^2 \left.\frac{\partial^2 g_i}{\partial x_i^2}\right|_{x_i=x_i^{(0)}}$$

$$= a_i + b_i x_i - \frac{1}{2} c_i x_i^2 \tag{1.41}$$

where

$$a_i = g(x_i^{(0)}) - x_i^{(0)} \left.\frac{\partial \log\{\pi(y_i|x_i, \boldsymbol{\theta})\}}{\partial x_i}\right|_{x_i=x_i^{(0)}} - \frac{1}{2}(x_i^{(0)})^2 \left.\frac{\partial^2 \log\{\pi(y_i|x_i, \boldsymbol{\theta})\}}{\partial x_i^2}\right|_{x_i=x_i^{(0)}}$$

$$b_i = \left.\frac{\partial \log\{\pi(y_i|x_i, \boldsymbol{\theta})\}}{\partial x_i}\right|_{x_i=x_i^{(0)}} + x_i^{(0)} \left.\frac{\partial^2 \log\{\pi(y_i|x_i, \boldsymbol{\theta})\}}{\partial x_i^2}\right|_{x_i=x_i^{(0)}}$$

$$c_i = \left.\frac{\partial^2 \log\{\pi(y_i|x_i, \boldsymbol{\theta})\}}{\partial x_i^2}\right|_{x_i=x_i^{(0)}}$$

Let

$$\mathbf{a}^0 = \sum_{i=1}^{2n} a_i, \quad \mathbf{b}^0 = (b_1, \ldots, b_{2n}), \quad \text{and} \quad \mathbf{C}^0 = \text{diag}(c_1, \ldots, c_{2n}),$$

where $\text{diag}(c_1, \ldots, c_{2n})$ denotes the diagonal matrix with diagonal entries given by $(c_1, \ldots, c_{2n})$. From (1.34),

$$\pi(\mathbf{x}|\boldsymbol{\theta}, \mathbf{y}) \propto \exp\left\{-\frac{1}{2}\mathbf{x}^T \mathbf{Q}(\boldsymbol{\theta})\mathbf{x} + g(\mathbf{x})\right\}$$

$$\approx \exp\left\{-\frac{1}{2}\mathbf{x}^T \mathbf{Q}(\boldsymbol{\theta})\mathbf{x} + \mathbf{a}^0 + (\mathbf{b}^0)^T\mathbf{x} - \frac{1}{2}c_i x_i^2\right\}$$

$$\propto \exp\left\{-\frac{1}{2}\mathbf{x}^T (\mathbf{Q}(\boldsymbol{\theta}) + \mathbf{C}^0)\,\mathbf{x} + (\mathbf{b}^0)^T\mathbf{x}\right\}. \tag{1.42}$$

We obtain in Eq. (1.42) the kernel of a Gaussian density function with precision matrix $\mathbf{Q}(\boldsymbol{\theta}) + \mathbf{C}^0$, and mode $\mathbf{x}^{(1)}$ given by the solution of

$$(\mathbf{Q}(\boldsymbol{\theta}) + \mathbf{C}^0)\mathbf{x}^{(1)} = \mathbf{b}^0 . \tag{1.43}$$

Next, we approximate each $g(x_i)$ around $x_i^{(1)}$, where $x_i^{(1)}$ is the ith entry of $x^{(1)}$, and obtain a new kernel of a Gaussian distribution, with precision matrix $\mathbf{Q}(\boldsymbol{\theta}) + \mathbf{C}^1$ and mode given by the solution of

$$(\mathbf{Q}(\boldsymbol{\theta}) + \mathbf{C}^1)\mathbf{x}^{(2)} = \mathbf{b}^1 . \tag{1.44}$$

1.3.4 Gaussian Observations

In the case where we consider Gaussian observations, the conditional distribution of the GMRF $\mathbf{x}$, $\pi(\mathbf{x}|\boldsymbol{\theta}, \mathbf{y})$, will be Gaussian. To show that, consider the following distribution for the observations $\mathbf{y}$, written using matricial form,

$$\pi(\mathbf{y}|\mathbf{x}, \boldsymbol{\theta}) = (2\pi)^{-n}|\mathbf{Q}_e|^{1/2}\exp\left\{-\frac{1}{2}(\mathbf{y} - (\boldsymbol{\alpha} + \mathbf{A}\mathbf{x}))^T\mathbf{Q}_e(\mathbf{y} - (\boldsymbol{\alpha} + \mathbf{A}\mathbf{x}))\right\} \tag{1.45}$$

where $\mathbf{Q}_e$ is the precision matrix for the vector of observations $\mathbf{y}$,

$$\mathbf{Q}_e = \begin{bmatrix} \tau_1\mathbf{I} & \mathbf{0} \\ \mathbf{0} & \tau_2\mathbf{I} \end{bmatrix}, \tag{1.46}$$

$\mathbf{I}$ is the identity matrix of order n, $\mathbf{1}$ is a n dimensional vector of 1's, $\boldsymbol{\alpha} = (\alpha_1\mathbf{1}, \alpha_2\mathbf{1})$ stands for the intercept of the model, $\tau_1\mathbf{I}$ is the precision matrix for $\mathbf{y_1}$, and $\tau_2\mathbf{I}$ is the precision matrix for $\mathbf{y_2}$. Moreover, the matrix $\mathbf{A}$ is the $2n \times 2m$ projector matrix, which links the GMRF $\mathbf{x}$ to the observations $\mathbf{y}$,

$$\mathbf{A} = \begin{bmatrix} A_{\text{proj}} & \mathbf{0} \\ \mathbf{0} & A_{\text{proj}} \end{bmatrix}$$

where A_{proj} is defined in (1.27). Following Hu et al. (2013), and given (1.30) and (1.45),

$$\begin{aligned}
\pi(\mathbf{x}|\boldsymbol{\theta}, \mathbf{y}) &\propto \pi(\mathbf{y}|\mathbf{x}, \boldsymbol{\theta})\pi(\mathbf{x}|\boldsymbol{\theta})\pi(\boldsymbol{\theta}) \\
&\propto \exp\left\{-\frac{1}{2}\mathbf{x}^T\mathbf{Q}(\boldsymbol{\theta})\mathbf{x} - \frac{1}{2}(\mathbf{y} - (\boldsymbol{\alpha} + \mathbf{A}\mathbf{x}))^T\mathbf{Q}_e(\mathbf{y} - (\boldsymbol{\alpha} + \mathbf{A}\mathbf{x}))\right\} \\
&\propto \exp\left\{-\frac{1}{2}\mathbf{x}^T(\mathbf{Q}(\boldsymbol{\theta}) + \mathbf{A}^T\mathbf{Q}_e\mathbf{A})\mathbf{x} + (\mathbf{y} - \boldsymbol{\alpha})^T\mathbf{Q}_e\mathbf{A}\mathbf{x}\right\} .
\end{aligned} \tag{1.47}$$

Let $\tilde{\mathbf{Q}} = \mathbf{Q}(\boldsymbol{\theta}) + \mathbf{A}^T\mathbf{Q}_e\mathbf{A}$. In order to obtain the mode $\mathbf{x}_0$ of $\pi(\mathbf{x}|\boldsymbol{\theta}, \mathbf{y})$, we maximize, for instance, $\log(\pi(\mathbf{x}|\boldsymbol{\theta}, \mathbf{y}))$, which can be done deriving the logarithm of (1.47),

$$\frac{\partial}{\partial \mathbf{x}}\left(-\frac{1}{2}\mathbf{x}^T\tilde{\mathbf{Q}}\mathbf{x} + (\mathbf{y}-\boldsymbol{\alpha})^T\mathbf{Q}_e\mathbf{A}\mathbf{x}\right) = 0$$

$$\Leftrightarrow -\mathbf{x}^T\tilde{\mathbf{Q}} + (\mathbf{y}-\boldsymbol{\alpha})^T\mathbf{Q}_e\mathbf{A} = \mathbf{0}^T$$

$$\Leftrightarrow \mathbf{x}_0 = \tilde{\mathbf{Q}}^{-1}\mathbf{A}^T\mathbf{Q}_e(\mathbf{y}-\boldsymbol{\alpha})\,. \tag{1.48}$$

Moreover, from (1.47),

$$\pi(\mathbf{x}|\boldsymbol{\theta},\mathbf{y}) \propto \exp\left\{-\frac{1}{2}(\mathbf{x}-\mathbf{x}_0)^T\tilde{\mathbf{Q}}(\mathbf{x}-\mathbf{x}_0)\right\}$$

which is the kernel of a Gaussian distribution with precision matrix

$$\mathbf{Q}_{\mathbf{x}|\boldsymbol{\theta},\mathbf{y}} = \mathbf{Q}(\boldsymbol{\theta}) + \mathbf{A}^T\mathbf{Q}_e\mathbf{A}$$

and mode given by

$$\mathbf{x}_0 = \mathbf{Q}_{\mathbf{x}|\boldsymbol{\theta},\mathbf{y}}^{-1}\mathbf{A}^T\mathbf{Q}_e(\mathbf{y}-\boldsymbol{\alpha})$$

that is,

$$\pi(\mathbf{x}|\boldsymbol{\theta},\mathbf{y}) \sim MNV(\mathbf{x}_0, \mathbf{Q}_{\mathbf{x}|\boldsymbol{\theta},\mathbf{y}})\,. \tag{1.49}$$

To obtain an estimate for $\boldsymbol{\theta}$, we need to maximize $\pi(\boldsymbol{\theta}|\mathbf{y})$. From (1.32), we have the following,

$$\pi(\boldsymbol{\theta}|\mathbf{y}) \propto \pi(\mathbf{y}|\boldsymbol{\theta})\pi(\boldsymbol{\theta}) = \frac{\pi(\mathbf{y}|\mathbf{x},\boldsymbol{\theta})\pi(\mathbf{x}|\boldsymbol{\theta})\pi(\boldsymbol{\theta})}{\pi(\mathbf{x}|\boldsymbol{\theta},\mathbf{y})}$$

so, applying (1.30) (1.45), and (1.49), we obtain,

$$\begin{aligned}\pi(\boldsymbol{\theta}|\mathbf{y}) \propto |\mathbf{Q}_e|^{1/2}\exp&\left\{-\frac{1}{2}(\mathbf{y}-(\boldsymbol{\alpha}+\mathbf{A}\mathbf{x}))^T\mathbf{Q}_e(\mathbf{y}-(\boldsymbol{\alpha}+\mathbf{A}\mathbf{x}))\right\}\cdot\\ &\cdot|\mathbf{Q}(\boldsymbol{\theta})|^{1/2}\exp\left\{-\frac{1}{2}\mathbf{x}^T\mathbf{Q}(\boldsymbol{\theta})\mathbf{x}\right\}\cdot\\ &\cdot|\mathbf{Q}_{\mathbf{x}|\boldsymbol{\theta},\mathbf{y}}|^{-1/2}\exp\left\{\frac{1}{2}(\mathbf{x}-\mathbf{x}_0)^T\mathbf{Q}_{\mathbf{x}|\boldsymbol{\theta},\mathbf{y}}(\mathbf{x}-\mathbf{x}_0)\right\}\pi(\boldsymbol{\theta})\,.\end{aligned} \tag{1.50}$$

The left hand side of (1.50) does not depend on $\mathbf{x}$, so the right side also does not. This means that we can choose any suitable value for $\mathbf{x}$. From (1.48), we conclude that the mode satisfies

$$\mathbf{x}_0^T \mathbf{Q}_{\mathbf{x}|\boldsymbol{\theta},\mathbf{y}} = (\mathbf{y} - \boldsymbol{\alpha})^T \mathbf{Q}_e \mathbf{A} \,,$$

additionally, choosing $\mathbf{x} = 0$, we obtain

$$\pi(\boldsymbol{\theta}|\mathbf{y}) \propto |\mathbf{Q}_e|^{1/2} \exp\left\{-\frac{1}{2}(\mathbf{y} - \boldsymbol{\alpha})^T \mathbf{Q}_e (\mathbf{y} - \boldsymbol{\alpha})\right\} .$$
$$\cdot |\mathbf{Q}(\boldsymbol{\theta})|^{1/2} |\mathbf{Q}_{\mathbf{x}|\boldsymbol{\theta},\mathbf{y}}|^{-1/2} \exp\left\{\frac{1}{2}(\mathbf{y} - \boldsymbol{\alpha})^T \mathbf{Q}_e \mathbf{A} \mathbf{x}_0\right\} \pi(\boldsymbol{\theta}) \,. \tag{1.51}$$

Applying the logarithm to (1.51), we obtain

$$\log(\pi(\boldsymbol{\theta}|\mathbf{y})) = C + \frac{1}{2} \log |\mathbf{Q}_e| - \frac{1}{2}(\mathbf{y} - \boldsymbol{\alpha})^T \mathbf{Q}_e (\mathbf{y} - \boldsymbol{\alpha}) +$$
$$+ \frac{1}{2} \log|\mathbf{Q}(\boldsymbol{\theta})| - \frac{1}{2} \log|\mathbf{Q}_{\mathbf{x}|\boldsymbol{\theta},\mathbf{y}}| + \frac{1}{2}(\mathbf{y} - \boldsymbol{\alpha})^T \mathbf{Q}_e \mathbf{A} \mathbf{x}_0 + \log(\pi(\boldsymbol{\theta})) \,. \tag{1.52}$$

where C is a constant. The next step is to maximize (1.52) in order to obtain the estimated parameters $\boldsymbol{\theta}^*$, and consequently the estimate for the GMRF, which is the mode evaluated at $\boldsymbol{\theta}^*$,

$$\mathbf{x}^* = \mathbf{x}|_{\boldsymbol{\theta}^*} \,.$$

The fitted values $\mathbf{y}$ are obtain by (1.28), which can be written as

$$\hat{\mathbf{y}} = \boldsymbol{\alpha}^* + \mathbf{A}\mathbf{x}^* \,.$$

1.4 Discussion

Consider for instance a vector field $v(s) = (v_1(s), v_2(s))$ and a vector of observations of each component $\mathbf{y} = (y_{1,1}, \ldots, y_{1,n}, y_{2,1}, \ldots, y_{2,n}) = (\mathbf{y}_1, \mathbf{y}_2)$, taken at the same locations $\{s_i \in D \subset \mathbb{R}^2 : i = 1, \ldots, n\}$.

In our case, we considered a data set with wind velocity from the data set Hersbach et al. (2023), on the 1st of January 2021, at UTC+00, downloaded from the Copernnicus Climate Change Service (C3S) Climate Data Store. It should be mentioned that the European Commission and the European Center for Medium-Range Weather Forecasts are not responsible for any use that may be made of the Copernicus information or data it contains.

Let us point out that, taking into account (1.3), (1.28) and (1.46), the parameters to be estimated are $\boldsymbol{\theta} = (\alpha_1, \tau_1, \alpha_2, \tau_2, b_{11}, b_{21}, b_{22}, k_{11}, k_{21}, k_{22})$. Nevertheless, according to Rue et al. (2009), the number of parameters to estimate should be less or equal to 6, otherwise, one may loose the computational benefits of their approach. In that sense, one should be able to reduce the number of parameters to estimate. That could be done, for instance, applying the R-INLA package to the components $v_1(s)$ and $v_2(s)$ separately, meaning that we consider $\mathcal{L}_{12} = 0$, in order to obtain estimated values for the intercepts, α_1 and α_2, as well as for the precision parameters τ_1 and τ_2. In fact, we chose randomly 1000 observations from a total of 4015 observations, located mainly over the Iberian Peninsula, in a squared domain D defined by the longitude interval $[-15°, 5°]$ and the latitude interval $[30°, 45°]$. We also use the R-INLA package to create a mesh in that domain, with 1572 nodes. The estimates for the intercept and precision were the following,

$$\alpha_1^* = 2.23, \quad \tau_1^* = 11.40, \quad \alpha_2^* = -2.88, \quad \tau_2^* = 15.16\,.$$

The summary statistics can be found in Table 1.1. Moreover, Fig. 1.1 presents observed wind velocity versus predicted wind velocity, considering independent modeling. If needed, this independent modelation also provides estimates and summary statistics for b_{11} and b_{22}.

The next step is to obtain the estimates for the remaining parameters,

$$(b_{11}, b_{21}, b_{22}, k_{11}, k_{21}, k_{22})\,.$$

The function to be minimized is $-\log(\pi(\boldsymbol{\theta}|\mathbf{y}))$, and as we can see in (1.51), this function involves sparse matrices, whose inverse are usually dense (Lindgren et al.,

Table 1.1 Summary statistics for the independent modelling of the wind components v_1 and v_2

	Mean	St dev	2.5%quant	50%quant	97.5%quant	Mode
α_1	2.23	0.71	0.82	2.24	3.64	2.24
τ_1	11.40	2.45	7.88	10.94	17.36	9.94
α_2	−2.88	1.12	−5.10	−2.88	−0.65	−2.88
τ_2	15.16	2.81	10.27	14.95	21.26	14.57

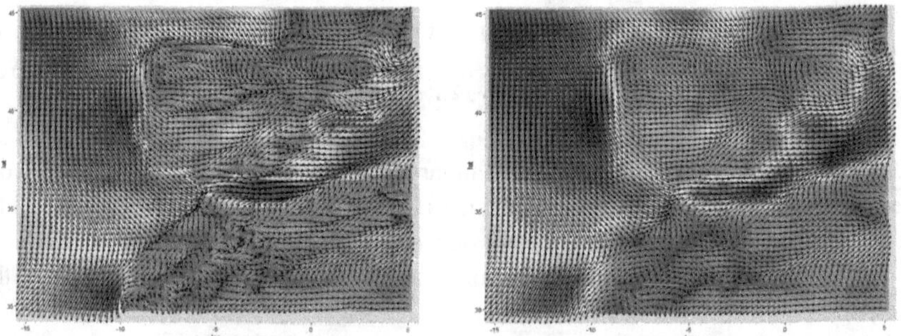

Fig. 1.1 (**a**) Observed wind velocity. (**b**) Predicted wind velocity, with independent modeling

2011), and also determinants of matrices that can easily have hundreds of entries, depending on the number of observations. Because of that, standard methods of optimization, taking into account first and second order approximations to the objective function, can easily become computationally very expensive, or even not conclusive. In fact, neither `optim` function was able to converge to an optimum, nor the `lbfgs` routine. Another difficulty was to choose a starting point, or a range of values, for the optimization routines. Although the initial values for b_{ij} and k_{ij} were chosen in a certain range arround b_{11}^* and b_{22}^* estimates given by the separated modelation, no satisfying optimum was found. In fact, the fitted values were very different from the actual data.

Another possibility is to apply non derivative optimization methods to the objective function, $-\log(\pi(\boldsymbol{\theta}|\mathbf{y}))$, in order to obtain estimates for the set of parameters. Still, choosing a range of initial values for the optimization, as well as the routine's performance, may be an issue that we look forward to overcome.

Acknowledgments This work is funded by national funds through the FCT—Fundação para a Ciência e a Tecnologia, I.P., under the scope of the projects: UIDB/00297/2020 (https://doi.org/10.54499/UIDB/00297/2020) and UIDP/00297/2020 (https://doi.org/10.54499/UIDP/00297/2020) (Center for Mathematics and Applications).

We would like to thank Doctor Ricardo Tomé, from Instituto Dom Luiz, for the wind data set: the Hersbach et al. (2023).

Appendix

Let us present the results that allow us to obtain (1.11) and (1.17). Let $D \subset \mathbb{R}^2$. Consider $x(s), \in H^2(D)$ and let $\varphi(s) \in H^2(D)$ be a scalar base element function with compact support. We have the following,

$$\langle b(k^2 - \Delta)x(s), \varphi(s)\rangle = k^2 b \int_D x(s)\varphi(s)\, ds + b \int_D \Delta x(s)\varphi(s)\, ds\,. \tag{1.53}$$

Let $\mathbf{n}$ be the normal vector on the boundary ∂D. The Green's first identity states the following.

Theorem 1.1 (Green's First Identity)

$$\int_D \Delta x(s)\varphi(s)\, ds + \int_D \nabla x(s)\nabla\varphi(s)\, ds = \int_{\partial D} \varphi(s)\nabla x(s)\cdot \mathbf{n}\, ds$$

For a proof see Evans (2022). Applying the Green's first identity to (1.53) we obtain

$$\begin{aligned}\langle b(k^2 - \Delta)x(s), \varphi(s)\rangle = k^2 b \int_D x(s)\varphi(s)\, ds + b \int_D \nabla x(s)\nabla\varphi(s)\, ds \\ - b \int_{\partial D} \varphi(s)\nabla x(s)\cdot \mathbf{n}\, ds\,.\end{aligned} \tag{1.54}$$

Taking into account the Neumann boundary condition, which states that

$$\left.\frac{\partial x(s)}{\partial \mathbf{n}}\right|_{\partial D} = \nabla x(s) \cdot \mathbf{n} = 0 \, ,$$

then we can rewrite (1.54) in the following way,

$$\langle b(k^2 - \Delta)x(s), \varphi(s)\rangle = k^2 b \int_D x(s)\varphi(s)\, ds + b \int_D \nabla x(s) \nabla \varphi(s)\, ds \, .$$

References

Matérn, B. (1960). *Spatial variation* (2nd ed.). Springer-Verlag. Reprinted 1986.

Blangiardo, M., & Cameletti, M. (2015). *Spatial and spatio-temporal Bayesian models with R-INLA*. John Wiley & Sons.

Evans, L. C. (2022). *Partial differential equations* (Vol. 19). American Mathematical Society.

Hersbach, H., Bell, B., Berrisford, P., Biavati, G., Horányi, A., Muñoz Sabater, J., Nicolas, J., Peubey, C., Radu, R., Rozum, I., Schepers, D., Simmons, A., Soci, C., Dee, D., & Thépaut, J-N. (2023). ERA5 hourly data on single levels from 1940 to present. *Copernicus Climate Change Service (C3S) Climate Data Store (CDS)* (2023)

Hu, X., & Steinsland, I. (2016). Spatial modeling with system of stochastic partial differential equations. *Wiley Interdisciplinary Reviews: Computational Statistics*, *8*(2), 112–125.

Hu, X., Simpson, D., Lindgren, F., & Rue, H. (2013). Multivariate Gaussian random fields using systems of stochastic partial differential equations. Preprint, arXiv:1307.1379.

Krainski, E., Gómez-Rubio, V., Bakka, H., Lenzi, A., Castro-Camilo, D., Simpson, D., Lindgren, F., & Rue, H. (2018). *Advanced spatial modeling with stochastic partial differential equations using R and INLA*. Chapman and Hall/CRC.

Lindgren, F., Rue, H., & Lindström, J. (2011). An explicit link between Gaussian fields and Gaussian Markov random fields: The stochastic partial differential equation approach. *Journal of the Royal Statistical Society Series B: Statistical Methodology*, *73*(4), 423–498.

Gelfand, A. E., Diggle, P., Guttorp, P., & Fuentes, M. (2010). *Handbook of spatial statistics*. In Handbooks of modern statistical methods. Chapman and Hall/CRC. https://books.google.pt/books?id=Xf4leslPDzsC.

Guerra, S. (2022). Stochastic partial differential equations and applications in Bayesian inference. Ph.D. Thesis. NOVA School of Science and Technology, NOVA University Lisbon.

Rozanov, Y. (1998). *Random fields and stochastic partial differential equations* (Vol. 438). Springer Science & Business Media.

Rue, H., & Held, L. (2005). *Gaussian Markov random fields: Theory and applications*. Chapman and Hall/CRC.

Rue, H., Martino, S., & Chopin, N. (2009). Approximate Bayesian inference for latent Gaussian models by using integrated nested Laplace approximations. *Journal of the Royal Statistical Society Series B: Statistical Methodology*, *71*(2), 319–392.

Whittle, P. (1954). On stationary processes in the plane. *Biometrika*, *41*, 434–449.

Whittle, P. (1963). Stochastic processes in several dimensions. *Bulletin de L'Institut International de Statistique*, *40*, 974–994 (1963)

Chapter 2
A Poly-Cylindrical Bayesian Network for Clustering Oceanographic Data

Thasmika Mohan and Najmeh Nakhaei Rad

2.1 Introduction

Directional statistics constitutes a specialized branch of statistical analysis focused on the study of observations that are angles or directions. The most common form of directional data is circular data (Bekker et al., 2022; Rad et al., 2022), while other forms include toroidal (Arashi et al., 2021), spherical (Nakhaei Rad et al., 2022), and cylindrical data (Ramkilawon et al., 2023). Directional data analysis is challenging because conventional statistics are not suitable for handling periodicity (see Ley and Verdebout 2017 and Pewsey and García-Portugués 2021). When circular variables are observed along with linear variables, we refer to them as cylindrical or circular-linear in the bivariate case, and as poly-cylindrical when extending to higher dimensions. An example of a cylindrical observation in ecology can be seen in animal movements, which can be described using step length, as a linear variable, and turning angle, as a circular variable (Pewsey & García-Portugués, 2021; Ramkilawon et al., 2023). Another example can be seen in meteorology, where wind direction and wind speed can be jointly modeled through cylindrical distributions (D'elia, 2001). An emerging field of application that has encouraged research into cylindrical statistics is oceanography. Oceans are of paramount importance to the well-being of the planet, and understanding the characteristics of ocean climate is crucial for further quantitative studies and investigations into ocean dynamics. To this effect, we can use the poly-cylindrical statistics to analyse oceanographic phenomena when there is more than one directional and one linear variable.

In this chapter, we focus on the clustering of cylindrical data. To uncover patterns in the data, clustering is a common tool that helps separate data into homogeneous

T. Mohan · N. Nakhaei Rad (✉)
Department of Statistics, University of Pretoria, Pretoria, South Africa
e-mail: u18145541@tuks.co.za; najmeh.nakhaeirad@up.ac.za

C. A. Coelho, D.-G. Chen (eds.), *Statistical Modeling and Applications*, Emerging Topics in Statistics and Biostatistics, https://doi.org/10.1007/978-3-031-69622-0_2

groups. If we assume that the data follows a mixture of probability distributions, then model-based clustering would be applicable (Fraley & Raftery, 2002). Each cluster represents a distribution in a finite mixture model (McLachlan et al., 2019). Due to the periodic patterns of cylindrical data, the available clustering methods on the real line are not applicable and new approaches should be explored. For example, mixtures of circular distributions are useful in text categorization, gene expression analysis, and characterization of the structure of proteins (Mardia & Jupp, 2000). However, due to a lack of efficient density estimation methods and identifiability problems, clustering of directional-linear data remains a challenge. Carta et al. (2008) explored a distribution based on copulas; however, it cannot be extended to accommodate multivariate data. Lagona and Picone (2011) and Lagona et al. (2011) used a latent-class approach for clustering; however, independence assumptions are used. Mastrantonio et al. (2015) considered a multivariate hidden Markov model to cluster directional-linear data, although the use of multiple circular variables is limited. Roy et al. (2017) constructed a directional-linear distribution with correlation terms, but it is limited to correlations between one circular and several linear variables only. In later years, Bulla et al. (2012) used a hidden Markov approach; nevertheless, there are independence assumptions in the conditional linear and directional distributions. Subsequently, Luengo-Sanchez et al. (2016), explored a hybrid model; however, correlations between linear variables were only accounted for. Furthermore, Roy et al. (2016) used a copula approach for clustering of automatic disease localization; however, it could not be extended to accommodate multivariate variables.

Evidently, clustering of multivariate correlated directional-linear data is very complex. This is due to the fact that cylindrical distributions are cumbersome to extend to higher dimensions. Additionally, the normalization constants in multivariate cylindrical distributions are sometimes intractable. In this chapter, we apply a finite mixture model of the extended Mardia-Sutton (EMS) distribution (Luengo-Sanchez et al., 2019) based on a Bayesian network (BN) approach for clustering of poly-cylindrical data.

The rest of the chapter is organized as follows: In Sect. 2.2, the EMS distribution is introduced. In Sect. 2.3, BN theory is introduced along with a clustering algorithm based on a poly-cylindrical Bayesian network using a mixture of EMS distributions. To assess the performance of the proposed clustering approach, a simulation study is conducted in Sect. 2.4. Lastly, in Sect. 2.5, to demonstrate the application of this approach in oceanography, two real datasets are analysed.

2.2 Extended Mardia-Sutton Model

Mardia and Sutton (1978) were the pioneers in constructing the earliest cylindrical probability distribution and derived a six-parameter cylindrical distribution. In the case of independence, the linear and circular parts are the normal and von Mises distributions, respectively. Let X be the linear variable and Θ be the directional variable then the probability distribution function (pdf) is given by:

$$\begin{aligned} f(x,\theta) &= \{2\pi I_0(\kappa)\}^{-1}\exp\{\kappa\cos(\theta-\mu_0)\}\,(2\pi\sigma_c^2)^{-1/2} \\ &\times \exp\left[-\left\{(x-\mu_c)^2/2\sigma_c^2\right\}\right], \end{aligned} \tag{2.1}$$

where $-\infty < x < \infty$, $0 < \theta \leq 2\pi$, $\kappa > 0$, $0 < \mu_0 \leq 2\pi$, $\sigma_c^2 = \sigma^2(1-\rho^2)$ and $\mu_c = \mu + \sigma\sqrt{\kappa}\,\{\rho_1(\cos\theta - \cos\mu_0) + \rho_2(\sin\theta - \sin\mu_0)\}$, with $\rho^2 = \rho_1^2 + \rho_2^2$, $0 \leq \rho \leq 1$ and $\sigma \geq 0$. $I_0(\kappa)$ is the modified Bessel function of the first kind and order zero.

To model poly-cylindrical data, an extension of (2.1) was proposed by Luengo-Sanchez et al. (2019). Suppose $\boldsymbol{X} = (X_1, ..., X_L)$ are L linear variables with vector mean $\boldsymbol{\mu}_X$ and covariance matrix $\boldsymbol{\Sigma}_X$, and $\boldsymbol{\Theta} = (\Theta_1, ..., \Theta_D)$ are D independent circular variables with circular mean vector $\boldsymbol{\mu}_\Theta = (\mu_1, ..., \mu_D)^\top$ and concentration vector $\boldsymbol{\kappa}_\Theta = (\kappa_1, ..., \kappa_D)^\top$. Then to obtain the multivariate Mardia-Sutton distribution we define $\boldsymbol{Y} = \begin{pmatrix}\boldsymbol{Y}_a \\ \boldsymbol{Y}_b\end{pmatrix}$, such that $\boldsymbol{Y}_a \in \mathbb{R}^L$ and $\boldsymbol{Y}_b \in \mathbb{R}^{2D}$ and $\boldsymbol{Y}$ has a $L \times 2D$-variate normal (MVN) distribution. Therefore,

$$\begin{pmatrix}\boldsymbol{Y}_a \\ \boldsymbol{Y}_b\end{pmatrix} \sim N\left[\begin{pmatrix}\boldsymbol{\mu}_a \\ \boldsymbol{\mu}_b\end{pmatrix}, \begin{pmatrix}\boldsymbol{\Sigma}_{aa} & \boldsymbol{\Sigma}_{ab} \\ \boldsymbol{\Sigma}_{ba} & \boldsymbol{\Sigma}_{bb}\end{pmatrix}\right],$$

such that $\boldsymbol{\mu}_a \in \mathbb{R}^L$ and $\boldsymbol{\mu}_b \in \mathbb{R}^{2D}$ correspond to the means of $\boldsymbol{Y}_a$ and $\boldsymbol{Y}_b$, respectively, and

$$\boldsymbol{\mu}_b' = (\mu_1, ..., \mu_D, \mu_1, ..., \mu_D)^\top,$$

$\boldsymbol{\Sigma}_{aa}$ is a matrix of dimension $L \times L$, $\boldsymbol{\Sigma}_{ab}$ is a matrix of dimension $L \times 2D$, $\boldsymbol{\Sigma}_{ab} = \boldsymbol{\Sigma}_{ba}^\top$ and $\boldsymbol{\Sigma}_{bb}$ is a $2D \times 2D$ matrix. Let $\boldsymbol{Y}_a = \boldsymbol{X}$ so that $\boldsymbol{\mu}_a = \boldsymbol{\mu}_X$ and $\boldsymbol{\Sigma}_{aa} = \boldsymbol{\Sigma}_X$. From the definition of a conditional distribution we know that

$$f(\boldsymbol{x}, \boldsymbol{y}_b) - f(\boldsymbol{y}_b) f(\boldsymbol{x}|\boldsymbol{Y}_b = \boldsymbol{y}_b),$$

where $\boldsymbol{Y}_b \sim N_{2D}(\boldsymbol{\mu}_b, \boldsymbol{\Sigma}_{bb})$ and $\boldsymbol{X}|\boldsymbol{Y}_b = \boldsymbol{y}_b \sim N_L(\boldsymbol{\mu}, \boldsymbol{Q})$, in which

$$\begin{aligned} \boldsymbol{\mu} &= \boldsymbol{\mu}_X + \boldsymbol{\Sigma}_{ab}\boldsymbol{\Sigma}_{bb}^{-1}(\boldsymbol{y}_b - \boldsymbol{\mu}_b), \\ \boldsymbol{Q} &= \boldsymbol{\Sigma}_X - \boldsymbol{\Sigma}_{ab}\boldsymbol{\Sigma}_{bb}^{-1}\boldsymbol{\Sigma}_{ba}. \end{aligned}$$

Consider the following transformation on $\boldsymbol{Y}_b$

$$\boldsymbol{Y}_b = \begin{pmatrix} \boldsymbol{Y}_{b_1} \\ \boldsymbol{Y}_{b_2} \end{pmatrix} = \begin{pmatrix} r_1 \cos\theta_1 \\ \vdots \\ r_D \cos\theta_D \\ r_1 \sin\theta_1 \\ \vdots \\ r_D \sin\theta_D \end{pmatrix}, \quad \boldsymbol{\mu}_b = \begin{pmatrix} \boldsymbol{\mu}_{b_1} \\ \boldsymbol{\mu}_{b_2} \end{pmatrix} = \begin{pmatrix} r_1 \cos\mu_1 \\ \vdots \\ r_D \cos\mu_D \\ r_1 \sin\mu_1 \\ \vdots \\ r_D \sin\mu_D \end{pmatrix},$$

and assume $\sin\theta_d, \cos\theta_d, \forall d = 1, ..., D$ have the same variance $\frac{1}{\kappa_d}$. Because of the assumption of independence between the circular variables and since $\sin\theta_d$ and $\cos\theta_d$ are orthogonal we have

$$\boldsymbol{\Sigma}_{b_1 b_2} = \boldsymbol{\Sigma}_{b_2 b_1} = \boldsymbol{0},$$

and

$$\boldsymbol{\Sigma}_{b_1 b_1} = \boldsymbol{\Sigma}_{b_2 b_2} = \begin{pmatrix} \frac{1}{\kappa_1} & 0 & \dots & 0 \\ 0 & \frac{d}{\kappa_d} & \dots & 0 \\ \vdots & \vdots & \ddots & \vdots \\ 0 & 0 & \dots & \frac{1}{\kappa_d} \end{pmatrix}.$$

By following the Jacobian approach for a multivariate transformation (Ruben, 1966), the EMS distribution is obtained by

$$f(\boldsymbol{x}, \boldsymbol{\theta}) = \prod_{d=1}^{D} \frac{\exp(k_d \cos(\theta_d - \mu_d))}{2\pi I_0(k_d)} \cdot f_{\mathrm{MVN}}(\boldsymbol{x}; \boldsymbol{\mu}, \boldsymbol{Q}), \tag{2.2}$$

where $f_{\mathrm{MVN}}(\cdot)$ denotes the density of a L-variate normal distribution with mean $\boldsymbol{\mu}$ and covariance $\boldsymbol{Q}$, and

$$\begin{aligned}
\boldsymbol{\mu} &= \boldsymbol{\beta}_0 + \boldsymbol{\beta}_1^\top \cos\boldsymbol{\Theta} + \boldsymbol{\beta}_2^\top \sin\boldsymbol{\Theta}, \\
\boldsymbol{\beta}_0 &= \boldsymbol{\mu}_X - \boldsymbol{\beta}_1^\top - \cos\boldsymbol{\mu}_{\boldsymbol{\Theta}} - \boldsymbol{\beta}_2{}^\top \sin\boldsymbol{\mu}_{\boldsymbol{\Theta}}, \\
\boldsymbol{\beta}_1^\top &= \boldsymbol{\Sigma}_{ab_1} \boldsymbol{\Sigma}_{b_1 b_1}^{-1}, \\
\boldsymbol{\beta}_2^\top &= \boldsymbol{\Sigma}_{ab_2} \boldsymbol{\Sigma}_{b_2 b_2}^{-1}, \\
\boldsymbol{\Sigma}_{ab_1} &= cov(\boldsymbol{X}, \boldsymbol{Y}_{b_1}^\top), \\
\boldsymbol{\Sigma}_{ab_2} &= cov(\boldsymbol{X}, \boldsymbol{Y}_{b_2}^\top).
\end{aligned}$$

As can be seen in (2), the EMS distribution is the product of D von Mises (VM) distributions and a L-variate normal distribution.

2.3 Bayesian Network Approach Using a Mixture Model of the Extended Mardia-Sutton Distribution

BNs are powerful graphical models since they utilize information from various sources and handle missing data and uncertainty (Cheon et al., 2009). To utilize the capabilities of a BN, the data must be complete; however, this is not true in model-based clustering as the mixing coefficients (ie. latent variable Z) are unobserved. Hence we need to utilize the Expectation-Maximization (EM) algorithm. The EM algorithm has two steps, namely the "E-Step" where parameter estimation occurs, and the "M-Step" where the log-likelihood is maximized to find the optimum set of parameters. By calculating the expected values of Z for each cluster in the E-Step, we attain a complete dataset. Let us denote a parent of a variable X as Pa_X, such that an edge points in the direction of the node X. Let G be the structure of a Bayesian network together with a set of parameters, η. Hence, using a Bayesian network approach, the pdf of a mixture of EMS distributions using BN theory is given by

$$f(\boldsymbol{x}, \boldsymbol{\theta}; \boldsymbol{\eta}) = \sum_{k=1}^{K} p(z; \boldsymbol{\eta}^k) \prod_{d=1}^{D} f_{VM}(\theta_d | z; \boldsymbol{\eta}^k) \prod_{l=1}^{L} f_N(x_l | Pa_{x_l}^G; \boldsymbol{\eta}^k), \tag{2.3}$$

where $Pa_{X_l}^G \subset \{X, \boldsymbol{\Theta}, Z\}$ and $Z \in Pa_{X_l}^G$ and the conditional normal distribution from (3) is

$$\begin{aligned} f_N(x_l | Pa_{x_l}^G; \boldsymbol{\eta}^k) &= f_N(x_l | Pa_{x_l}^G; \boldsymbol{\beta}_l^k, \sigma_l^{2,k}) \\ &= f_N(\beta_{0l}^k + \boldsymbol{\beta}_{1l}^{k\,\mathsf{T}} \boldsymbol{x} + \boldsymbol{\beta}_{2l}^{k\,\mathsf{T}} \cos \boldsymbol{\Theta} + \boldsymbol{\beta}_{3l}^{k\,\mathsf{T}} \sin \boldsymbol{\Theta}, \sigma_l^{2,k}) \\ &= f_N\left(\beta_{0l}^k + \sum_{j=1}^{J} \beta_{jl}^{k\,\mathsf{T}} u_{jl}, \sigma_l^{2,k} \right), \end{aligned}$$

where β_{jl} are the regression coefficients, $\sigma_l^{2,k}$ corresponds to the variance of X_l in cluster k and $(U_{1l}, \ldots, U_{Jl}) = (X_1, \ldots, X_l, \sin \Theta_1, \sin \Theta_d, \cos \Theta_1, \ldots, \cos \Theta_d)$. The product of VM and conditional normal distributions above are due to the local Markov property, i.e., for each variable, given its parent variables, it is independent of any non-descendant variables.

To find the best structure and set of parameters, we need to utilize the Structural Expectation-Maximization (SEM) algorithm (Friedman et al., 1997). This allows the EM algorithm to learn $\hat{G}$ and $\hat{\boldsymbol{\eta}}$ from incomplete data. For a given initial structure G and a set of parameters $\boldsymbol{\eta}$, the algorithm iterates between a pair of steps. Firstly, the parameters are maximized according to the EM algorithm. Secondly, when it converges, the algorithm searches for the best structure with the output of the expectation step. Both steps are repeated until convergence. To search for the best structure, a common score-search metric is the Bayesian Information Criterion

(BIC) score; see Algorithm 2.1 for more details. For each latent state ($k = 1, \ldots, K$) of Z, the maximum likelihood estimates (MLEs) for each directional parameter of cluster k is calculated as

$$\hat{\mu}_d^k = \arctan\left(\frac{\sum_{i=1}^N p(z^i|\boldsymbol{x}^i, \boldsymbol{\theta}^i; \boldsymbol{\eta}^k)\sin\theta_d^i}{\sum_{i=1}^N p(z^i|\boldsymbol{x}^i, \boldsymbol{\theta}^i; \boldsymbol{\eta}^k)\cos\theta_d^i}\right),$$

$$\hat{\kappa}_d^k = A^{-1}\left(\frac{\sum_{i=1}^N p(z^i|\boldsymbol{x}^i, \boldsymbol{\theta}^i; \boldsymbol{\eta}^k)\cos(\theta_d^i - \hat{\mu}_d^k)}{\sum_{i=1}^N p(z^i|\boldsymbol{x}^i, \boldsymbol{\theta}^i; \boldsymbol{\eta}^k)}\right), \tag{2.4}$$

where $A(\hat{\kappa}_d^k) = \frac{I_1(\hat{\kappa}_d^k)}{I_0(\hat{\kappa}_d^k)}$. For an approximation of $A^{-1}(.)$ refer to Best and Fisher (1981).

For X_l each regression coefficient in (2.4) is obtained by solving a system of coefficients given below. Let the weighted expectation of a random variable $\boldsymbol{X}$ be

$$\mathbb{E}_D(\boldsymbol{X}) = \sum_{i=1}^N P(Z^i|\boldsymbol{X}^i = \boldsymbol{x}^i, \boldsymbol{\Theta}^i = \boldsymbol{\theta}^i; \boldsymbol{\eta}^k)\boldsymbol{X}^i. \tag{2.5}$$

The regression coefficients can be obtained by solving the system of equations,

$$\begin{aligned}
\mathbb{E}_D(X_l) &= \hat{\beta}_{0l}^k\mathbb{E}_D(\mathbf{1}) + \ldots + \hat{\beta}_{Jl}^k\mathbb{E}_D(U_{Jl}),\\
\mathbb{E}_D(X_l \cdot U_{1l}) &= \hat{\beta}_{0l}^k\mathbb{E}_D(U_{1l}) + \ldots + \hat{\beta}_{Jl}^k\mathbb{E}_D(U_{1l} \cdot U_{Jl}),\\
&\vdots\\
\mathbb{E}_D(X_l \cdot U_{Jl}) &= \hat{\beta}_{0l}^k\mathbb{E}_D(U_{Jl}) + \ldots + \hat{\beta}_{Jl}^k\mathbb{E}_D(U_{Jl} \cdot U_{Jl}).
\end{aligned}$$

After the coefficients have been obtained, the variance of X_l is calculated as

$$\hat{\sigma}_l^{2,k} = \frac{\sum_{i=1}^N p(z^i|\boldsymbol{x}^i, \boldsymbol{\theta}^i; \boldsymbol{\eta}^k)(x_l^i - \hat{\beta}_{0l}^k - \sum_{j=1}^J \beta_{jl}^k u_{jl}^i)^2}{\sum_{i=1}^N p(z^i|\boldsymbol{x}^i, \boldsymbol{\theta}^i; \boldsymbol{\eta}^k)},$$

such that u_{jl}^i is the ith observation of the jth random variable in $\boldsymbol{U}_l$. Lastly, we can calculate prior probabilities for cluster k

$$p(z; \boldsymbol{\theta}^k) = \frac{1}{N}\sum_{i=1}^N p(z^i|\boldsymbol{x}^i, \boldsymbol{\theta}^i; \boldsymbol{\eta}^k).$$

Algorithm 2.1 The Structural Expectation-Maximization (SEM) algorithm to implement Bayesian network clustering.

1. Input a Dataset.
   2.1 E-Step: Estimate membership probability of each observation to each cluster.
   2.2 M-Step: Locally optimize each variable until convergence.
3. Structure learning takes place to discover independence relationships between variables in the Bayesian network.
4. Repeat E-Step, M-Step and structure learning until convergence.
5. Model selection takes place where different number of clusters are tried to see best solution based on a chosen criterion (BIC).

The heterogeneity between cluster solutions are indicative of the quality of a clustering algorithm. Patterns and trends can be found when homogeneity is evident within a specific cluster. Furthermore, similarity measures are used between distributions for model-based clustering, to evaluate the overlap of groups. Among them, one of the most widely used measure is relative entropy or Kullback Leibler divergence (KL) (Kullback & Leibler, 1951), which is defined as a measure of the difference between two distributions. The formula is given as

$$D_{KL}\left(f(x)||g(x)\right) = \int f(x)\log\frac{f(x)}{g(x)}dx.$$

This measure is used to quantify the difference between two probability distributions. A lower KL divergence value indicates that the two distributions are closer or more similar to each other, while a higher value indicates greater dissimilarity or divergence. The KL divergence between two EMS distributions, f and g, with $\boldsymbol{X}$ being the vector of linear variables and $\boldsymbol{\Theta}$ being the vector of directional variables, is defined as

$$\begin{aligned} D_{KL}\left(f(\boldsymbol{x},\boldsymbol{\theta})||g(\boldsymbol{x},\boldsymbol{\theta})\right) &= \sum_{d=1}^{D} D_{KL}\left(f(\theta_d)||g(\theta_d)\right) \\ &+ D_{KL}\left(f(\boldsymbol{x}|Pa_{\boldsymbol{x}}^G)||g(\boldsymbol{x}|Pa_{\boldsymbol{x}}^G)\right). \end{aligned}$$

2.4 Simulation

In this section by generating synthetic data, we can systematically evaluate the accuracy and robustness of the BN approach for model-based clustering using a mixture of the EMS distributions. Using the procedure described by Kalisch and Bühlman (2007), the artificial BNs were randomly generated via the Peter Cark algorithm to obtain a structure G. For each mixture component, the parameters

Table 2.1 Simulation results from artificially generated Bayesian networks

BIC score	KL divergence	Silhouette coefficient
(−30432.11, −26116.75)	(90.6544, 383.0195)	(0.1442, 0.4676)

were randomly generated such that each directional variable Θ_d follows a von Mises distribution with mean direction uniformly distributed around the circle, $f(\mu_d) = \frac{1}{2\pi}$ and concentration parameter $\kappa_d \sim U(1, 5)$, where U denotes the uniform distribution. Each linear variable X_l follows a normal distribution with mean coefficients $\boldsymbol{\beta} = (\beta_1, \beta_2) \sim U(-10, 10)$ and standard deviation $\sigma_l \sim U(1, 5)$. If there are no parents of X_l, then $\beta_0 \sim U(-10, 10)$ and lastly, for each mixture component $k = 1, \ldots, K$, $p(z; \boldsymbol{\theta}^k) = \frac{1}{K}$. We consider three directional and eight linear variables. The maximum number of parent nodes is limited to five. To this effect, we randomly generate ten artificial BNs and 100 observations for each cluster ($K = 3$). This data was used as the input for the SEM algorithm (Algorithm 2.1), and the performance metrics chosen were the BIC, KL divergence, and Silhouette coefficient. The Silhouette coefficient is a metric used to measure the quality of clustering by evaluating the similarity of points in each cluster, while the KL divergence is used to evaluate the heterogeneity between the clusters. The Silhouette coefficient ranges from −1 to 1, where higher scores indicate well-separated clusters.

To reduce the probability of convergence to non-optimal solutions, the SEM algorithm was initialized with 30 different random starting points. From the complete set of solutions provided by all the restarts, we selected the solution that minimized the BIC score. Table 2.1 shows the results from the simulation study, which are presented in 95% confidence intervals for each performance measure.

From Table 2.1, the confidence interval for the KL divergence is indicative of the heterogeneity between cluster solutions. Furthermore, the silhouette coefficient indicates the similarity within each cluster solution. Therefore, the SEM algorithm for implementing BN model-based clustering approach provides accurate results due to the large values of KL divergence and Silhouette coefficient.

2.5 Oceanographic Application

In this section, we will apply the poly-cylindrical BN clustering approach to two dataset to illustrate the performance of this technique in practice. We capitalize the potential of Bayesian networks along with the SEM algorithm to learn the finite mixture model for clustering of oceanographic data. As performance metrics, we use the BIC score, the Kullback Liebler (KL) divergence, and the Silhouette coefficient.

2.5.1 Atlantic Ocean, Southern Africa

This dataset contains information pertaining to the climatology of the South Atlantic Ocean in Southern Africa obtained via the NASA research website (https://power.larc.nasa.gov/). Taken from the coast near the border of South Africa (−28.6205°S, 15.0374°E) and Namibia, this dataset contains 456 daily climatology recordings from 2020 to 2021. The variables of interest are wind direction and wind speed at 10 and 50 meters above ground (degrees), precipitation (mm/day), temperature (degrees), dew point (degrees), humidity (g/kg), and surface pressure (kPa). The climatic characteristics of this region were reported using nine variables: two directional and seven linear variables.

Table 2.2 shows the BIC values for the BN clustering and the results in Table 2.2 indicate that a two-cluster solution is the best for the poly-cylindrical BN. The KL Divergence and Silhouette coefficient of the best solution for the Atlantic Ocean datasets are 708.336 and 0.1880, respectively. The KL divergence shows that the heterogeneity is very high between the two clusters and the Silhouette coefficient confirms the similarity within each cluster. Using the membership probability for each observation to each cluster (see Appendix, Table 2.6), the observations are allocated to a cluster for which the probability is the largest. In Fig. 2.1, the global cluster allocation is given and 72% of observations are allocated to cluster one while 28% of observations are allocated to cluster two. Figure 2.2 shows the pairwise scatter plots of the observations and Fig. 2.3 displays the stacked relative frequency plots of the observations for each variable in two colours, corresponding to the two-cluster solution.

We employ a t-test to assess mean differences between the two-cluster solutions across each variable and the results are summarized in Table 2.3. The means of wind direction (10 and 50 m), temperature, dew point, precipitation, humidity and

Table 2.2 BIC values from the poly-cylindrical BN fitted to the Atlantic Ocean dataset

Number of clusters	BIC value
2	**−2207.710**
3	−1948.312
4	−1695.243

The bold values indicate the best-fitting model in terms of the number of clusters

Fig. 2.1 Global cluster allocation bar chart for the Atlantic Ocean dataset

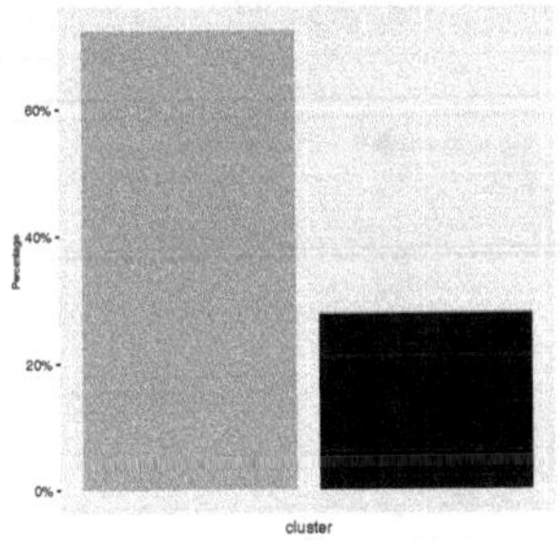

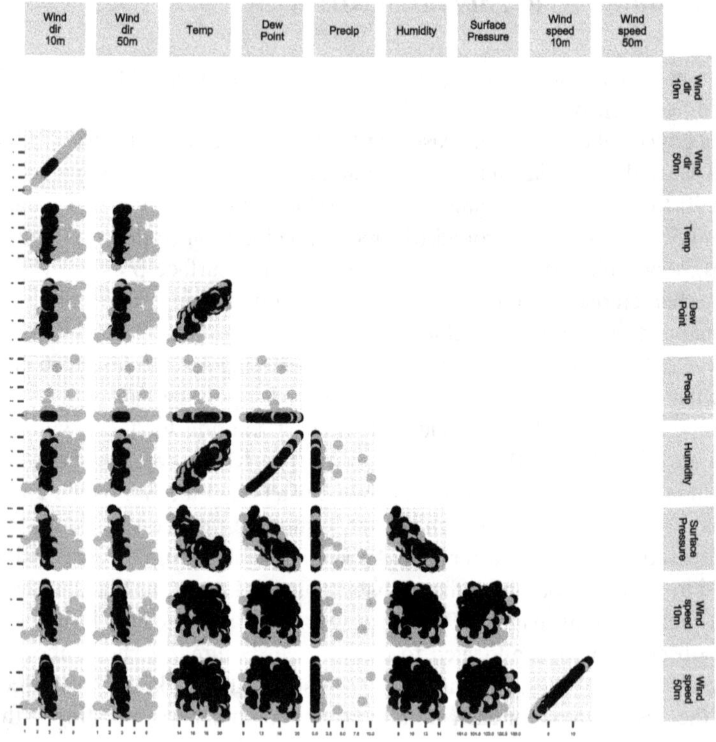

Fig. 2.2 Pairwise plots of the Atlantic Ocean dataset

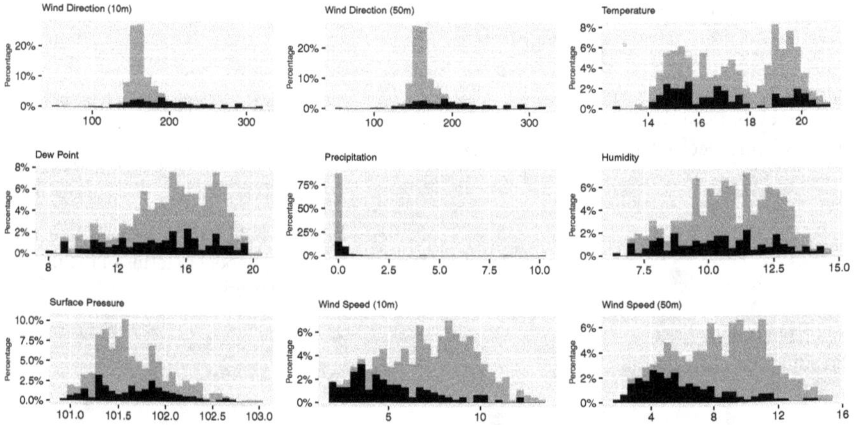

Fig. 2.3 Stacked relative frequency plots of each variable in the Atlantic Ocean dataset

Table 2.3 Summary of P-values from t-tests

Variable	P-value of t-test
Wind direction (10 m)	< 0.001***
Wind direction (50 m)	< 0.001***
Temperature	0.0024***
Dew point	0.0019***
Precipitation	0.0004***
Humidity	0.0037***
Surface pressure	0.8908
Wind speed (10 m)	< 0.001***
Wind speed (50 m)	< 0.001***

Note: The cosine of the directional variables was used to accommodate the t-tests. Significance codes: 0.01 "***", 0.05 "**", 0.1 "*"

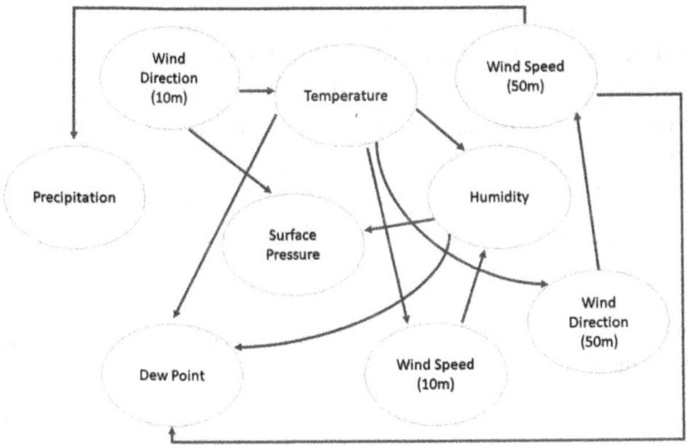

Fig. 2.4 The BN structure learned by the Structural Expectation-Maximization algorithm obtained from the clustering of the Atlantic Ocean dataset

wind speed (10 and 50 m) show significant difference between two clusters. We can determine characteristics for each cluster to identify a distinct set of oceanographic conditions. These characteristics can be stated as

- Cluster 1: 140°< Wind Direction (10 m/50 m) <170° (with probability of 83%) , Precipitation < 0.018 mm/day (with probability of 66%), Wind Speed (10 m)>6.098 m (81%), Wind Speed (50 m)>8 m (with probability of 69%).
- Cluster 2: 150° < Wind Direction (10 m/50 m)< 300°, (86%), Precipitation>0.018 mm/day (with probability of 81%), Wind Speed (10 m)<6.098 m (with probability of 75%), Wind Speed (50 m)<8 m (with probability of 80%).

Figure 2.4 represents the BN structure learned by the SEM algorithm and from this, it can be seen that temperature is the most influential metaocean parameter with the highest node degree. A node degree is a measure of how many children a node has.

2.5.2 *Arctic Ocean, Russia*

This dataset contains information describing the hourly meteorology observations at a Russian polar station in Nowaja Semlja in an archipelago (72.38°N, 52.7°E). The dataset contains hourly recordings from September 1882 to July 1883. The data can be accessed via the Earth and Environmental Science website (https://doi.pangaea.de/10.1594/PANGAEA.549034). There are 120 hourly readings of station pressure (hPa), temperature (degrees Celsius), vapour tension (hPa), humidity (%), wind direction (degrees), wind speed (m/s), cloud coverage (%) and cloud direction (degrees). Hence we can describe the climatic conditions in this area with eight variables: two directional and six linear variables.

The results in Table 2.4 indicate that a three-cluster solution is the best for the poly-cylindrical BN. The KL Divergence between each pair of clusters are $KL_{1,2} = 29.836$, $KL_{1,3} = 11.71$, $KL_{2,3} = 7.395$ and the Silhouette coefficient is 0.2629. The KL divergence shows that the heterogeneity between the three clusters and the Silhouette coefficient shows the similarity within each cluster. Hence, the clusters with the most similarity were cluster two and cluster three, with a KL divergence of 7.395 and the least similar clusters were cluster one and cluster two, with a KL divergence of 29.836.

In Fig. 2.5, the global cluster allocation is given. This is based on the probability of allocation for each observation which can be found in the Appendix in Table 2.7.

Table 2.4 BIC values from the poly-cylindrical BN fitted to the Arctic Ocean dataset

Number of clusters	BIC value
2	−475.539
3	**−647.060**
4	−497.616

The bold values indicate the best-fitting model in terms of the number of clusters

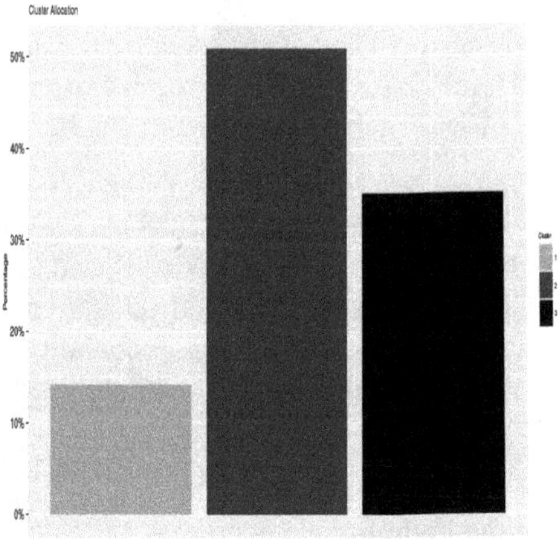

Fig. 2.5 Global cluster allocation bar chart for the Arctic Ocean dataset

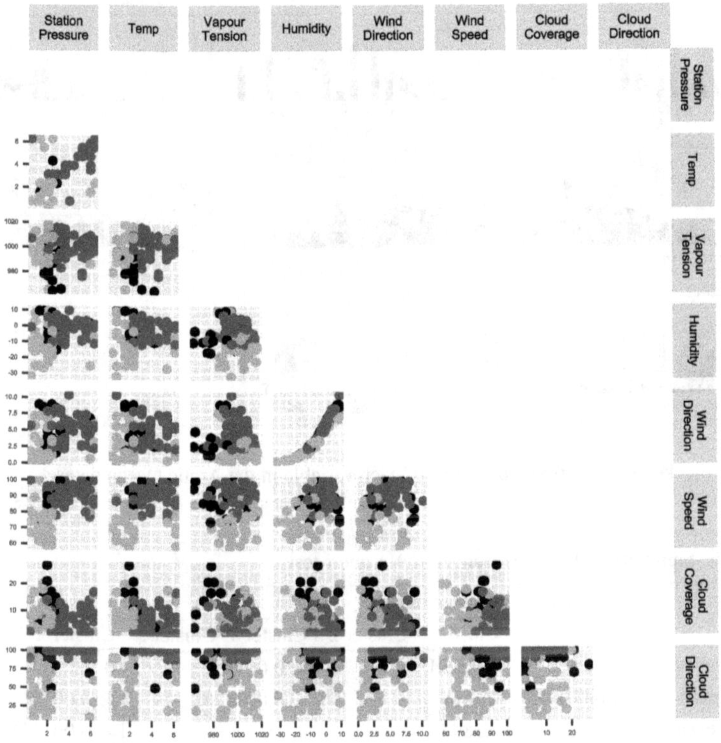

Fig. 2.6 Pairwise plots of the Atlantic Ocean dataset

It should be noted that 14, 50 and 35% of observations are allocated to cluster one, two and three, respectively. Figure 2.6 shows the pairwise scatter plots of the observations and Fig. 2.7 displays the stacked relative frequency plots of the observations for each variable in three colours, corresponding to the three-cluster solution.

The analysis of variance (ANOVA) test is used to analyze differences amongst the means of variables across the three-cluster solution. The ANOVA test is used to analyse differences amongst the means of variables across the three-cluster solution. The results are reported in Table 2.5 together with post-hoc t-tests for each cluster pair. From Table 2.5, the means of all variables were significantly different at a 10% level of significance between three clusters. Some characteristics of the clusters are as follows,

- Cluster 1: Wind is blowing in a northeastern direction (with probability of 85%), station pressure < 1000 hPa (with probability of 80%).
- Cluster 2: Wind is blowing in a southwestern to northwestern direction (with probability of 75%) with warmer temperatures around 1° C (with probability of 85%) and humidity > 90% (with probability of 95%).

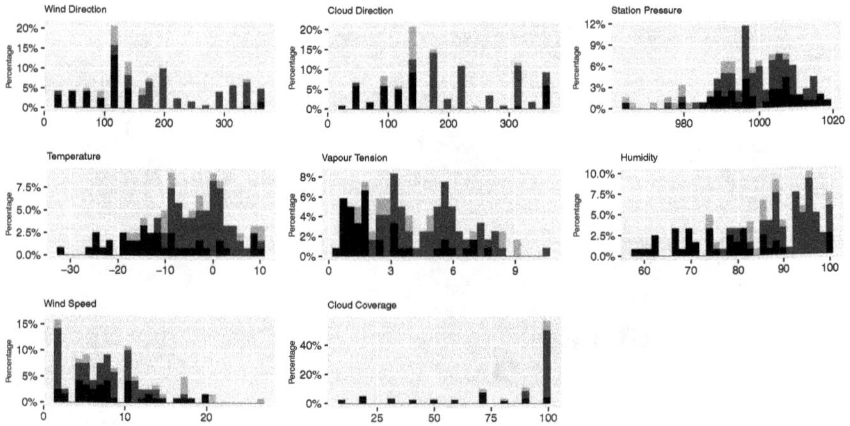

Fig. 2.7 Stacked relative frequency plots of each variable in the Arctic Ocean dataset

Table 2.5 Summary of P-values from ANOVA and t-tests

Variable	P-value of ANOVA	P-value of t-test Cluster 1–Cluster 2	Cluster 1–Cluster 3	Cluster 2–Cluster 3
Station pressure	< 0.001***	< 0.001***	< 0.001***	0.7
Temperature	< 0.001***	0.3668	< 0.001***	< 0.001***
Vapour tension	< 0.001***	0.3895	0.0037***	< 0.001***
Humidity	< 0.001***	0.01**	< 0.001***	< 0.001***
Wind direction	0.0233**	0.29	0.1*	0.22
Wind speed	0.0771*	< 0.001***	0.0071 ***	0.1414
Cloud coverage	< 0.001***	0.025 **	< 0.001***	< 0.001***
Cloud direction	< 0.001***	0.275	0.014**	0.019**

Note: The cosine of the directional variables were used in order to accommodate these tests. Significance codes: 0.01 "***", 0.05 "**", 0.1 "*"

- Cluster 3: Extremely cold temperatures around −30° C (with probability of 92%) with station pressure > 1000 hPa (with probability of 72%) and humidity < 90% (with probability of 96%).

Figure 2.8 represents the BN structure and from this, it can be seen that wind direction is the most influential metaocean parameter with the highest node degree and cloud direction is the least influential parameter.

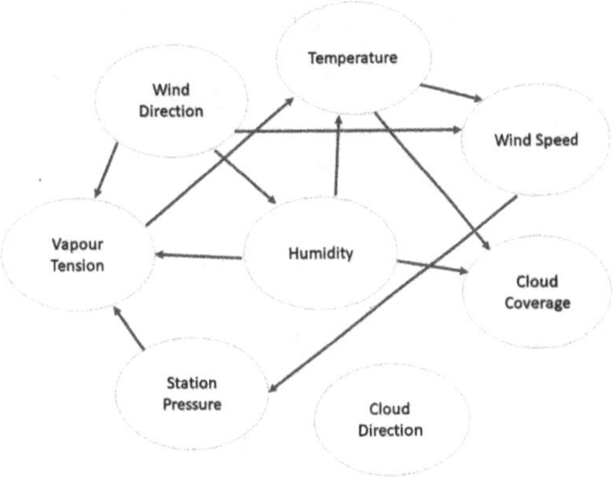

Fig. 2.8 The BN structure learned by the Structural Expectation-Maximization algorithm obtained from the clustering of the Arctic Ocean dataset

2.6 Conclusion

In this chapter, we used a finite mixture model of the EMS distribution based on a Bayesian network approach for clustering poly-cylindrical data. This method has the advantage of capturing directional-linear correlations. Furthermore, by using the EM algorithm to estimate the parameters, we avoided computationally costly methods. Moreover, by employing the SEM algorithm, we can cluster data and learn the structure of the Bayesian network. Additionally, the simulation study demonstrates the good performance of the clustering approach. To show the practical usefulness of this approach, we analysed two different ocean datasets from two different geographical locations. The results emphasize the capability of the Bayesian network clustering approach in application. We can identify distinct characteristics of climatology in specific oceanographic regions, as illustrated in the application. Clustering in oceanography is important to identify hidden patterns in vast spaces of water, which aids in understanding ecological and environmental dynamics. Secondly, Bayesian networks allow us to take a probabilistic approach that recognizes the uncertainty in a dynamic and powerful ocean environment. Furthermore, by clustering ocean data, distinct ecological zones are identified, which helps marine scientists make informed decisions about conservation, resource management, and disaster response. These insights can enable better predictive models, which facilitate the early detection of environmental changes.

Acknowledgments This research is supported in part by the RDP grant at University of Pretoria and the National Research Foundation (NRF) of South Africa, Ref.: RA210106581084, grant No. 150170. The opinions expressed and conclusions arrived at are those of the authors and are not necessarily to be attributed to the NRF.

Appendix

Table 2.6 Probabilities of allocation for each cluster

Observation	Cluster 1	Cluster 2	Observation	Cluster 1	Cluster 2	Observation	Cluster 1	Cluster 2	Observation	Cluster 1	Cluster 2	Observation	Cluster 1	Cluster 2	Observation	Cluster 1	Cluster 2
1	1	0	41	1	0	81	1	0	121	0	1	161	0	1	201	0.9884	0.0116
2	1	0	42	1	0	82	1	0	122	0.9997	0.0003	162	0	1	202	0.8944	0.1056
3	1	0	43	0.9297	0.0703	83	1	0	123	0.9989	0.0011	163	0.0087	0.9913	203	0.9793	0.0207
4	1	0	44	0.9999	0.0001	84	1	0	124	0.9987	0.0013	164	0.9986	0.0014	204	0.9999	0.0001
5	0.9998	0.0002	45	0.9998	0.0002	85	0.998	0.002	125	1	0	165	0.9963	0.0037	205	0.9985	0.0015
6	0.9999	0.0001	46	0	1	86	0.0003	0.9997	126	1	0	166	1	0	206	0.0011	0.9989
7	0.9987	0.0013	47	0	1	87	1	0	127	0.9996	0.0004	167	1	0	207	0.9998	0.0002
8	0	1	48	0.9999	0.0001	88	0.9999	0.0001	128	0.9964	0.0036	168	0	1	208	1	0
9	0.9997	0.0003	49	0.9999	0.0001	89	0.9999	0.0001	129	0.9965	0.0035	169	0	1	209	1	0
10	1	0	50	0.9842	0.0158	90	1	0	130	0.9905	0.0095	170	0	1	210	0.9997	0.0003
11	1	0	51	0.9999	0.0001	91	0.9999	0.0001	131	0.3592	0.6408	171	0	1	211	1	0
12	1	0	52	1	0	92	1	0	132	0.9944	0.0056	172	0.0008	0.9992	212	1	0
13	1	0	53	1	0	93	0.681	0.319	133	0.2496	0.7504	173	0.9968	0.0032	213	0.9935	0.0065
14	1	0	54	1	0	94	0	1	134	1	0	174	0.9965	0.0035	214	0	1
15	1	0	55	0.5823	0.4177	95	0	1	135	1	0	175	0.9616	0.0384	215	0	1
16	0.9977	0.0023	56	0	1	96	0.9859	0.0141	136	0.9999	0.0001	176	0.9996	0.0004	216	0.828	0.172
17	0.9998	0.0002	57	0	1	97	0	1	137	1	0	177	0	1	217	0.9998	0.0002
18	1	0	58	1	0	98	0	1	138	0.9999	0.0001	178	0	1	218	0.9978	0.0022
19	0.9979	0.0021	59	1	0	99	1	0	139	0.8413	0.1587	179	0	1	219	1	0
20	0	1	60	0	1	100	0.9995	0.0005	140	0	1	180	0.9971	0.0029	220	0	1

21	0	1	61	0	1	101	0.9997	0.0003	141	0	1	181	1	0	221	0	1
22	0	1	62	0	1	102	0.8642	0.1358	142	0	1	182	0.2177	0.7823	222	0	1
23	0.1468	0.8532	63	0.9911	0.0089	103	0	1	143	0.9999	0.0001	183	0.0038	0.9962	223	0.9995	0.0005
24	0	1	64	0.9178	0.0822	104	1	0	144	1	0	184	1	0	224	0	1
25	0.9204	0.0796	65	0.9999	0.0001	105	1	0	145	0.6676	0.3324	185	1	0	225	0.9997	0.0003
26	0.999	0.001	66	0.9979	0.0021	106	1	0	146	0	1	186	0.9999	0.0001	226	0.9999	0.0001
27	1	0	67	0.9993	0.0007	107	1	0	147	0.997	0.003	187	0.9996	0.0004	227	0.8954	0.1046
28	1	0	68	1	0	108	1	0	148	0.9998	0.0002	188	0.0007	0.9993	228	0	1
29	1	0	69	1	0	109	0.9454	0.0546	149	1	0	189	0.9997	0.0003	229	0.9899	0.0101
30	0.9998	0.0002	70	1	0	110	0	1	150	0.9909	0.0091	190	0	1	230	0.6044	0.3956
31	0.9998	0.0002	71	0	1	111	0.0034	0.9966	151	0	1	191	0	1	231	0.9942	0.0058
32	0.9999	0.0001	72	0.4269	0.5731	112	0	1	152	0	1	192	0	1	232	0.9998	0.0002
33	0.9997	0.0003	73	0.0035	0.9965	113	0	1	153	0	1	193	0.997	0.003	233	1	0
34	0.9997	0.0003	74	0.9999	0.0001	114	0	1	154	0	1	194	0	1	234	1	0
35	0.8255	0.1745	75	1	0	115	0	1	155	0	1	195	0	1	235	0.0012	0.9988
36	0.9439	0.0561	76	0.9999	0.0001	116	0	1	156	0.9904	0.0096	196	0.9992	0.0008	236	0.9984	0.0016
37	1	0	77	0.9978	0.0022	117	0	1	157	0	1	197	0.0223	0.9777	237	0.9684	0.0316
38	0.9993	0.0007	78	0.9999	0.0001	118	1	0	158	0.9995	0.0005	198	0	1	238	0.1476	0.8524
39	0.9908	0.0092	79	0.9999	0.0001	119	0.9937	0.0063	159	0.9986	0.0014	199	0	1	239	0	1
40	1	0	80	0.9999	0.0001	120	0.0001	0.9999	160	0	1	200	0	1	240	0	1

(continued)

Table 2.6 (continued)

Observation	Cluster 1	Cluster 2	Observation	Cluster 1	Cluster 2	Observation	Cluster 1	Cluster 2	Observation	Cluster 1	Cluster 2	Observation	Cluster 1	Cluster 2	Observation	Cluster 1	Cluster 2
241	0	1	281	0.9999	0.0001	321	1	0	361	0.9992	0.0008	401	1	0	441	1	0
242	0	1	282	0.9999	0.0001	322	0.9999	0.0001	362	0.9999	0.0001	402	1	0	442	1	0
243	0.9895	0.0105	283	0.9999	0.0001	323	1	0	363	1	0	403	1	0	443	1	0
244	0.9999	0.0001	284	0.9999	0.0001	324	1	0	364	1	0	404	1	0	444	1	0
245	0	1	285	1	0	325	0.0001	0.9999	365	1	0	405	1	0	445	1	0
246	0	1	286	0.9999	0.0001	326	0.9729	0.0271	366	1	0	406	1	0	446	1	0
247	0.9981	0.0019	287	1	0	327	0.8048	0.1952	367	1	0	407	0.9999	0.0001	447	0.996	0.004
248	0	1	288	1	0	328	0	1	368	1	0	408	0.9996	0.0004	448	0.9924	0.0076
249	0	1	289	1	0	329	0.9867	0.0133	369	0.9578	0.0422	409	0.9997	0.0003	449	0.9809	0.0191
250	1	0	290	1	0	330	1	0	370	0	1	410	0.9986	0.0014	450	0.1659	0.8341
251	0.9997	0.0003	291	0.9998	0.0002	331	1	0	371	0	1	411	0.9946	0.0054	451	0	1
252	0.0001	0.9999	292	0.9999	0.0001	332	0.9999	0.0001	372	1	0	412	1	0	452	0.9999	0.0001
253	1	0	293	0.9999	0.0001	333	1	0	373	1	0	413	1	0	453	1	0
254	0.9998	0.0002	294	1	0	334	1	0	374	1	0	414	0.9983	0.0017	454	0.9874	0.0126
255	0.9327	0.0673	295	1	0	335	0.9997	0.0003	375	1	0	415	1	0	455	0.9997	0.0003
256	0	1	296	0.9679	0.0321	336	1	0	376	1	0	416	1	0	456	1	0
257	0.0706	0.9294	297	0	1	337	1	0	377	1	0	417	0.9992	0.0008			
258	0.9847	0.0153	298	0.9941	0.0059	338	1	0	378	0	1	418	1	0			
259	0.9991	0.0009	299	1	0	339	1	0	379	0	1	419	1	0			
260	0.9995	0.0005	300	1	0	340	0.999	0.001	380	0	1	420	1	0			

261	1	0	301	0.9998	0.0002	341	0.999	0.001	381	0	1	421	1	0		
262	0.9999	0.0001	302	0.9662	0.0338	342	0.0056	0.9944	382	0.9993	0.0007	422	1	0		
263	1	0	303	0	1	343	0	1	383	0.9999	0.0001	423	1	0		
264	0.9999	0.0001	304	1	0	344	0.9995	0.0005	384	0.9994	0.0006	424	0.5754	0.4246		
265	1	0	305	1	0	345	0.9995	0.0005	385	1	0	425	0	1		
266	1	0	306	0.9999	0.0001	346	0.373	0.627	386	1	0	426	0	1		
267	0.9999	0.0001	307	0	1	347	1	0	387	1	0	427	0.9997	0.0003		
268	0	1	308	0	1	348	0.9999	0.0001	388	1	0	428	0.9998	0.0002		
269	0	1	309	0	1	349	1	0	389	0	1	429	0.9999	0.0001		
270	0.9992	0.0008	310	0	1	350	1	0	390	0	1	430	0.9999	0.0001		
271	0.9979	0.0021	311	0	1	351	1	0	391	1	0	431	0.9998	0.0002		
272	0.9999	0.0001	312	0	1	352	1	0	392	1	0	432	0.9999	0.0001		
273	1	0	313	0	1	353	0.9999	0.0001	393	1	0	433	0	1		
274	0.9993	0.0007	314	1	0	354	0.9999	0.0001	394	1	0	434	0	1		
275	0	1	315	1	0	355	0.9999	0.0001	395	1	0	435	0	1		
276	0	1	316	1	0	356	0.9999	0.0001	396	1	0	436	0.2542	0.7458		
277	0.9999	0.0001	317	0.9983	0.0017	357	0.9969	0.0031	397	0.9999	0.0001	437	0.9995	0.0005		
278	0.9991	0.0009	318	0	1	358	1	0	398	0.9999	0.0001	438	0.8494	0.1506		
279	0.9996	0.0004	319	0.9998	0.0002	359	1	0	399	0.9993	0.0007	439	0.9308	0.0692		
280	0.9997	0.0003	320	1	0	360	1	0	400	0.9999	0.0001	456	0.0007	0.9993		

Table 2.7 Probabilities of allocation to each cluster

Obser-vation	Cluster 1	Cluster 2	Cluster 3	Obser-vation	Cluster 1	Cluster 2	Cluster 3	Obser-vation	Cluster 1	Cluster 2	Cluster 3	Obser-vation	Cluster 1	Cluster 2	Cluster 3
1	0	0.0024	0.9976	31	0	0.9999	0.0001	61	0.9999	0	0	91	0.9993	0	0.0007
2	0.9989	0	0.0011	32	0.0133	0.0003	0.9864	62	0.2942	0.7055	0.0003	92	0	0	1
3	0	0.8262	0.1738	33	0	0.9999	0	63	0.9655	0	0.0345	93	0.3653	0.6346	0
4	0	0.0649	0.9351	34	1	0	0	64	0.9971	0	0.0029	94	0.9802	0	0.0198
5	1	0	0	35	0	0.0391	0.9609	65	0	0.0057	0.9942	95	0	0.0044	0.9956
6	0	0	1	36	0	0.9999	0	66	0	0.9934	0.0066	96	0.9826	0.0174	0
7	0	0.9963	0.0037	37	0.0061	0.9916	0.0023	67	0	1	0	97	0.9998	0	0.0002
8	0	0.9122	0.0877	38	0	0.9904	0.0096	68	0.0293	0.0001	0.9707	98	0	0.7898	0.2102
9	0.9784	0.0002	0.0214	39	0	0.0023	0.9977	69	0	0.0004	0.9996	99	0.9416	0.0584	0
10	0.4909	0	0.5091	40	0	0.0635	0.9365	70	0.1235	0.8747	0.0018	100	0.0001	0.9992	0.0007
11	0.9998	0	0.0002	41	0.0011	0.9989	0.0001	71	0.9402	0.0582	0.0016	101	1	0	0
12	0	0.0056	0.9944	42	0	0.0009	0.9991	72	0.1637	0	0.8363	102	0.0002	0.9995	0.0003
13	1	0	0	43	0	1	0	73	0	0.618	0.382	103	0.9822	0	0.0178
14	0.9998	0.0002	0	44	0.9987	0	0.0013	74	0	0.9995	0.0005	104	0.0001	0.9996	0.0004
15	0	0.9991	0.0009	45	0	0.0004	0.9996	75	0.0005	0	0.9995	105	0	0.0003	0.9997
16	0	0.0003	0.9997	46	0.9999	0.0001	0	76	0.9875	0.0125	0	106	0.9999	0	0.0001
17	0	0.0002	0.9998	47	0.0036	0.9955	0.0008	77	0.9995	0	0.0005	107	0.5302	0.4697	0.0001
18	1	0	0	48	0	0.9934	0.0066	78	0.9999	0	0.0001	108	0	0	1
19	0	0.0027	0.9973	49	0	0.8355	0.1645	79	0.9999	0	0.0001	109	0.0005	0.7706	0.2289
20	0.9999	0	0.0001	50	0.0011	0.9971	0.0018	80	0	0.9888	0.0112	110	0	1	0

21	0	0.989	0.011	51	0.9997	0	0.0003	81	0.9746	0	0.0254	111	0.9999	0.0001	0.0001
22	0.0007	0.9991	0.0002	52	0.8773	0.0001	0.1226	82	0.0041	0	0.9959	112	0.8805	0.1069	0.0125
23	0	0.9994	0.0006	53	0.9934	0.0058	0.0008	83	0.9996	0.0003	0.0001	113	0	0	1
24	0	0.0001	0.9999	54	1	0	0	84	0	0.9997	0.0003	114	0	0.9931	0.0069
25	0.0257	0.0003	0.974	55	0.0004	0.9982	0.0014	85	0.0049	0.9888	0.0063	115	0.004	0.9956	0.0004
26	0.9999	0	0.0001	56	0.9998	0	0.0002	86	0	0.9838	0.0162	116	0.896	0	0.104
27	0	0.0028	0.9972	57	0.0003	0.0001	0.9997	87	0.0002	0.9971	0.0027	117	0.0638	0.9359	0.0002
28	0.3105	0.6895	0	58	0.9999	0	0	88	0	0.9999	0.0001	118	0	0.1267	0.8733
29	0	0.9986	0.0014	59	0	1	0	89	0	0.0008	0.9992	119	0.9998	0	0.0002
30	0	0	1	60	0	0.9999	0.0001	90	0	0.9351	0.0649	120	0	0.8105	0.1895

References

Arashi, M., Nakhaei Rad, N., Bekker, A., & Schubert, W.-D. (2021). Möbius transformation-induced distributions provide better modelling for protein architecture. *Mathematics*, *9*(21), 2749.

Bekker, A., Nakhaei Rad, N., Arashi, M., & Ley, C. (2022). Generalized skew-symmetric circular and toroidal distributions. In *Directional statistics for innovative applications: A bicentennial tribute to Florence nightingale* (pp. 161–186). Springer.

Best, D. J., & Fisher, N. I. (1981). The bias of the maximum likelihood estimators of the von mises-fisher concentration parameters: The bias of the maximum likelihood estimators. *Communications in Statistics-Simulation and Computation*, *10*(5), 493–502.

Bulla, J., Lagona, F., Maruotti, A., & Picone, M. (2012). A multivariate hidden markov model for the identification of sea regimes from incomplete skewed and circular time series. *Journal of Agricultural, Biological, and Environmental Statistics*, *17*, 544–567.

Carta, J. A., Ramirez, P., & Bueno, C. (2008). A joint probability density function of wind speed and direction for wind energy analysis. *Energy Conversion and Management*, *49*(6), 1309–1320.

Cheon, S.-P., Kim, S., Lee, S.-Y., & Lee, C.-B. (2009). Bayesian networks based rare event prediction with sensor data. *Knowledge-Based Systems*, *22*(5), 336–343.

D'elia, A. (2001). A statistical model for orientation mechanism. *Statistical Methods and Applications*, *10*, 157–174.

Fraley, C., & Raftery, A. E. (2002). Model-based clustering, discriminant analysis, and density estimation. *Journal of the American statistical Association*, *97*(458), 611–631.

Friedman, N., et al. (1997). Learning belief networks in the presence of missing values and hidden variables. In *Icml, Berkeley, CA* (Vol. 97, pp. 125–133).

Kalisch, M., & Bühlman, P. (2007). Estimating high-dimensional directed acyclic graphs with the pc-algorithm. *Journal of Machine Learning Research*, *8*(3).

Kullback, S., & Leibler, R. A. (1951). On information and sufficiency. *The Annals of Mathematical Statistics*, *22*(1), 79–86.

Lagona, F., Maruotti, A., & Picone, M. (2011). A non-homogeneous hidden markov model for the analysis of multi-pollutant exceedances data. *Hidden Markov models, theory and applications* (pp. 207–222). InTech. doi: 10.5772/601

Lagona, F., Picone, M. (2011). A latent-class model for clustering incomplete linear and circular data in marine studies. *Journal of Data Science*, *9*(4), 585–605.

Ley, C., & Verdebout, T. (2017). *Modern directional statistics*. Chapman and Hall/CRC.

Luengo-Sanchez, S., Bielza, C., & Larranaga, P. (2016). Hybrid gaussian and von mises model-based clustering. In *Proceedings of the twenty-second European conference on artificial intelligence* (pp. 855–862).

Luengo-Sanchez, S., Larranaga, P., & Bielza, C. (2019). A directional-linear bayesian network and its application for clustering and simulation of neural somas. *IEEE Access*, *7*, 69907–69921.

Mardia, K., & Sutton, T. (1978). A model for cylindrical variables with applications. *Journal of the Royal Statistical Society Series B: Statistical Methodology*, *40*(2), 229–233.

Mardia, K. V., & Jupp, P. E. (2000). *Directional statistics*. John Wiley & Sons.

Mastrantonio, G., Maruotti, A., & Jona-Lasinio, G. (2015). Bayesian hidden markov modelling using circular-linear general projected normal distribution. *Environmetrics*, *26*(2), 145–158.

McLachlan, G. J., Lee, S. X., & Rathnayake, S. I. (2019). Finite mixture models. *Annual Review of Statistics and Its Application*, *6*, 355–378.

Nakhaei Rad, N., Bekker, A., Arashi, M., & Ley, C. (2022). Coming together of bayesian inference and skew spherical data. *Frontiers in Big Data*, *4*, 769726.

Pewsey, A., & García-Portugués, E. (2021). Recent advances in directional statistics. *Test*, *30*(1), 1–58.

Rad, N. N., Bekker, A., & Arashi, M. (2022). Enhancing wind direction prediction of south africa wind energy hotspots with bayesian mixture modeling. *Scientific Reports*, *12*(1), 11442.

Ramkilawon, G. D., Ferreira, J., & Nakhaeirad, N. (2023). Sine-skewed von mises-and lindley/gumbel models as candidates for direction and distance in modelling animal movement. *Brazilian Journal of Biometrics*, *41*(2), 175–190.

Roy, A., Parui, S. K., & Roy, U. (2016). Swgmm: a semi-wrapped gaussian mixture model for clustering of circular–linear data. *Pattern Analysis and Applications*, *19*, 631–645.

Roy, A., Pal, A., & Garain, U. (2017). Jclmm: A finite mixture model for clustering of circular-linear data and its application to psoriatic plaque segmentation. *Pattern Recognition*, *66*, 160–173.

Ruben, H. (1966). *Multidimensional Gaussian distributions.* John Wiley and Sons.

Chapter 3
A Copula-Based Approach to Statistical Modelling of Solar Irradiance

Francois von Holtzhausen, Najmeh Nakhaei Rad, Priyanka Nagar, and Andriette Bekker

3.1 Introduction

3.1.1 Problem Contextualisation

As the world's population is constantly growing and more corporations are expanding on a local and global scale is the world's electricity demand at an all-time high. This increase in electricity demand is resulting in the depletion of fossil fuels at a staggering rate. Not only has this continuous consumption of fossil fuels had a damaging effect on our ecosystems and leads to environmental problems such as global warming. But has it also resulted in numerous countries having an unstable and unreliable power supply. One such country is South Africa which has had for the majority of the twenty-first century the regular implementation of load-shedding to lower the total electricity demand on the national energy grid. This is due to the fact that 70% of South Africa's electricity is produced through the consumption of coal resulting in the emission of 433.57 metric tons of Carbon Dioxide (CO2) which

F. von Holtzhausen · N. Nakhaei Rad (✉)
Department of Statistics, University of Pretoria, Pretoria, South Africa
e-mail: u18023691@tuks.co.za; najmeh.nakhaeirad@up.ac.za

P. Nagar
Department of Statistics and Actuarial Science, Stellenbosch University, Stellenbosch, South Africa
e-mail: pnagar@sun.ac.za

A. Bekker
Department of Statistics, University of Pretoria, Pretoria, South Africa

DSI-NRF Centre of Excellence in Mathematical and Statistical Sciences (CoE-MaSS), Centre for Environmental Studies, Department of Geography, Geoinformatics and Meteorology, University of Pretoria, Pretoria, South Africa
e-mail: andriette.bekker@up.ac.za

C. A. Coelho, D.-G. Chen (eds.), *Statistical Modeling and Applications*, Emerging Topics in Statistics and Biostatistics, https://doi.org/10.1007/978-3-031-69622-0_3

amounts to approximately 42% of the total greenhouse gas emissions of South Africa.[1] We definitely need a viable substitute, particularly renewable energy. South Africa has a plentiful renewable energy resource in the form of ample sunshine, with an average of over 2500 hours per year, showing tremendous potential for solar energy production.[2]

Estimating and predicting solar irradiance on the Earth's surface is important for photovoltaic power generation. Solar irradiance variation has a significant impact on the performance of instantaneous non-linear devices such as solar energy converters Assuncao et al. (2003). Cloud cover is the primary element for solar irradiance reduction, making it a valuable tool for forecasting solar radiation globally. Clouds are quite unpredictable in their behavior. Factors affecting the duration of sunlight obstruction by clouds include cloud type, cloud thickness, and the reflection of solar irradiance between cloud layers. All of this greatly affects the solar irradiation reaching the Earth's surface. Cloud transmission is regarded as the most unpredictable variable affecting surface irradiance (Smith et al., 2017; Wacker et al., 2015; Munkhammar & Widén, 2016). By quantifying the solar irradiance variability, it is possible to effectively manage numerous power systems that are used to inject the solar-generated power into the energy grid Mutavhatsindi et al. (2020). We achieve this by conducting statistical modelling on the main variables of solar energy, namely diffused horizontal irradiance (DHI), direct normal irradiance (DNI) and global horizontal irradiance (GHI), followed by computing the statistical relationship between these variables and the correlation between these specific components from different locations (Tovar-Pescador, 2008):

- the DHI is the radiation energy received on a unit area of a horizontal surface from all directions excluding radiation energy directly from the sun's rays (DNI). This is the radiation that is scattered by the atmosphere from molecules and particles in the air as well as the radiation bounced off surrounding surfaces (Pasari & Nandigama, 2020).
- the DNI is the direct beam of radiation from the sun excluding the diffused irradiance. DNI is the total amount of solar energy received per unit area of a surface that is always held perpendicular to the rays coming directly in a straight line from the sun. The DNI is important for solar installations that track the position of the sun and for concentrating solar technologies such as concentrating solar power plants (Pasari & Nandigama, 2020).
- the GHI is the total amount of short-wave radiation received by a surface horizontal to the ground. The GHI is jointly modelled by DHI and DNI $\times \cos\theta$ where θ is the solar zenith angle, and is defined as the angle between the sun's rays and the vertical direction. The GHI is important to anybody using photovoltaic technologies since they can make use of both the diffused and beam components of solar irradiance to generate electricity (Pasari & Nandigama, 2020).

[1] https://www.iea.org/countries/south-africa.

[2] https://www.energy.gov.za/files/esources/renewables/r_solar.html.

The clear-sky index (normalized surface solar irradiance) is a dimensionless form of solar irradiance (Pavanello et al., 2015) and estimates solar irradiance attenuation caused by clouds (Smith et al., 2017). This is achieved by calculating the ratio of the GHI to the total amount of solar irradiance that would have reached the surface under clear-sky conditions at a specific time. The clear-sky index can be used to analyse solar irradiance variability at a specific location (Lohmann et al., 2017), since the variability of solar irradiance comprises of the position of the sun and the variability of cloudiness (Munkhammar & Widén, 2016).

In 1960 the first study about the statistical distribution of the clear-sky index was conducted by Liu and Jordan (1960) in an attempt to study higher frequency variability in solar irradiance. They accomplished this by tracking the daily average clear-sky index of both the DNI and DHI variables at 98 locations in Canada and the United States. Subsequently, they computed the distribution functions for the monthly-average clear-sky indices (Liu & Jordan, 1960). For a few years it was believed that these functions were independent of location and the month considered (Liu & Jordan, 1960). It was later shown by Saunier et al. (1987) that these functions were not so universal since the generalized distribution curves of Liu and Jordan (1960) were not suitable for tropical conditions, whereas Bendt et al. (1981) probability density function was suitable. It was initially believed by Bendt et al. (1981), Hollands and Huget (1983) and Saunier et al. (1987) that the clear-sky index follow a unimodal probability distribution as it was a reasonably good fit to the cumulative distribution functions (cdf) of Liu and Jordan (1960). However, in 1988 a study by Suehrcke and McCormick (1988) noticed that the distribution curves from short-term data readings (1 and 5-minute intervals) differ significantly from the daily curves from Liu and Jordan (1960). It was concluded that the clear-sky index actually follows a bimodal distribution. This was achieved by using 1-minute intervals of solar irradiance readings in Western Australia and deriving the model from concepts of Boltzmann statistics (Suehrcke & McCormick, 1988). Multiple probability distribution approaches have since been developed, but to this day is there no universal preference for a specific distribution or distribution family to represent the clear-sky index (Munkhammar & Widén, 2016).

Numerous statistical models were created in an attempt to estimate and predict South Africa's solar irradiance. These models range from machine learning models (Mutavhatsindi et al., 2020) to regression models (Mulaudzi et al., 2013), however, a convolution model was not yet used to obtain the complete probability distribution model of South Africa's solar irradiance. Hence in this chapter the convolution model is used to obtain the complete probability distribution model for solar irradiance.

3.1.2 Data Description

In this chapter, we used data from the Southern Africa Universities Radiometric Network (SAURAN) database. The SAURAN database includes the minute-average

Table 3.1 Descriptive statistics of minute-average data for GHI(W/m^2)

	Min	Q1	Median	Mean	Q3	Max	Skewness	Kurtosis
UP	0.1174	251.4175	487.9658	509.3179	724.3266	1553.2980	0.3267	2.2181
SUN	0.7761	277.3994	528.6825	535.2764	780.1922	2367.2900	0.1325	2.0559

Table 3.2 Descriptive statistics of hour-average data for GHI(W/m^2)

	Min	Q1	Median	Mean	Q3	Max	Skewness	Kurtosis
UP	0	0	8.354	218.018	417.507	1145.991	1.179	3.0819
SU	0	0	8.704	218.15	402.706	1232.387	1.24	3.263

solar irradiance data (which include the GHI, DHI and DNI values) of two locations in South Africa. The two radiometric stations are situated at the University of Pretoria (UP) (latitude: −25.75308037, longitude: 28.22859001) in Gauteng and at Stellenbosh University (SUN) (latitude: −33.92810059, longitude:18.86540031) in the Western Cape. Kipp & Zonen CMP11 pyranometers were used to capture the data for the GHI and DHI variables, whereas SOLYS-2 trackers and CHP1 pyrheliometers were used to capture the data of the DNI variable. It is worth noting that the pyranometers at SUN and at UP were last calibrated 2016/09/06 and 2018/08/11 respectively. The instruments were regularly scheduled for maintenance and the windows/glass domes were cleaned on a weekly basis.[3]

Data from 1 January 2021 until 31 December 2021 were utilised. Only the solar irradiance during the daylight times in the presence of sunlight was considered. To mitigate the removal of observation the following margins for both of locations were chosen:

- For University of Pretoria's data a margin of between 7:00-17:00 was chosen since the latest the sun came up was on 2 July 2021 at 6:54 and the earliest it went down was on 9 June 2021 at 17:23.[4]
- For Stellenbosh University's data a margin of between 8:00-17:00 was chosen since the latest the sun came up was on 30 June 2021 at 7:50 and the earliest it went down was on 12 June 2021 at 17:42.[5]

Tables 3.1 and 3.2 show the descriptive statistics of minute-average and hour-average data for GHI(W/m^2) for two stations. From Table 3.1 is it observed that the minute-average has a positive skewness indicating that it is skewed to the right and that it has a heavy right tail. The minute-average data has a kurtosis less than 3 indicating that it is platykurtic. This means that it has a lower peak and lighter tails

[3] https://sauran.ac.za/,2022/03/14.

[4] https://weatherspark.com/h/y/95267/2021/Historical-Weather-during-2021-in-Pretoria-South-Africa#Figures-Daylight,2022/07/18.

[5] https://weatherspark.com/h/y/82954/2021/Historical-Weather-during-2021-in-Stellenbosch-South-Africa#Figures-Daylight,2022/07/18.

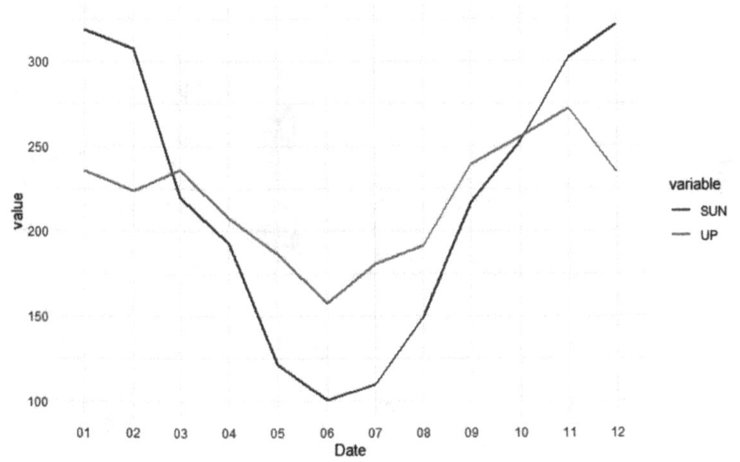

Fig. 3.1 Average GHI of University of Pretoria (Orange) and Stellenbosch University (Purple) for 2021

compared to a normal distribution. To ensure that we have independent observations in the datasets we filtered the dataset to only include every fifth observation.

Figure 3.1 shows that Stellenbosch's average GHI is higher than Pretoria's for the months of October through March. This is due to the fact that Pretoria's rainfall season is from September until May which directly results in it being the cloudier part of the year. In 2021, the rainy period of the year lasted 8.1 Months, from September 8 until May 11.[6] Whereas Stellenbosch's rainfall season is from March until November. In 2021, Stellenbosch's rainfall season lasted 7.9 months and started around 31 March and lasted until 25 November.[7] In Fig. 3.2 we see that for the first 5 days of January that the plot for the University of Pretoria's GHI jumps up and down frequently, whereas Stellenbosch's plot is more consistent, this indicates that there is a significant change in the amount of irradiance reaching the University of Pretoria's station, this is most likely caused by clouds in the sky. Additionally, the exact opposite is observed during the first 5 days of July, with Stellenbosch's plot showing more ups and downs and University of Pretoria's plot becoming more stable.

To calculate the clear-sky condition variable $G_c(t)$, the famous Ineichen-Perez model (Zhandire, 2017) is used. For this calculation consider a solar zenith angle of 0° as midday is theoretically the time of day that the maximum solar irradiance is observed (Zhandire, 2017). For the radiometric station at the University of Pretoria,

[6] https://weatherspark.com/y/95267/Average-Weather-in-Pretoria-South-Africa-Year-Round#Sections-Rain,2022/06/17.

[7] https://weatherspark.com/y/82954/Average-Weather-in-Stellenbosch-South-Africa-Year-Round,2022/06/17.

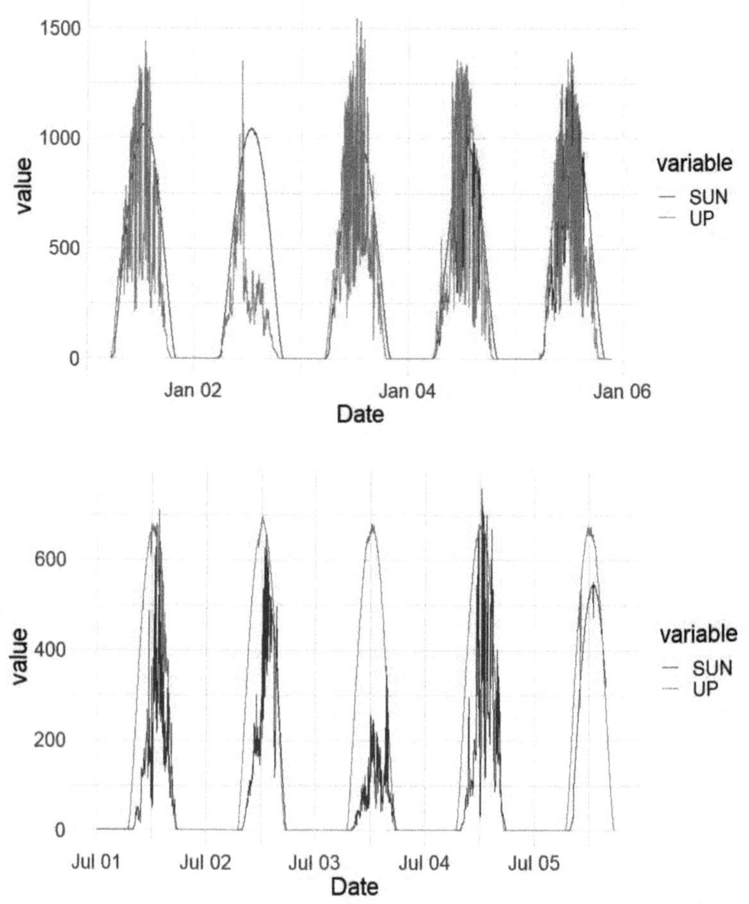

Fig. 3.2 GHI of University of Pretoria (Orange) and Stellenbosch University (Blue) for the first 5 days of January and July of 2021

Table 3.3 Descriptive statistics of clear-sky conditions for UP

Min	Q1	Median	Mean	Q3	Max	Skewness	Kurtosis
814.1	1154.2	1186.3	1157.7	1196.2	1215.4	−2.4847	9.099

which has an altitude of 1410m, the descriptive statistics for the $G_c(t)$ are shown in Table 3.3.

For the radiometric station at Stellenbosch University, which has an altitude of 119m, the descriptive statistics for the $G_c(t)$ are reported in Table 3.4.

The time-specific $G_c(t)$ values for each location were used to determine the clear-sky indices and $k(t)$ variable at a given time.

Table 3.4 Descriptive statistics of clear-sky conditions for SUN

Min	Q1	Median	Mean	Q3	Max	Skewness	Kurtosis
661.3	1047.0	1094.5	1063.7	1112.3	1139.4	−2.2877	8.5167

The rest of the chapter is as follows: in Sect. 3.2, the marginal distributions and the copula approach are discussed. Section 3.3 is dedicated to data application. Section 3.4 summarises and concludes this chapter.

3.2 Methodology

The instantaneous global horizontal irradiance $G(t)$ is jointly modelled by $G_d(t)$ and $G_b(t)$, where $G(t)$ represents the instantaneous GHI, $G_d(t)$ represents the instantaneous DHI and $G_b(t)$ represents the instantaneous $DNI * \cos(\theta)$. Furthermore, by normalising the GHI variable for clear-sky conditions, the global clear-sky index $k(t)$ can be jointly modelled by $k_d(t)$ and $k_b(t)$ where $k_d(t) = \frac{G_d(t)}{G_c(t)}$ and $k_b(t) = \frac{G_b(t)}{G_c(t)}$. The variable $k_d(t)$ is the diffused irradiance index and $k_b(t)$ is the direct irradiance index. The purpose of normalizing the instantaneous irradiance variables with $G_c(t)$ is to eliminate the deterministic variation and time dependency in both our irradiance variables resulting from the sun's movement in the sky throughout each day (Munkhammar & Widén, 2017).

To implement the convolution model to obtain the complete probability distribution model for solar irradiance, consider the following two steps:

- Step 1: Model both DHI and DNI clear-sky indices separately with suggested probability distributions. Distributions are then fitted to the data of both the clear-sky indices.
- Step 2: The Global clear-sky index is jointly modelled by the DHI and DNI indices. The correlation between the two indices is estimated from the data of clear-sky index variables and is modelled using a copula.

3.2.1 *Marginal Distributions for* $k_d(t)$ *and* $k_b(t)$

It is usually assumed that the distribution of $k_d(t)$ is unimodal and it follows the well-known gamma distribution (Munkhammar & Widén, 2016). The cdf of the gamma distribution is given by the following:

$$F(x;\theta;\kappa) = \frac{1}{\Gamma(\kappa)}\gamma(\kappa, \frac{x}{\theta}) \qquad x > 0, \kappa > 0 \text{ and } \theta > 0, \tag{3.1}$$

where $\Gamma(\kappa) = \int_0^\infty x^{\kappa-1}e^{-x}dx$ is the gamma function and $\gamma(\kappa, x) = \int_0^x t^{\kappa-1}e^{-t}dt$ is the lower incomplete gamma function.

For the $k_b(t)$ variable, it is usually assumed in literature that it is of bimodal nature consisting of two peaks, where one will be close to zero and the other between 0 and 1 (Munkhammar & Widén, 2016). This is due to fact that DNI will only reach the surface of the Earth when the sky is clear and there is no obstruction from clouds in the sky. The lower peak is indicative of clouds covering the sun and the second peak corresponds to direct bright sunlight. Since the $k_b(t)$ variable follows a bimodal distribution, in this chapter various distributions with the ability to capture this bimodal behaviour are considered. Since the probability density function (pdf) of the $k_b(t)$ variable has a compact support $S \in [a, b]$ where $a = 0$ and $b = 1$, the subsequent distributions have been applied:

- A mixture distribution of two mode-parameterised beta distributions, where
  - a general finite mixture distribution is given by Leisch (2004):

$$p(\boldsymbol{x}, \boldsymbol{\pi}, \boldsymbol{\nu}) = \sum_{j=1}^{k} \pi_j f(\boldsymbol{x}, \nu_j) \quad \pi_j \geq 0 \text{ such that} \sum_{j=1}^{k} \pi_j = 1, \tag{3.2}$$

    where k is the number of components, $\boldsymbol{\pi}$ is a vector of mixture weights π_j ($j = 1, 2, ..., k$), the sum of all π_j is equal to 1, ν_j is the component specific parameters of each of the underlying densities of the mixture distribution, and finally $\boldsymbol{\nu}$ is a vector containing all of the parameters of the various k underlying pdfs.
  - the mode-parameterised beta distribution is given by Bagnato and Punzo (2013)

$$f(x, m, \nu) = \frac{(x - a)^{\frac{m-a}{\nu(b-a)}} (b - x)^{\frac{b-m}{\nu(b-a)}}}{(b - a)^{\frac{\nu+1}{\nu}} B\left[\frac{m-a}{\nu(b-a)} + 1, \frac{b-m}{\nu(b-a)} + 1\right]}, \tag{3.3}$$

    where $m = \frac{\alpha-1}{\alpha+\beta-2}$, $\nu = \frac{1}{\alpha+\beta-2}$, $\alpha > 0$ and $\beta > 0$ are shape parameters such that $a \leq x \leq b \in S, m \in S$ and $\nu > 0$, where m is the mode of the distribution and $B(p, q) = \int_0^1 x^{p-1}(1 - x)^{q-1}dx$ is the beta function.
    As stated earlier $a = 0$ and $b = 1$, resulting in $m \in [0, 1]$ giving the following pdf (Bagnato & Punzo, 2013):

$$f(x, m, \nu) = \frac{x^{\frac{m}{\nu}} (1 - x)^{\frac{1-m}{\nu}}}{B\left[\frac{m-a}{\nu} + 1, \frac{1-m}{\nu} + 1\right]}. \tag{3.4}$$

- A unit inverse Gaussian (UIG) distribution (Ghitany et al., 2019) as

$$f(x, \lambda, \mu) = \sqrt{\frac{\lambda}{2\pi} \frac{1}{x^{3/2}}} exp\left[-\frac{\lambda}{2\mu^2 x}(x - \mu)^2\right] \quad x, \lambda, \mu > 0. \tag{3.5}$$

The UIG distribution was put forth in recent research in an effort to develop a unit interval distribution free from difficult estimation problems, such as having a large number of parameters to estimate or having their ranges depending on these factors (Ghitany et al., 2019). The UIG shares similar properties to other unit interval distributions such as the beta distribution. To obtain the parameters of the UIG one needs to transform it is data into the inverse Gaussian distribution. On the other hand, the maximum likelihood estimates (MLEs) for the UIG distribution are available in closed form (Ghitany et al., 2019), allowing us to compute the UIG's parameters directly without the need for iterative processes:

$$\hat{\mu} = \frac{1}{n} \sum_{i=1}^{n} x_{\mathrm{i}},$$

$$\hat{\lambda} = \frac{1}{\frac{1}{n} \sum_{i=1}^{n} (\frac{1}{x_{\mathrm{i}}} - \frac{1}{\bar{x}})}.$$

- A mixture of an exponential distribution with parameter ν and the unit inverse Gaussian distribution as

$$p(x, \mu, \lambda, \nu, \pi) = \pi \nu exp[-\nu x] + (1 - \pi) UIG(x, \mu, \lambda). \tag{3.6}$$

Estimation of the parameters for both of the proposed mixture distributions are performed using the maximum likelihood method and the function 'optim_sa' from the "optimization" package in R software.

3.2.2 *Copula Modelling*

The global clear-sky index, $k(t)$, is jointly modeled by the DHI and DNI indices using a copula approach. A copula is a function that expresses a multivariate cdf or a joint distribution function in terms of individual marginal cdfs (Nelsen, 2007). Copulas can be used to express the interdependence between variables explicitly and to model the correlation between these variables.

For a copula to capture and portray the correct properties of a joint probability function should they meet the following three requirements: (1) the copula function always returns a valid probability; (2) it is an increasing function of its marginal distribution inputs and (3) if all the marginal distribution inputs are equal to 1 except for one, then the copula is equal to the value of the marginal distribution which was not equal to 1 (Nelsen, 2007).

In this chapter, a bivariate copula is used to model the dependencies between two clear-sky variables $k_{\mathrm{d}}(t)$ and $k_{\mathrm{b}}(t)$ by generating a set of random values from each of two marginal distributions (Munkhammar & Widén, 2016). The bivariate copula is defined as:

$$F_{X_1,X_2}(x_1, x_2) = C(F_{X_1}(x_1), F_{X_2}(x_2)), \tag{3.7}$$

where $F_{X_i}(x_i)$ represents the cdf of X_i and $F_{X_i,X_j}(x_i, x_j)$ represents the cumulative joint distribution between variables X_i and X_j. The inversion method is applied to obtain the copula of the clear-sky variables. With the inversion method, it is assumed that if X_i is a random variable with a cdf F_{X_i} and U is a random uniformly distributed variable between 0 and 1, then $F_{X_i}^{-1}(U)$ have the same distribution as X_i. Using the inversion method, a random sample, x_i, from the distribution of X_i can easily be generated using a random value, u, from a Uniform distribution. In the bivariate case, Sklar's theorem states that if there are two continuous marginal distribution functions $F_{X_1}(x_1)$ and $F_{X_2}(x_2)$ and the joint distribution $F(X_1, X_2)$ exists, then a unique copula exists (Nelsen, 2007). By using the inversion method, the unique copula function for the clear-sky variable can be obtained and is given by:

$$C(X_1, X_2) = F_{X_1,X_2}(F_{X_1}^{-1}(U_1), F_{X_2}^{-1}(U_2)). \tag{3.8}$$

The copula enables us to determine the joint distribution of the two clear-sky variables as well as explain the interdependencies between them. We apply three different types of copula functions:

- The Clayton copula (Joe, 2014) as:

$$C(X, Y) = (X^{-\alpha} + Y^{-\alpha} - 1)^{\frac{-1}{\alpha}}, \tag{3.9}$$

  where α is the dependency parameter and is used to adjust the strength of the correlation between the two clear-sky variables. The Clayton copula is an explicit continuous bivariate copula meaning that it has a simple closed-form expression. The Clayton copula is used to explain a model where there exists a lower tail dependence and no upper tail dependence between two marginal distributions.
- The Gaussian copula (Joe, 2014) as:

$$C(X, Y) = \Phi_\rho[\Phi^{-1}(X), \Phi^{-1}(Y)], \tag{3.10}$$

  where Φ^{-1} is the inverse of the standard normal cdf, and Φ_ρ is the cdf of a bivariate standard normal distribution with a correlation parameter $\rho \in [-1, 1]$.
- The Student's t copula (Joe, 2014) as:

$$C(X, Y) = t_{\gamma,\rho}[t_\gamma^{-1}(X), t_\gamma^{-1}(Y)] \tag{3.11}$$

  where t_γ^{-1} is the inverse of the Student's t distribution function with γ degrees of freedom, and $t_{\gamma,\rho}$ is the cdf of a bivariate Student's t distribution with a correlation parameter $\rho \in [-1, 1]$.

The Gaussian copula and Student's t copula are both reflection symmetric implicit copulas and there does not exist a simple closed-form expression for either of them (Joe, 2014). The main differences between these two copula models are that with the Gaussian copula the marginal distributions are transformed normal distributions, whereas with the Student's t copula the marginal distributions are transformed to Student's t distributions. Additionally, the Gaussian copula is used to model the correlation of two marginal distributions which lack tail dependency, whereas the Student's t copula is used for marginal distributions where tail dependency exists (Joe, 2014). Both the Gaussian and Student's t copulas are based on the Pearson correlation coefficient:

$$\rho_{\mathrm{XY}} = \frac{COV(X,Y)}{\sqrt{var(X)var(Y)}}, \tag{3.12}$$

where $COV(X, Y)$ is the covariance between the X and Y variable, and $var(X)$ and $var(Y)$ is the variance of the X and Y variable, respectively.

To determine which copula best represents the correlation between the two clear-sky variables, the Akaike's Information Criterion (AIC) and the Bayesian Information Criterion (BIC) are used.

3.3 Application

In this section, the minute-average data from the two SAURAN radiometric stations situated at the University of Pretoria (UP) in Gauteng South Africa and at Stellenbosch University (SUN) in the Western Cape in South Africa are modeled.

3.3.1 *Results of the Distribution Fitting of the Clear-Sky Indices*

As stated in Sect. 3.2 a gamma distribution was assumed for $k_{\mathrm{d}}(t)$ at both of the locations. Table 3.5 shows the obtained MLEs of the gamma distribution for $k_{\mathrm{d}}(t)$ and Figs. 3.3 and 3.4 illustrate the histogram of $k_{\mathrm{d}}(t)$ and the fitted gamma density plots for UP and SUN, respectively.

The data for the $k_{\mathrm{b}}(t)$ variable has a bimodal pattern as previously mentioned. For this reason, three different distributions to model the $k_{\mathrm{b}}(t)$ variable were considered, namely (1) the UIG distribution (3.5), (2) a mixture of two re-parameterised beta

Table 3.5 MLEs of the parameters of the gamma distribution for clear-sky DHI variables

Station	Distribution	θ	κ
UP	Gamma	1.66302	0.09019
SUN	Gamma	1.64	0.0736

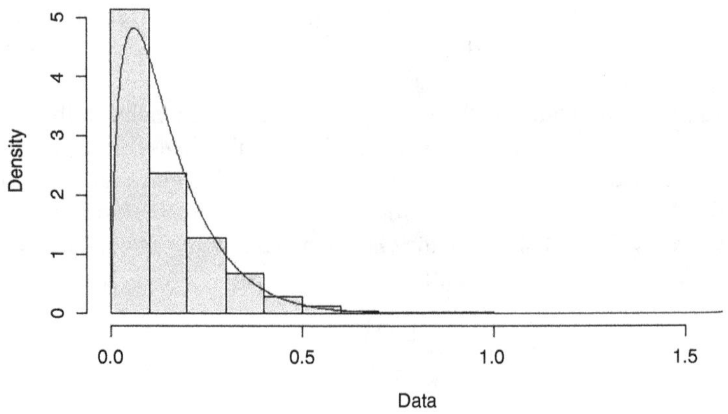

Fig. 3.3 Gamma density plots of UP's clear-sky DHI variable

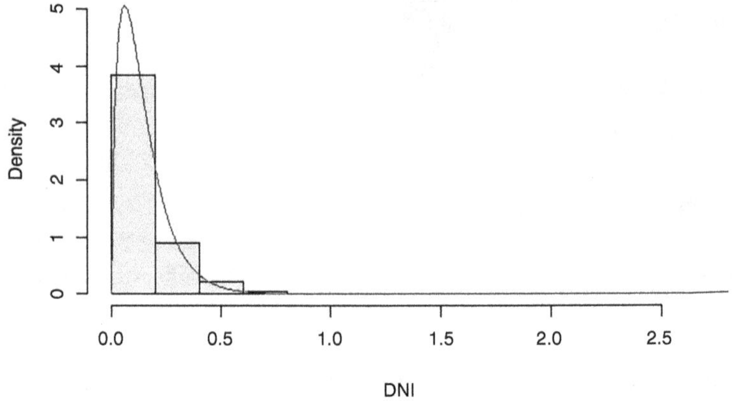

Fig. 3.4 Gamma density plots of SUN's clear-sky DHI variable

distributions (3.4) and (3) a mixture of an exponential distribution and the UIG distribution (3.6).

From Table 3.6 it can be concluded that the best distribution to model UP's $k_b(t)$ variable is the UIG distribution with parameters $\hat{\mu} = 3.07$ and $\hat{\lambda} = 0.526$ since it had the lowest AIC and BIC score. It can also be concluded that the mixture of two beta distributions was the worst-performing distribution having the highest AIC and BIC scores. Similar results were obtained for SUN's $k_b(t)$ variable, as the UIG with parameters $\hat{\mu}$=1.64 and $\hat{\lambda}$=0.323 was the best fit model. However, the mixture of an exponential distribution and the UIG with parameters $\hat{\mu}$=99.8, $\hat{\lambda}$=0.249, $\hat{\upsilon}$=100 and $\hat{\pi}$=0.1 could also have been used as a reasonable fit for SUN's $k_b(t)$ variable as its goodness-of-fit scores were quite close to that of the UIG. This is further emphasised

Table 3.6 Estimated parameters of various distributions for the clear-sky DNI variables and goodness-of-fit measure values

Station	Distibution	Parameters								Goodness-of-fit	
		μ	λ	ν	π	μ_1	ν_1	μ_2	ν_2	AIC	BIC
UP	UIG	6.73	0.287	–	–	–	–	–	–	−40,911	−40,894
	Mixture of exponential and UIG	89.5	0.382	36.2	0.0168	–	–	–	–	−36,617	−36,583
	Mixture of re-parameterised beta distributions	–	–	–	0.0392	0.00114	48.3	0.0138	6.64	−1074	−1031
SUN	UIG	1.64	0.323	–	–	–	–	–	–	−44,414	−44,397
	Mixture of exponential and UIG	99.8	0.249	100	0.1	–	–	–	–	−40,880	−40,846
	Mixture of re-parameterised beta distributions	–	–	–	0.995	0.00998	99.8	0.264	99.7	117	160

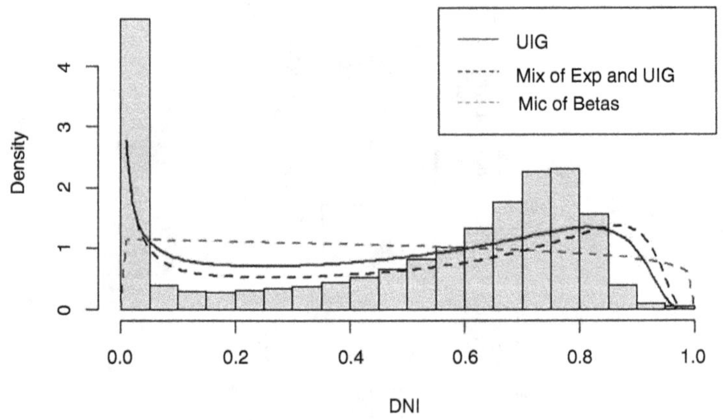

Fig. 3.5 Density plots of various distributions modelling UP's clear-sky DNI

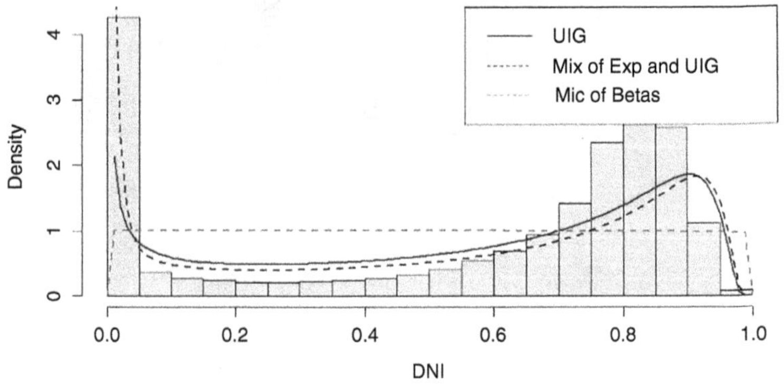

Fig. 3.6 Density plots of various distributions modelling SUN's clear-sky DNI

in Figs. 3.5 and 3.6 which plots a histogram of UP's and SUN's $k_{\text{d}}(t)$ data along with the three fitted distributions. From this it can be seen that both the UIG distribution and the mixture of the exponential distribution and UIG distribution are reasonable fits, whereas the mixture of the two re-parameterised beta distributions is not well suited.

3.3.2 Results for Joint Modelling

As discussed in Sect. 3.2 three different copula functions were considered, namely the bivariate Gaussian copula, bivariate Student's t copula and Clayton copula to model the joint distribution between the $k_{\text{d}}(t)$ and $k_{\text{b}}(t)$ variables for both locations.

Table 3.7 Goodness-of-fit measures of the joint model with different copula functions

Station	Copula	AIC	BIC
UP	Gaussian	−8698.26	−8689.67
	Student's t	−8846.38	−8829.21
	Clayton	−237.702	−229.114
SUN	Gaussian	−12,780.648	−12,772.163
	Student's t	−13,165.710	−13,148.740
	Clayton	−1820.252	−1811.766

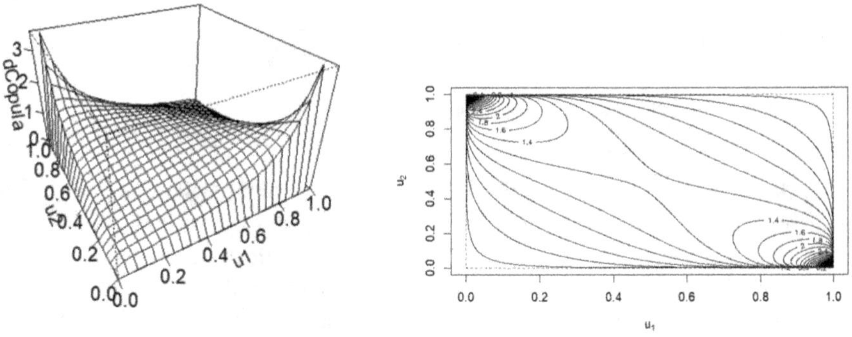

Fig. 3.7 Joint model of clear-sky DNI and clear-sky DHI of UP

The following marginal distributions were considered for both UP's and SUN's clear-sky indices: a gamma distribution for the $k_{\mathrm{d}}(t)$ variable and a UIG distribution for the $k_{\mathrm{b}}(t)$ variable. The correlation coefficients of the Gaussian copula and Student's t copula evaluated to $\rho = -0.485$ for UP's dataset and $\rho = -0.649$ for SUN's dataset.

The Clayton copula's correlation coefficient was equal to $\alpha = -0.0144$ for UP's dataset and $\alpha = -0.0875$ for SUN's dataset. The performance measures for the joint distribution when considering the three different copula functions is given in Table 3.7.

From Table 3.7 it is observed that for both UP's and SUN's clear-sky variables joint distribution function the bivariate Student's t copula was the most suitable as it has the lowest AIC and BIC scores, with a score of $AIC = -8846.38$, $BIC = -8829.21$ and $AIC = -13,165.710$, $BIC = -13,148.740$ respectively. This was closely followed by the bivariate Gaussian copula and the Clayton copula having significantly higher AIC and BIC scores. Figures 3.7 and 3.8 display the Student's t copula depicting the joint distribution of $k_{\mathrm{d}}(t)$ and $k_{\mathrm{b}}(t)$ variables of UP and SUN. Now since the Global clear-sky index $k(t)$ is jointly modelled of $k_{\mathrm{d}}(t)$ and $k_{\mathrm{b}}(t)$ variables we can conclude that the joint distribution function of $k_{\mathrm{d}}(t)$ and $k_{\mathrm{b}}(t)$ variables is the distribution of our $k(t)$ variable. Therefore, we can utilize the convolution method with the copula functions depicted in Figs. 3.7 and 3.8 for analysing $k(t)$ variable, which in turn can be used to model the GHI variable for both locations.

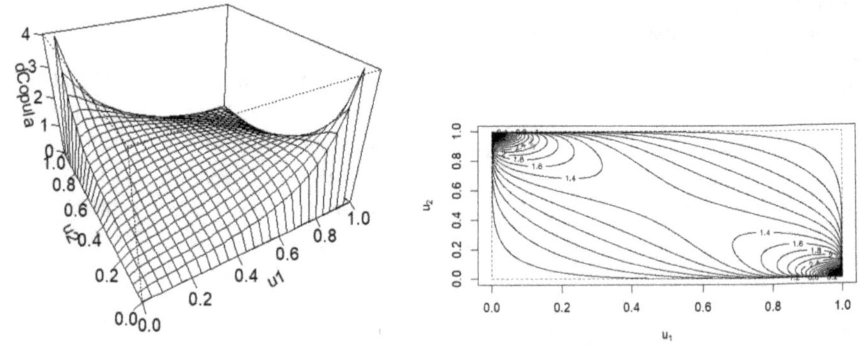

Fig. 3.8 Joint model of clear-sky DNI and clear-sky DHI of SUN

Furthermore, since $\rho = -0.485$ for UP and $\rho = -0.649$ for SUN we concluded that our $k_d(t)$ and $k_b(t)$ are negatively correlated for both locations.

In Figs. 3.7 and 3.8, the fitted t-copula function illustrates the dependency between the DHI and DNI variables. From Figs. 3.7 and 3.8, the negative correlation between the variables is noted. These results concur with practise in that if there is an increase in cloud cover then we will see an increase in the total amount of DHI as there will be an increase of radiation refracting of the clouds to a specific surface area. Furthermore, we observe a decrease of DNI since the clouds will obscure the sun and decrease the amount of radiation received in a straight line from the sun to the specific surface area. This is further emphasized by the important property of a Student's t copula, which is that this specific copula is used for marginal distributions where tail dependencies do exist. This indicates to us that the extreme values of the $k_d(t)$ and $k_b(t)$ variables are dependent on each other. Hence if $k_b(t)$ increases then the $k_d(t)$ decreases and vice versa.

3.4 Conclusion

Multiple efforts have been undertaken to model the Global Horizontal Irradiance required for photovoltaic solar energy generation in South Africa. Cloud cover is the primary component that reduces solar irradiance by restricting the quantity of radiation that reaches the Earth's surface, making it the main cause of fluctuations in solar irradiance. This chapter specifically concentrated on solar irradiance data generated by the University of Pretoria and Stellenbosch University.

We analyzed and modelled various variables of the global clear-sky index $k(t)$, specifically the direct irradiance index $k_b(t)$ and the diffused irradiance index $k_d(t)$. The variable $k_d(t)$ is distributed according to a gamma distribution, whereas the $k_b(t)$ variables follow a UIG distribution at both locations. These distributions were determined based on the values of AIC and BIC. We utilized the previously described distributions as marginal distributions to represent the combined distribution of the $k_b(t)$ and $k_d(t)$ variables by employing a copula. We utilized three distinct copulas for this phase of the investigation: the bivariate Gaussian copula, bivariate Student's t copula, and the Clayton copula. We determined that the Student's t copula was the most appropriate choice for modeling the joint distribution of the clear-sky indices, based on the AIC and BIC values. Additionally, while employing the Student's t copula to represent the joint model of the clear-sky index variables, we determined that the $k_b(t)$ and $k_d(t)$ variables exhibited a general negative correlation.

Ultimately, we determined that the clear-sky index variables exhibit a higher level of dependence at the extreme points of the joint distribution. If the $k_b(t)$ variable increases, then the $k_d(t)$ variable will decrease, and vice versa. Estimating the variable $k(t)$ can positively impact the estimation and optimization of solar power generation and aid in protecting the systems utilized for solar power generation. Understanding the relationship between the two clear-sky index variables is crucial for planning and maintaining solar energy installations.

Acknowledgments This work was based upon research supported in part by the RDP grant at University of Pretoria, the National Research Foundation (NRF) of South Africa, Ref.: RA210106581084, grant No. 150170; ref. SRUG2204203865 and DSI-NRF Centre of Excellence in Mathematical and Statistical Sciences (CoE-MaSS), South Africa. The opinions expressed and conclusions arrived at are those of the authors and are not necessarily to be attributed to the CoE-MaSS or the NRF.

References

Assuncao, H., Escobedo, J., & Oliveira, A. (2003). Modelling frequency distributions of 5 minute-averaged solar radiation indexes using beta probability functions. *Theoretical and Applied Climatology, 75*(3), 213–224.

Bagnato, L., & Punzo, A. (2013). Finite mixtures of unimodal beta and gamma densities and the k-bumps algorithm. *Computational Statistics, 28*(4), 1571–1597.

Bendt, P., Collares-Pereira, M., & Rabl, A. (1981). The frequency distribution of daily insolation values. *Solar Energy, 27*(1), 1–5.

Ghitany, M., Mazucheli, J., Menezes, A., & Alqallaf, F. (2019). The unit-inverse gaussian distribution: A new alternative to two-parameter distributions on the unit interval. *Communications in Statistics-Theory and Methods, 48*(14), 3423–3438.

Hollands, K. G. T., & Huget, R. (1983). A probability density function for the clearness index, with applications. *Solar Energy, 30*(3), 195–209.

Joe, H. (2014). *Dependence modeling with copulas*. CRC Press.

Leisch, F. (2004). Flexmix: A general framework for finite mixture models and latent glass regression in r. *Journal of Statistical Software, 11*(8), 1–18.

Liu, B. Y. H., & Jordan, R. C. (1960). The interrelationship and characteristic distribution of direct, diffuse and total solar radiation. *Solar Energy*, 4(3), 1–19.

Lohmann, G. M., Hammer, A., Monahan, A. H., Schmidt, T., & Heinemann, D. (2017). Simulating clear-sky index increment correlations under mixed sky conditions using a fractal cloud model. *Solar Energy, 150*, 255–264.

Mulaudzi, S. T., Sankaran, V., & Lysko, M. D. (2013). Solar radiation analysis and regression coefficients for the vhembe region, limpopo province, south africa. *Journal of Energy in Southern Africa, 24*(3), 02–07.

Munkhammar, J., & Widén, J. (2016). Correlation modeling of instantaneous solar irradiance with applications to solar engineering. *Solar Energy 133*, 14–23.

Munkhammar, J., & Widén, J. (2017). An autocorrelation-based copula model for producing realistic clear-sky index and photovoltaic power generation time-series. In *2017 IEEE 44th photovoltaic specialist conference (PVSC)* (pp. 3067–3072). IEEE.

Mutavhatsindi, T., Sigauke, C., & Mbuvha, R. (2020). Forecasting hourly global horizontal solar irradiance in south africa using machine learning models. *IEEE Access, 8*, 198872–198885.

Nelsen, R. B. (2007). *An introduction to copulas*. Springer Science & Business Media.

Pasari, S., & Nandigama, V. (2020). Statistical modeling of solar energy. In *Enhancing future skills and entrepreneurship* (pp. 157–165). Springer.

Pavanello, D., Zaaiman, W., Colli, A., Heiser, J., & Smith, S. (2015). Statistical functions and relevant correlation coefficients of clearness index. *Journal of Atmospheric and Solar-Terrestrial Physics, 130*, 142–150.

Saunier, G., Reddy, T. A., & Kumar, S. (1987). A monthly probability distribution function of daily global irradiation values appropriate for both tropical and temperate locations. *Solar Energy, 38*(3), 169–177.

Smith, C. J., Bright, J. M., & Crook, R. (2017). Cloud cover effect of clear-sky index distributions and differences between human and automatic cloud observations. *Solar Energy, 144*, 10–21.

Suehrcke, H., & McCormick, P. (1988). The frequency distribution of instantaneous insolation values. *Solar Energy, 40*(5), 413–422.

Tovar-Pescador, J. (2008). Modelling the statistical properties of solar radiation and proposal of a technique based on boltzmann statistics. In *Modelling solar radiation at the Earth's surface* (pp. 55–91). Springer.

Wacker, S., Groebner, J., Zysset, C., Diener, L., Tzoumanikas, P., Kazantzidis, A., Vuilleumier, L., Stoeckli, R., Nyeki, S., & Kaempfer, N. (2015). Cloud observations in switzerland using hemispherical sky cameras. *Journal of Geophysical Research: Atmospheres*, 120(2), 695–707.

Zhandire, E. (2017). Predicting clear-sky global horizontal irradiance at eight locations in south africa using four models. *Journal of Energy in Southern Africa, 28*(4), 77–86.

Chapter 4
Two-Sample Intraclass Correlation Coefficient Tests for Matrix-Valued Data

Yuli Liang, Chengcheng Hao, and Deliang Dai

4.1 Introduction

The study of the patterned covariance matrices with Kronecker product of two components, that is, $\boldsymbol{\Sigma} \otimes \boldsymbol{\Psi}$, where $\boldsymbol{\Sigma}(q \times q)$ and $\boldsymbol{\Psi}(p \times p)$, has raised much attention in recent years. Among others, this structure can be particularly useful to model spatial-temporal dependency simultaneously, where $\boldsymbol{\Sigma}$ is connected to temporal dependency and $\boldsymbol{\Psi}$ models the dependency over space (see Srivastava et al., 2009, for example). In the Kronecker product of two matrices there are only $\frac{1}{2}(p(p+1)+q(q+1))$ parameters to be estimated as compared to the general case where we need to estimate $pq(pq+1)/2$ parameters. Hence it is also preferable in many analysis when there is not enough data otherwise to estimate the model parameters (Dutilleul, 2018). From an inferential point of view, the Kronecker structure makes the estimation more complicated since the identification problem should be resolved and some restrictions have to be imposed on the parameter space. Then it results in non-explicit maximum likelihood estimates (MLEs) which depend on the choice of restrictions imposed on the covariance matrix (Srivastava et al., 2008).

One interesting extension is when there can be some patterns imposed on matrices $\boldsymbol{\Sigma}$ and $\boldsymbol{\Psi}$, the compound symmetry (CS) structure. Roy and Leiva (2008) provide an algorithm to compute the MLEs of the parameters when both $\boldsymbol{\Psi}$ and $\boldsymbol{\Sigma}$ have a CS structure or an autoregressive of order one structure. Worsley et al.

Y. Liang (✉) · D. Dai
Department of Economics and Statistics, Linnaeus University, Växjö, Sweden
e-mail: yuli.liang@lnu.se; deliang.dai@lnu.se

C. Hao
Departmet of Data Science, School of Statistics and Information, Shanghai University of International Business and Economics, Shanghai, China
e-mail: chengcheng.hao@suibe.edu.cn

C. A. Coelho, D.-G. Chen (eds.), *Statistical Modeling and Applications*, Emerging Topics in Statistics and Biostatistics, https://doi.org/10.1007/978-3-031-69622-0_4

(1991) considered a trial where glucose measurements were taken from two groups of 10 subjects ($n_1 = n_2 = 10$) which were analyzed by two different experimental techniques. There are 15 different regions of these subjects' brains paired in both hemispheres ($q = 2$). It was of interest to test whether three specific pairs of the regions ($p = 3$) with equal distance for two different groups are equally correlated. This requires the task of the equality test of two correlation parameters and this application will be used as an illustrative example later in this article.

In this paper, we also consider another case where one Kronecker component is of CS structure and the one is of symmetric circular Toeplitz (CT) structure. Circular dependence is common in many applications, especially in signal processing (Olkin & Press, 1969), medical (Louden & Roy, 2010) and educational studies (Steinmetz et al., 2012). A CT matrix means that the correlation between two measurements depends only on the circular distance between them, in other words, on the number of positions between them on a circle.

Testing Kronecker product covariance matrix is often challenging especially regarding the estimating procedure. Among others, this type of structure has a nonidentifiability issue which means that the estimators will not be unique without imposing a restriction on the parameters (Dai et al., 2023). Hao et al. (2016) was one of only a few studies (Lu & Zimmerman, 2005; Srivastava et al., 2008) which have proposed some testing procedures and among others it has been noted that two independent exact F tests exist for testing hypothesis concerning the correlations. Moreover, the methodologies for combining the two F tests have been explored.

Motivated by the above arguments, the aim of this paper is to study two-sample intraclass correlation coefficient tests under a model having a Kronecker product covariance structure with compound symmetry or circular symmetry. We shall use one of the higher order asymptotic procedures that modifies the likelihood ratio test so as to achieve accurate performance in small samples; see Brazzale and Davison (2008). By using those independent minimal sufficient statistics, it will also be noted that different independent tests exist by using the ratio of F distributions.

The rest of this paper is organized as follows. In Sect. 4.2 the model and the hypothesis are presented. Several test procedures are investigated in Sect. 4.3 and it will also be noted that different independent exact tests exist for testing the equality of the two correlations. The performance of all the tests will be evaluated in Sect. 4.4 based on simulated type I error probabilities and power. Finally, the example mentioned earlier will be used to illustrate presented methods in Sect. 4.5. The article is summarized in the Discussion section.

4.2 Model and Hypothesis

Let $\mathbf{X}_{k1}, \mathbf{X}_{k2}, \ldots, \mathbf{X}_{kn_k}$ be a random sample from the distribution of $\mathbf{X}_k$, so that each $\mathbf{X}_{ki}$ is a $p \times q$ matrix. We assume a matrix normal model having the following structure:

$$\mathbf{X}_{ki} \sim N_{p,q}(\boldsymbol{\mu}_k, \boldsymbol{\Psi}_k, \boldsymbol{\Sigma}_k), \quad i = 1, \ldots, n_k, \quad k = 1, 2, \tag{4.1}$$

where $\boldsymbol{\mu}_k : p \times q$ is an unstructured mean, $\boldsymbol{\Psi}_k : p \times p$ describes the covariance between the rows of $\mathbf{X}_{ki}$ at any given column, and $\boldsymbol{\Sigma}_k : q \times q$ is the corresponding covariance matrix between the columns at any given row. The matrices $\boldsymbol{\mu}_k$, $\boldsymbol{\Sigma}_k$ and $\boldsymbol{\Psi}_k$ are unknown. Let $\boldsymbol{I}_p$ be the $p \times p$ identity matrix, $\mathbf{1}_p$ be the $p \times 1$ vector with all elements equal to 1 and $\boldsymbol{J}_p = \mathbf{1}_p\mathbf{1}_p'$, $\mathbf{P}_{1_p} = \frac{1}{p}\boldsymbol{J}_p$, $\mathbf{Q}_{1_p} = \boldsymbol{I}_p - \frac{1}{p}\boldsymbol{J}_p$, $\mathbf{X}_{kic} = \mathbf{X}_{ki} - \bar{\mathbf{X}}_k$, where $\bar{\mathbf{X}}_k$ are the means of the $\mathbf{X}_{ki}$, $k = 1, 2$. The CS structured matrix $\boldsymbol{\Psi}_k$ can be expressed as

$$\boldsymbol{\Psi}_k = \psi_k\left[(1 - r_k)\mathbf{I}_p + r_k\mathbf{J}_p\right], \tag{4.2}$$

where ψ_k and r_k are unknown parameters. For the purpose of identifiability, we shall assume that $\psi_k = 1$ in (4.2). Furthermore, the following two specific structures of $\boldsymbol{\Sigma}_k$ are of interest:

(i) $\boldsymbol{\Sigma}_k$ is a CS matrix, i.e., $\boldsymbol{\Sigma}_k = \sigma_k\left[(1 - \rho_k)\mathbf{I}_q + \rho_k\mathbf{J}_q\right]$, where σ_k and ρ_k are unknown parameters.
(ii) $\boldsymbol{\Sigma}_k = (\sigma_{hl})$ is a CT matrix which depends on $\omega = [q/2] + 1$ parameters, and for $h, l = 1, \ldots, q$,

$$\sigma_{hl} = \begin{cases} \tau_{|l-h|}, & \text{if } |l-h| \le \omega - 1, \\ \tau_{q-|l-h|}, & \text{otherwise,} \end{cases}$$

where the τ_j's are unknown parameters for $j = 0, \ldots, \omega - 1$. We shall consider the following hypothesis testing problem concerning the correlation coefficients:

$$H_0 : r_1 = r_2 = r \text{ versus } H_a : -\tfrac{1}{p-1} < r_1 \neq r_2 < 1. \tag{4.3}$$

The null hypothesis in (4.3) means that the correlation between the rows of $\mathbf{X}_{1i}$ is equal to the correlation between the rows of $\mathbf{X}_{2i}$ at any column, and the covariance between the columns of $\mathbf{X}_{1i}$ and $\mathbf{X}_{2i}$ at any row, are characterized by $\boldsymbol{\Sigma}_1$ and $\boldsymbol{\Sigma}_2$, respectively.

4.3 Test Procedures

In this section we shall consider several possible tests for the hypothesis in (4.3). First we give the eigenvalues and eigenvectors of a CS and CT matrix respectively, since they are important for the subsequent inference. Let $\boldsymbol{\Psi}_k^*$ denote the matrix $\boldsymbol{\Psi}_k$ under the restriction $\psi_k = 1$. The distinct eigenvalues of $\boldsymbol{\Sigma}_k$ and $\boldsymbol{\Psi}_k^*$ in (i) are

$$\begin{cases} \lambda_{k1} = \sigma_k(1 + (q-1)\rho_k) \\ \lambda_{k2} = \sigma_k(1 - \rho_k) \end{cases} \text{and} \begin{cases} \lambda_{k3} = 1 + (p-1)r_k \\ \lambda_{k4} = 1 - r_k \end{cases}. \tag{4.4}$$

The distinct eigenvalues of the CT structured $\boldsymbol{\Sigma}_k$ in (ii), i.e., ξ_{kj}, $j = 0, \ldots, [\frac{q}{2}]$, are given in the following lemma. The spectral properties of CT matrices can, for example, be found in Basilevsky (1983), Nahtman and von Rosen (2008), and Liang et al. (2021). It is worth noting that a CS matrix is a special case of a CT matrix so the following lemma can also be applied to a CS matrix.

Lemma 4.1 *Let $\boldsymbol{\Sigma} : q \times q$ be a CT matrix and let $\lambda_h, h = 1, \ldots, q$, be an eigenvalue of $\boldsymbol{\Sigma}$.*

(i) *If q is odd, then*

$$\lambda_h = \tau_0 + 2 \sum_{j=1}^{[q/2]} \tau_j cos(2\pi hj/q).$$

It follows that, $\lambda_h = \lambda_{q-h}$, $h = 1, \ldots, q-1$, and there is only one eigenvalue, λ_q, which has multiplicity 1, all other eigenvalues are of multiplicity 2.
If q is even, then

$$\lambda_h = \tau_0 + 2 \sum_{j=1}^{q/2-1} \tau_j cos(2\pi hj/q) + \tau_{q/2} cos(\pi h).$$

It follows that, for $h \neq q, q/2 : \lambda_h = \lambda_{q-h}$, there are only two eigenvalues, λ_q and $\lambda_{q/2}$, which have multiplicity 1, and all other eigenvalues are of multiplicity 2.

(ii) *The number of distinct eigenvalues for CT matrix is $[\frac{q}{2}] + 1$.*

(iii) *A set of eigenvectors $(\boldsymbol{v}_1, \ldots, \boldsymbol{v}_q)$ corresponding to the eigenvalues $\lambda_1, \ldots, \lambda_q$, is defined by*

$$v_{hi} = \frac{1}{\sqrt{q}} \left(cos(2\pi ih/q) + sin(2\pi ih/q)\right), \ i, h = 1, \ldots, q. \tag{4.5}$$

The eigenvalues and eigenvectors of a CT matrix in Lemma 4.1 will be utilized to obtain the minimum sufficient statistics in the next subsection, which in turn can be used to constitute the likelihood function in canonical form (see Sect. 4.3.2).

4.3.1 Minimal Sufficient Statistics

Let $\boldsymbol{\eta}$ a vector containing all unknown parameters in the covariance matrices $(\boldsymbol{\Psi}_1^*, \boldsymbol{\Sigma}_1)$ and $(\boldsymbol{\Psi}_2^*, \boldsymbol{\Sigma}_2)$, i.e., $\boldsymbol{\eta}_{CS} = (r_1, r_2, \sigma_1, \sigma_2, \rho_1, \rho_2)'$ and $\boldsymbol{\eta}_{CT} = (r_1, r_2, \tau_{10}, \ldots, \tau_{1,\omega-1}, \tau_{20}, \ldots, \tau_{2,\omega-1})'$. Since the hypothesis of interest concerns two components of $\boldsymbol{\eta}$, the mean parameters $\boldsymbol{\mu}_1$ and $\boldsymbol{\mu}_2$ are the nuisance parameters. Therefore we only consider the restricted log-likelihood function of $\boldsymbol{\eta}$, say $\ell_c(\boldsymbol{\eta})$.

With the CS structured $\boldsymbol{\Psi}_k^*$ and $\boldsymbol{\Sigma}_k$ in model (4.1), the minimal sufficient statistics and the corresponding distributions are given in the following theorem by Hao et al. (2016).

Theorem 4.1 *When $\boldsymbol{\Sigma}_k$ is assumed to be a CS matrix in model* (4.1), $k = 1, 2$, *the minimal sufficient statistics of $\boldsymbol{\eta}_{CS}$ are $\boldsymbol{t}_k = (t_{k1}, \ldots, t_{k4})'$, where*

$$
\begin{aligned}
t_{k1} &= \sum_{i=1}^{n_k} tr\left\{\mathbf{P}_{1_q}\mathbf{X}'_{kic}\mathbf{P}_{1_p}\mathbf{X}_{kic}\right\} \sim (-2\theta_{k1})^{-1}\chi^2(n_k - 1),\\
t_{k2} &= \sum_{i=1}^{n_k} tr\left\{\mathbf{P}_{1_q}\mathbf{X}'_{kic}\mathbf{Q}_{1_p}\mathbf{X}_{kic}\right\} \sim (-2\theta_{k2})^{-1}\chi^2((n_k - 1)(p-1)),\\
t_{k3} &= \sum_{i=1}^{n_k} tr\left\{\mathbf{Q}_{1_q}\mathbf{X}'_{kic}\mathbf{P}_{1_p}\mathbf{X}_{kic}\right\} \sim (-2\theta_{k3})^{-1}\chi^2((n_k - 1)(q-1)),\\
t_{k4} &= \sum_{i=1}^{n_k} tr\left\{\mathbf{Q}_{1_q}\mathbf{X}'_{kic}\mathbf{Q}_{1_p}\mathbf{X}_{kic}\right\} \sim (-2\theta_{k4})^{-1}\chi^2((n_k - 1)(p-1)(q-1)),
\end{aligned} \tag{4.6}
$$

and

$$
\begin{aligned}
\boldsymbol{\theta}_k(\boldsymbol{\eta}_{CS}) &= (\theta_{k1}, \ldots, \theta_{k4})'\\
&= -\frac{1}{2}\left((\lambda_{k1}\lambda_{k3})^{-1}, (\lambda_{k1}\lambda_{k4})^{-1}, (\lambda_{k2}\lambda_{k3})^{-1}, (\lambda_{k2}\lambda_{k4})^{-1}\right)'.
\end{aligned} \tag{4.7}
$$

Furthermore, $t_{k1}, \ldots, t_{k4}$ are independent.

Next, we present the minimal sufficient statistics and the corresponding distributions when $\boldsymbol{\Sigma}_k$ is a CT matrix in model (4.1).

Theorem 4.2 *When $\boldsymbol{\Sigma}_k$ is assumed to be a CT matrix in model* (4.1), $k = 1, 2$, *the minimal sufficient statistics of $\boldsymbol{\eta}_{CT}$ are given by*

$$
\begin{aligned}
t_{k1j} &= \sum_{i=1}^{n_k} \left\{\boldsymbol{E}_j\mathbf{X}'_{ic}\mathbf{P}_{1_p}\mathbf{X}_{ic}\right\} \sim \lambda_{k1}\,\xi_{kj}\,\chi^2((n_k - 1)m_j),\\
t_{k2j} &= \sum_{i=1}^{n_k} \left\{\boldsymbol{E}_j\mathbf{X}'_{ic}\mathbf{Q}_{1_p}\mathbf{X}_{ic}\right\} \sim \lambda_{k2}\,\xi_{kj}\,\chi^2((n_k - 1)(p-1)m_j),
\end{aligned}
$$

where $\boldsymbol{E}_j$ is the orthogonal projection onto the corresponding eigenspace of a CT matrix and m_j is the multiplicity of ξ_{kj}, for $j = 0, \ldots, [q/2]$.

Proof Based on the distribution

$$\sum_{i=1}^{n_k} \operatorname{vec}\mathbf{X}_{kic}\operatorname{vec}'\mathbf{X}_{kic} \sim W_{pq}(n_k - 1, \boldsymbol{\Sigma}_k \otimes \boldsymbol{\Psi}_k^*),$$

the restricted log-likelihood function for $\boldsymbol{\mu}_k$ is obtained as follows:

$$\ell_c(\eta_{CT}) = -\frac{n_k - 1}{2}\ln\left|\boldsymbol{\Sigma} \otimes \boldsymbol{\Psi}^*\right| - \frac{1}{2}\mathrm{tr}\left\{\left(\sum_{i=1}^{n_k} \operatorname{vec}\mathbf{X}_{ic}\operatorname{vec}'\mathbf{X}_{ic}\right)(\boldsymbol{\Sigma}^{-1} \otimes \boldsymbol{\Psi}^{*-1})\right\}. \tag{4.8}$$

By using the spectral decomposition $\boldsymbol{\Sigma}_k = \sum_{j=0}^{\omega-1} \xi_{kj}\boldsymbol{E}_j$, and $\boldsymbol{\Psi}_k^* = \lambda_{k3}\mathbf{P}_{1_p} + \lambda_{k4}\mathbf{Q}_{1_p}$, where ξ_{kj}, $j = 0, \ldots, \omega - 1$ are the eigenvalues given by Lemma 4.1 and λ_{k3} and λ_{k4} are given in equation (4.4). Equation (4.8) can be simplified as

$$\begin{aligned}
\ell_c(\eta_{CT}) = &-\frac{n_k - 1}{2}\ln(\xi_{k0}\lambda_{k3}) - \frac{n_k - 1}{2}\sum_{j=1}^{\omega-1} m_j \ln(\xi_{kj}\lambda_{k3}) \\
&- \frac{(n_k - 1)(p - 1)}{2}\ln(\xi_{k0}\lambda_{k4}) - \frac{(n_k - 1)(p - 1)}{2}\sum_{j=1}^{\omega-1} m_j \ln(\xi_{kj}\lambda_{k4}) \\
&- \frac{1}{2}\xi_{k0}^{-1}\lambda_{k3}^{-1}\sum_{i=1}^{n_k}\mathrm{tr}(\boldsymbol{P}_{1_q}\boldsymbol{X}'_{ic}\boldsymbol{P}_{1_p}\boldsymbol{X}_{ic}) \\
&- \frac{1}{2}\lambda_{k3}^{-1}\sum_{j=1}^{\omega-1}\xi_{kj}^{-1}\left\{\sum_{i=1}^{n_k}\mathrm{tr}(\boldsymbol{E}_j\boldsymbol{X}'_{ic}\boldsymbol{P}_{1_p}\boldsymbol{X}_{ic})\right\} \\
&- \frac{1}{2}\xi_{k0}^{-1}\lambda_{k4}^{-1}\sum_{i=1}^{n_k}\mathrm{tr}(\boldsymbol{P}_{1_q}\boldsymbol{X}'_{ic}\boldsymbol{Q}_{1_p}\boldsymbol{X}_{ic}) \\
&- \frac{1}{2}\lambda_{k4}^{-1}\sum_{j=1}^{\omega-1}\xi_{kj}^{-1}\left\{\sum_{i=1}^{n_k}\mathrm{tr}(\boldsymbol{E}_j\boldsymbol{X}'_{ic}\boldsymbol{Q}_{1_p}\boldsymbol{X}_{ic})\right\}.
\end{aligned} \tag{4.9}$$

The log-likelihood (4.9) belongs to a curved exponential family, and the minimal sufficient statistic for the unknown covariance parameters is $\boldsymbol{t} = (t_{k10}, \ldots, t_{k1,\omega-1}, t_{k20}, \ldots, t_{k2,\omega-1})'$, where

$$t_{k1j} = \sum_{i=1}^{n_k}\left\{\boldsymbol{E}_j\mathbf{X}'_{ic}\mathbf{P}_{1_p}\mathbf{X}_{ic}\right\}, \quad t_{k2j} = \sum_{i=1}^{n_k}\left\{\boldsymbol{E}_j\mathbf{X}'_{ic}\mathbf{Q}_{1_p}\mathbf{X}_{ic}\right\}, \tag{4.10}$$

where m_j is the multiplicity of ξ_{kj}, for $j = 0, \ldots, \omega - 1$. It is readily seen that t_{k1j} and t_{k2j} are independent and

$$t_{k1j} \sim \lambda_{k3}\, \xi_{kj}\, \chi^2((n_k - 1)m_j), \quad t_{k2j} \sim \lambda_{k4}\, \xi_{kj}\, \chi^2((n_k - 1)(p - 1)m_j).$$

4.3.2 Likelihood Ratio Test

With the CS structured $\boldsymbol{\Sigma}_k$, the restricted log-likelihood function $\ell_c(\eta_{CS})$ can be written in canonical form as

$$\ell_c(\eta_{CS}) = [(n_1-1)\, c(\boldsymbol{\theta}_1(\eta_{CS})) + \boldsymbol{t}_1' \boldsymbol{\theta}_1(\eta_{CS})] \times [(n_2-1)\, c(\boldsymbol{\theta}_2(\eta_{CS})) + \boldsymbol{t}_2' \boldsymbol{\theta}_2(\eta_{CS})],$$

where $\boldsymbol{\theta}_k(\eta_{CS})$ is given in (4.7) and is the so called canonical parameterization of η_{CS}, and

$$c(\boldsymbol{\theta}_k(\eta_{CS})) = -\frac{1}{2}[\ln(-\theta_{k1}) + (p-1)\ln(-\theta_{k2}) + (q-1)\ln(-\theta_{k3}) + (p-1)(q-1)\ln(-\theta_{k4})].$$

With the CT structured $\boldsymbol{\Sigma}_k$, the restricted log-likelihood function $\ell_c(\eta_{CT})$ is given in (4.9). The REML estimators of ξ_{kj}, λ_{k3} and λ_{k4} are given by

$$\hat{\xi}_{kj} = \frac{1}{p(n_k - 1)m_j}\left(\frac{t_{k1j}}{\lambda_{k3}} + \frac{t_{k2j}}{\lambda_{k4}}\right),$$

$$\hat{\lambda}_{k3} = \frac{1}{q(n_k - 1)} \sum_{j=0}^{\omega-1} \frac{t_{k1j}}{\xi_{kj}},$$

$$\hat{\lambda}_{k4} = \frac{1}{q(p-1)(n_k - 1)} \sum_{j=0}^{\omega-1} \frac{t_{k2j}}{\xi_{kj}},$$

where the statistics t_{k1j}, t_{k2j} and m_j are given in Theorem 4.2.

Now let $\hat{r}_1$ and $\hat{r}_2$ be the restricted maximum likelihood estimator (REML) of r_1 and r_2. Based on the restrictive likelihood function, the classical signed log-likelihood ratio statistic can be written as

$$LD_c(r) = \text{sign}(\hat{r}_1 - \hat{r}_2)\left[2\left\{\sup_{H_0} \ell_c(\eta) - \sup_{H_1} \ell_c(\eta)\right\}\right]^{1/2}, \qquad (4.11)$$

which sup is the supremum function and sign(x) is $+1$ or -1 depending on whether $x > 0$ or $x < 0$, respectively.

It is known that the test statistic given in (4.11) is asymptotically standard normal distributed under regularity conditions, with the tail area approximations having error $O(n^{-1/2})$. One can show the log-likelihoood belongs to a curved exponential family. The MLEs of η_{CS} and η_{CT} do not have closed forms under both H_0 and H_1 and are obtained iteratively.

4.3.3 Higher Order Procedure

We shall now consider one higher order modification of $LD_c(r)$ in order to have better type I error performance in small samples. The modification that we shall consider is due to DiCiccio et al. (2001). The modified statistic, say $LD_{c_{mv}}(r)$, is given by

$$LD_{c_{mv}}(r) = \frac{LD_c(r) - m(\hat{\eta}_{c0})}{\sqrt{1 + v(\hat{\eta}_{c0})}}, \tag{4.12}$$

where $\hat{\eta}_{c0}$ is the estimator of η under H_0, $m(\hat{\eta}_{c0})$ and $1 + v(\hat{\eta}_{c0})$ are the mean and variance of $LD_c(r)$, respectively, which can be estimated from Monte Carlo simulation of t_k at the parameter value $\eta = \hat{\eta}_{c0}$. Subject to regularity conditions, $LD_{c_{mv}}(r)$ has an asymptotic standard normal distribution, and tail area approximation based on $LD_c(r)$ has relative error $O(n^{-3/2})$.

4.3.4 Tests Based on Ratio of Independent F Distributions

We note that the eigenvalues λ_{k1} and λ_{k2} in (4.4) are free of r_k, and

$$\lambda_{k4}/\lambda_{k3} = (1 - r_k)/[1 + (p - 1)r_k].$$

Then we have the following distributions:

$$c_k\, t_{k1}/t_{k2} \sim F_{n_k-1,\ (n_k-1)(p-1)},$$
$$c_k\, t_{k3}/t_{k4} \sim F_{(n_k-1)(q-1),\ (n_k-1)(p-1)(q-1)},$$

where $c_k = (p-1)\frac{1-r_k}{1+(p-1)r_k}$. Using this observation, two exact tests can be constructed for testing the null hypothesis in (4.3), based on the distributions in (4.6). Construct the following independent test statistics:

$$F_1^{CS}(r) = c_1\left(\frac{t_{11}}{t_{12}}\right)\Big/ c_2\left(\frac{t_{21}}{t_{22}}\right),$$
$$F_2^{CS}(r) = c_1\left(\frac{t_{13}}{t_{14}}\right)\Big/ c_2\left(\frac{t_{23}}{t_{24}}\right). \tag{4.13}$$

Under the null hypothesis $H_0 : r_1 = r_2 = r$, we have the distributions $F_1^{CS}(r) \sim \frac{F_{n_1-1,(n_1-1)(p-1)}}{F_{n_2-1,(n_2-1)(p-1)}}$ and $F_2^{CS}(r) \sim \frac{F_{(n_1-1)(q-1),(n_1-1)(p-1)(q-1)}}{F_{(n_2-1)(q-1),(n_2-1)(p-1)(q-1)}}$. Under the alternative hypothesis $H_a : -\frac{1}{p-1} < r_1 \neq r_2 < 1$, we have $F_1^{CS}(r) \sim \frac{c_1}{c_2}\frac{F_{n_1-1,(n_1-1)(p-1)}}{F_{n_2-1,(n_2-1)(p-1)}}$ and $F_2^{CS}(r) \sim \frac{c_1}{c_2}\frac{F_{(n_1-1)(q-1),(n_1-1)(p-1)(q-1)}}{F_{(n_2-1)(q-1),(n_2-1)(p-1)(q-1)}}$.

When $\boldsymbol{\Sigma}_k$ is a CT matrix, the eigenvalues ξ_{kj} are functions of r_k given in (4.4). Thus,

$$t_{k1j}/t_{k2j} \sim (\lambda_{k3}/\lambda_{k4})F_{(n_k-1)m_j,(n_k-1)(p-1)m_j}.$$

Thus $\omega = [q/2] + 1$ exact tests can be constructed for testing hypothesis in (4.3):

$$F_j^{CT}(r) = c_1\left(\frac{t_{11j}}{t_{12j}}\right)\Big/ c_2\left(\frac{t_{21j}}{t_{22j}}\right), \tag{4.14}$$

where $c_k = (p-1)\frac{1-r_k}{1+(p-1)r_k}$, t_{k1j} and t_{k2j} are given in Theorem 4.2, $k = 1, 2$, $j = 0, \ldots, \omega - 1$. In the next example the proposed tests in (4.14) will be illustrated when $q = 4$.

Example 4.1 When $\boldsymbol{\Sigma}_k$ has a 4×4 CT structure, we have the following distributions:

$$c_k\, t_{k10}/t_{k20} \sim F_{n_k-1,\,(n_k-1)(p-1)},$$
$$c_k\, t_{k11}/t_{k21} \sim F_{2(n_k-1),\,2(n_k-1)(p-1)},$$
$$c_k\, t_{k12}/t_{k22} \sim F_{(n_k-1),\,(n_k-1)(p-1)},$$

then we can construct 3 exact tests for testing the hypothesis in (4.3):

$$F_0^{CT}(r) = c_1\left(\frac{t_{110}}{t_{120}}\right)\Big/ c_2\left(\frac{t_{210}}{t_{220}}\right),$$
$$F_1^{CT}(r) = c_1\left(\frac{t_{111}}{t_{121}}\right)\Big/ c_2\left(\frac{t_{211}}{t_{221}}\right), \tag{4.15}$$
$$F_2^{CT}(r) = c_1\left(\frac{t_{112}}{t_{122}}\right)\Big/ c_2\left(\frac{t_{212}}{t_{222}}\right).$$

Under the null hypothesis $H_0 : r_1 = r_2 = r$, we have the distributions:

$$F_0^{CT}(r) \sim \frac{F_{n_1-1,\,(n_1-1)(p-1)}}{F_{n_2-1,\,(n_2-1)(p-1)}},$$

$$F_1^{CT}(r) \sim \frac{F_{2(n_1-1),\,2(n_1-1)(p-1)}}{F_{2(n_2-1),\,2(n_2-1)(p-1)}},$$

$$F_2^{CT}(r) \sim \frac{F_{(n_1-1),\,(n_1-1)(p-1)}}{F_{(n_2-1),\,(n_2-1)(p-1)}}.$$

Under the alternative hypothesis $H_a : -\frac{1}{p-1} < r_1 \neq r_2 < 1$, the distributions are given by $F_0^{CT}(r) \sim \frac{c_1}{c_2}\frac{F_{n_1-1,(n_1-1)(p-1)}}{F_{n_2-1,(n_2-1)(p-1)}}$, $F_1^{CT}(r) \sim \frac{c_1}{c_2}\frac{F_{2(n_1-1),2(n_1-1)(p-1)}}{F_{2(n_2-1),2(n_2-1)(p-1)}}$ and $F_2^{CT}(r) \sim \frac{c_1}{c_2}\frac{F_{(n_1-1),(n_1-1)(p-1)}}{F_{(n_2-1),(n_2-1)(p-1)}}$.

The exact distributions of the test statistics in (4.13) are non-standard distributions and the determinations are relatively complex. The exact distributions can be specified by their characteristic functions and therefore can be computed with the use of R package *CharFunToolR* (Gajdoš, 2018) based on numerical inversion of the characteristic functions.

4.4 Simulation Studies

To investigate the performance of the proposed tests in Sect. 4.3, we now evaluate their performance in terms of type I error probabilities and power. The results of testing H_0 are in the context of $r_1 = r_2 = 0.9$ and based on 10,000 simulations and we put $\boldsymbol{\mu}_1 = \boldsymbol{\mu}_2 = 0$, $\sigma_1 = \sigma_2 = 30$ and $\rho_1 = \rho_2 = 0.5$. In addition, both balanced and unbalanced cases are included in the simulations. For a 5% significance level, Table 4.1 gives the estimated type I error probabilities of the tests based on the likelihood ratio test (LRT) given in (4.11), the mean-variance modified statistics (m-v) given in (4.12), and the two independent test statistics $F_1^{CS}(r)$ and $F_2^{CS}(r)$ given in (4.13), respectively.

From Tables 4.1 and 4.2, we draw the following conclusions. It is clear that m-v, $F_1^{CS}(r)$, $F_2^{CS}(r)$ and $F_j^{CT}(r)$ $(j = 0, \ldots, 2)$ are very accurate under all settings and they provide better control of the type I probability, compared to LRT. For the unbalanced cases, $F_1^{CS}(r)$, $F_2^{CS}(r)$ and $F_j^{CT}(r)$ $(j = 1, \ldots, 3)$ are superior to the other two testing procedures in terms of accuracy while both LRT and m-v produce relatively higher type I error probabilities than the nominal significance level.

We now compare the power performance of the different tests and the results are given in Table 4.3. The alternatives used are $(r_1, r_2) = (0.1, 0.2)$ and $(r_1, r_2) = (0.7, 0.8)$. Seen from Tables 4.3 and 4.4, for the balanced cases, LRT, m-v, $F_2^{CS}(r)$

Table 4.1 Type I error probabilities of testing $H_0 : r_1 = r_2 = 0.9$ for a 5% significance level when $\boldsymbol{\Sigma}_k$ are CS matrices

	(p, q, n)			
	(3, 2, 5)	(3, 2, 100)	(24, 12, 5)	(24, 12, 100)
LRT	0.0607	0.0496	0.0487	0.0440
m-v	0.0484	0.0532	0.0520	0.0503
$F_1^{CS}(r)$	0.0463	0.0501	0.0458	0.0500
$F_2^{CS}(r)$	0.0480	0.0501	0.0500	0.0499
	(p, q, n_1, n_2)			
	(3, 2, 5, 100)		(24, 12, 5, 100)	
LRT	0.0547		0.0520	
m-v	0.0505		0.0500	
$F_1^{CS}(r)$	0.0494		0.0500	
$F_2^{CS}(r)$	0.0488		0.0500	

Table 4.2 Type I error probabilities of testing $H_0 : r_1 = r_2 = 0.9$ for a 5% significance level when $\boldsymbol{\Sigma}_k$ are CT matrices

	(p, q, n)			
	(3, 4, 5)	(3, 4, 100)	(24, 5, 5)	(24, 5, 100)
LRT	0.0687	0.0556	0.0564	0.0520
m-v	0.0440	0.0450	0.0470	0.0501
$F_0^{CT}(r)$	0.0491	0.0500	0.0493	0.0500
$F_1^{CT}(r)$	0.0495	0.0500	0.0499	0.0500
$F_2^{CT}(r)$	0.0487	0.0501	0.0498	0.0500
	(p, q, n_1, n_2)			
	(3, 4, 5, 100)		(24, 5, 5, 100)	
LRT	0.0507		0.0468	
m-v	0.0535		0.0582	
$F_0^{CT}(r)$	0.0495		0.0500	
$F_1^{CT}(r)$	0.0500		0.0500	
$F_2^{CT}(r)$	0.0497		0.0500	

and $F_1^{CT}(r)$ are all competitive except when p, q and n are all small. In Table 4.3 all the tests have significantly higher powers when $(r_1, r_2) = (0.7, 0.8)$ than $(r_1, r_2) = (0.1, 0.2)$. For the unbalanced cases, except $F_1^{CS}(r)$ all the tests are competitively powerful when both p and q are relatively large and however the same conclusion cannot be made when both p and q are small. It is also worthwhile to note that $F_1^{CS}(r)$ has lower power than $F_2^{CS}(r)$ when $(p, q, n) = (24, 12, 5)$ and $(p, q, n_1, n_2) = (24, 12, 5, 100)$. Similarly, $F_0^{CT}(r)$ has lower power than $F_1^{C}T(r)$ and $F_2^{C}T(r)$ when $(p, q, n) = (24, 5, 5)$ and $(p, q, n_1, n_2) = (24, 5, 5, 100)$, even though the loss of power is not as drastic as that for $F_1^{CS}(r)$. It is because when $q > 2$, both of the degrees of freedom in the ratio $\frac{c_1}{c_2}\frac{F_{(n_1-1)(q-1),\,(n_1-1)(p-1)(q-1)}}{F_{(n_2-1)(q-1),\,(n_2-1)(p-1)(q-1)}}$

Table 4.3 Powers of the testings $H_0 : r_1 = r_2$ at $\alpha = 0.05$ when $\boldsymbol{\Sigma}_k$ are CS matrices

		(p, q, n)			
(r_1, r_2)		(3, 2, 5)	(3, 2, 100)	(24, 12, 5)	(24, 12, 100)
(0.1, 0.2)	LRT	0.0708	0.3386	0.5756	1.0000
	m-v	0.1007	0.2792	0.6150	1.0000
	$F_1^{CS}(r)$	0.0701	0.2912	0.1344	0.9303
	$F_2^{CS}(r)$	0.0739	0.3067	0.6644	1.0000
(0.7, 0.8)	LRT	0.1376	0.9943	0.7903	1.0000
	m-v	0.1138	0.9840	0.7902	1.0000
	$F_1^{CS}(r)$	0.0890	0.6182	0.1152	0.8253
	$F_2^{CS}(r)$	0.0984	0.6350	0.5218	1.0000
		(p, q, n_1, n_2)			
(r_1, r_2)		(3, 2, 5, 100)		(24, 12, 5, 100)	
(0.1, 0.2)	LRT	0.0731		0.8482	
	m-v	0.1005		0.8791	
	$F_1^{CS}(r)$	0.0947		0.2822	
	$F_2^{CS}(r)$	0.0888		0.8968	
(0.7, 0.8)	LRT	0.2201		0.9618	
	m-v	0.2419		0.9611	
	$F_1^{CS}(r)$	0.1404		0.2229	
	$F_2^{CS}(r)$	0.1297		0.7835	

are $(q - 1)$ times larger than the ones in the ratio $\frac{c_1}{c_2}\frac{F_{n_1-1,\,(n_1-1)(p-1)}}{F_{n_2-1,\,(n_2-1)(p-1)}}$, that's why $F_2^{CS}(r)$ has much higher power than $F_1^{CS}(r)$ when q is large and the sample size is small. Similarly, in Table 4.4, when $q = 4$, both $F_0^{CT}(r)$ and $F_1^{CT}(r)$ have the same distribution under H_1. So they perform equally in terms of power when $(p, q, n) = (3, 4, 5)$ and $(3, 4, 5, 100)$. And $F_2^{CT}(r) \sim \frac{c_1}{c_2}\frac{F_{2(n_1-1),\,2(n_1-1)(p-1)}}{F_{2(n_2-1),\,2(n_2-1)(p-1)}}$ has slightly higher power than the other two due to the larger degrees of freedom of the F-ratio. However, when $q = 5$, $F_1^{CT}(r)$ and $F_2^{CT}(r)$ have the same distribution under H_1, i.e., $\frac{c_1}{c_2}\frac{F_{2(n_1-1),\,2(n_1-1)(p-1)}}{F_{2(n_2-1),\,2(n_2-1)(p-1)}}$ while $F_0^{CT}(r) \sim \frac{c_1}{c_2}\frac{F_{n_1-1,\,(n_1-1)(p-1)}}{F_{n_2-1,\,(n_2-1)(p-1)}}$. So when the sample size is small (or one of the sample size is small for the unbalanced case), $F_0^{CT}(r)$ is more difficult to reject the null hypothesis than $F_1^{CT}(r)$ and $F_2^{CT}(r)$, see the case $(p, q, n) = (24, 4, 5)$ and $(24, 5, 5, 100)$.

The overall conclusions based on the numerical results in Tables 4.1 and 4.2 and Tables 4.3 and 4.4 are that the tests based on m-v, $F_2^{CS}(r)$ and $F_1^{CT}(r)$ can be recommended for practical use when we have balanced data on hand. For the unbalanced data we can choose LRT, m-v, $F_2^{CS}(r)$ and $F_1^{CT}(r)$. Since the quantities $m(\hat{\boldsymbol{\eta}}_{c0})$ and $v(\hat{\boldsymbol{\eta}}_{c0})$ in (4.12) are computed numerically, the test procedure m-v is very computer-intensive even through it provides satisfactory performance in terms of Type I error probabilities and power. For both cases both m-v, $F_2^{CS}(r)$ and $F_1^{CT}(r)$ are accurate in terms of type I error performance, and have reasonable power.

Table 4.4 Powers of the testings $H_0 : r_1 = r_2$ at $\alpha = 0.05$ when $\boldsymbol{\Sigma}_k$ are CT matrices

		(p, q, n)			
(r_1, r_2)		(3, 4, 5)	(3, 4, 100)	(24, 5, 5)	(24, 5, 100)
(0.1, 0.2)	LRT	0.0228	0.3050	0.1479	1.0000
	m-v	0.0800	0.6167	0.3329	1.0000
	$F_0^{CT}(r)$	0.0736	0.3073	0.1326	0.9364
	$F_1^{CT}(r)$	0.0894	0.4747	0.2132	0.9965
	$F_2^{CT}(r)$	0.0746	0.2917	0.2096	0.9985
(0.7, 0.8)	LRT	0.0634	0.8859	0.1220	0.9992
	m-v	0.1323	0.9761	0.2534	1.0000
	$F_0^{CT}(r)$	0.0958	0.6427	0.1130	0.8333
	$F_1^{CT}(r)$	0.1312	0.8753	0.1701	0.9747
	$F_2^{CT}(r)$	0.0996	0.6302	0.1702	0.9792
		(p, q, n_1, n_2)			
(r_1, r_2)		(3, 4, 5, 100)		(24, 5, 5, 100)	
(0.1, 0.2)	LRT	0.0850		0.4371	
	m-v	0.0866		0.4186	
	$F_0^{CT}(r)$	0.0904		0.2641	
	$F_1^{CT}(r)$	0.1153		0.3963	
	$F_2^{CT}(r)$	0.0918		0.3840	
(0.7, 0.8)	LRT	0.1709		0.3143	
	m-v	0.1752		0.2954	
	$F_0^{CT}(r)$	0.1347		0.2106	
	$F_1^{CT}(r)$	0.1994		0.3130	
	$F_2^{CT}(r)$	0.1391		0.3016	

4.5 Applications of the Results: An Example

In this section we use the positron emission tomography data given in Worsley et al. (1991) to illustrate our proposed testing methods in Sect. 4.3. Researchers are often interested in several selected regions of the brain with closed locations, which tend to be functionally related, such as the Anterior temporal (AT), Midtemporal (MT) and Posttemporal (PT) regions. The locations are defined as the centers of the regions in one hemisphere of the brain, measured in centimeters, taken from the brain atlas. We recall from Sect. 4.1 that the sample size here is $n = 10$ for each group. Furthermore, we also have $p = 3$ and $q = 2$. We are interested in testing whether the three specific pairs (AT, MT and PT) of the regions ($p = 3$) with equal distance for two groups (with two different experimental techniques) are equally correlated.

In this article, different regions, as well as different hemispheres are considered exchangeable (equicorrelated) among themselves. Thus, it is reasonable to assume that both vec $\mathbf{X}_{1i}$ and vec $\mathbf{X}_{2i}$, $i, \ldots, 10$, in the samples have the covariance matrices of the form $\boldsymbol{\Sigma}_1 \otimes \boldsymbol{\Psi}_1$ and $\boldsymbol{\Sigma}_2 \otimes \boldsymbol{\Psi}_2$, respectively, where $(\boldsymbol{\Psi}_1, \boldsymbol{\Sigma}_1)$ and $(\boldsymbol{\Psi}_2, \boldsymbol{\Sigma}_2)$

Table 4.5 Test statistics and p-values for the positron emission tomography data ($\alpha = 0.05$). Test statistics and p-value are reported for the hypotheses $r_1 = r_2$ against $r_1 \neq r_2$ given CS structures

	LRT	m-v	$F_1^{CS}(r)$	$F_2^{CS}(r)$
Stat	0.3219	0.4037	1.0453	1.7568
P-val	0.7475	0.6864	0.5267	0.7414

have the structures given in (4.2). Note, when $q = 2$ CT and CS become the same pattern. We shall test the following hypothesis:

$$H_0 : r_1 = r_2 = r \text{ against } H_a : -\frac{1}{p-1} < r_1 \neq r_2 < 1 \text{ given CS structures.}$$

We consider the test statistics LRT, m-v, $F_1^{CS}(r)$ and $F_2^{CS}(r)$. The test statistics and the p-values are given in Table 4.5. Note, the quantiles (and thus also p-value) of $F_1^{CS}(r)$ and $F_2^{CS}(r)$ exact distributions are computed with the use of R package *CharFunToolR*. The results indicate that the null hypothesis $H_0 : r_1 = r_2$ cannot be rejected with $\alpha = 0.05$.

4.6 Discussion

The tests $F_1^{CS}(r)$ and $F_2^{CS}(r)$ as well as $F_j^{CT}(r)$, $j = 0, \ldots, \omega - 1$, proposed in this paper provide a valid alternative to tests for the equality of two correlation parameters in multivariate models with Kronecker product covariance structure with either compound symmetric or circular Toeplitz patterns. In some cases these tests are more accurate and powerful than other presented tests although $F_1^{CS}(r)$ has lower power than $F_2^{CS}(r)$. Concerning the test statistics $F_j^{CT}(r)$, $j = 0, \ldots, \omega - 1$, we may incorporate the information of different statistics and combine them. Another advantage of the tests based on the ratio of independent F distributions is that the explicit null distributions of test statistics exist and therefore it is easy to implement in practice.

References

Basilevsky, A. (1983). *Applied matrix algebra in the statistical sciences*. North-Holland.
Brazzale, A. R., & Davison, A. C. (2008). Accurate parametric inference for small samples. *Statistical Science, 23*, 465–484.
Dai, D., Hao, C., Jin, S., & Liang, Y. (2023). Regularized estimation of kronecker structured covariance matrix using modified cholesky decomposition. *Journal of Statistical Computation and Simulation*. https://doi.org/10.1080/00949655.2023.2291536.
DiCiccio, T. J., Martin, M. A., & Stern, S. E. (2001). Simple and accurate one-sided inference from signed roots of likelihood ratios. *Canadian Journal of Statistics, 29*, 67–76.

Dutilleul, P. (2018). Estimation and testing for separable variance–covariance structures. *Wiley Interdisciplinary Reviews: Computational Statistics*, *10*, e1432.

Gajdoš, A. (2018). Charfuntoolr: The characteristic functions toolbox (R). Available from: https://github.com/gajdosandrej/CharFunToolR.

Hao, C., Liang, Y., & Mathew, T. (2016). Testing variance parameters in models with a kronecker product covariance structure. *Statistics & Probability Letters*, *118*, 182–189.

Liang, Y., Von Rosen, D., & von Rosen, T. (2021). On properties of toeplitz-type covariance matrices in models with nested random effects. *Statistical Papers*, *62*, 2509–2528.

Louden, C., & Roy, A. (2010). Classification rules under autoregressive and general circulant covariance. *Communications in Statistics–Theory and Methods*, *39*, 3294–3315.

Lu, N., & Zimmerman, D. L. (2005). The likelihood ratio test for a separable covariance matrix. *Statistics & Probability Letters*, *73*, 449–457.

Nahtman, T., & von Rosen, D. (2008). Shift permutation invariance in linear random factor models. *Mathematical Methods of Statistics*, *17*, 173–185.

Olkin, I., & Press, S. (1969). Testing and estimation for a circular stationary model. *The Annals of Mathematical Statistics*, *40*, 1358–1373.

Roy, A., & Leiva, R. (2008). Likelihood ratio tests for triply multivariate data with structured correlation on spatial repeated measurements. *Statistics & Probability Letters*, *78*, 1971–1980.

Srivastava, M. S., von Rosen, T., & von Rosen, D. (2008). Models with a Kronecker product covariance structure: Estimation and testing. *Mathematical Methods of Statistics*, *17*, 357–370.

Srivastava, M. S., von Rosen, T., & von Rosen, D. (2009). Estimation and testing in general multivariate linear models with Kronecker product covariance structure. *Sankhyā: The Indian Journal of Statistics, Series A*, 137–163.

Steinmetz, H., Isidor, R., & Baeuerle, N. (2012). Testing the circular structure of human values: A meta-analytical structural equation modelling approach. *Survey Research Methods*, *6*, 61–75.

Worsley, K., Evans, A., Strother, S., & Tyler, J. (1991). A linear spatial correlation model, with applications to positron emission tomography. *Journal of the American statistical association*, *86*, 55–67.

Chapter 5
Evolution on the Generation and Analysis of Single Imputation Synthetic Datasets in Statistical Disclosure Control

Ricardo Moura, Carlos A. Coelho, and Bimal Sinha

5.1 Introduction

In today's world, simply using a smartphone leads to the creation of a vast array of data. This data is automatically saved, and more and more, various organizations, businesses, and institutions are insisting on accessing this data for research and examination purposes. However, if this data is shared without proper control, it could threaten the privacy of the individuals or groups to whom the data pertains.

To adhere to the principle of statistical confidentiality besides the physical safeguarding of data—meaning data that is stored and accessible only to authorized personnel—many national and international bodies commonly implement Statistical Disclosure Control (SDC) methods. The purpose of these methods is to protect the confidentiality of existing data deemed sensitive, minimizing the risk of individual

R. Moura (✉)
Mathematics Department, Lisbon School of Economics & Management (ISEG), Lisbon, Portugal

NOVA Math – Center for Mathematics and Applications, Caparica, Portugal

Portuguese Navy Research Center (CINAV), Alfeite, Portugal
e-mail: rp.moura@fct.unl.pt

C. A. Coelho
Mathematics Department, NOVA FCT- NOVA University of Lisbon, Caparica, Portugal

NOVA Math – Center for Mathematics and Applications, Caparica, Portugal
e-mail: cmac@fct.unl.pt

B. Sinha
Department of Mathematics and Statistics, University of Maryland, Baltimore County, Baltimore, MD, USA

Center for Statistical Research and Methodology, U.S. Census Bureau, Washington, DC, USA
e-mail: sinha@umbc.edu

C. A. Coelho, D.-G. Chen (eds.), *Statistical Modeling and Applications*, Emerging Topics in Statistics and Biostatistics, https://doi.org/10.1007/978-3-031-69622-0_5

identification.[1] This allows for the public release of such data while ensuring privacy protection.

Noise addition, data rounding, selective data suppression, and the creation of synthetic data are a few methods of Statistical Disclosure Control (SDC) employed by organizations like EUROSTAT and the US CENSUS BUREAU before releasing data to the public (Hundepool et al., 2010; Abowd et al., 2020). This text will delve into the synthetic data generation techniques, which in brief, involve replacing the actual data with its synthetic counterparts. This technique is not only relatively new but also has the significant advantage of maintaining the statistical characteristics of the original model assumed for the data, a feature not commonly found in other SDC methods (Drechsler, 2011). Consequently, this approach is being actively promoted for research by government institutions worldwide.

Little (1993) and Rubin (1993), are often credited as being the forerunners in developing this technique, being the first to propose the generation of synthetic data through multiple imputation (Rubin, 1987) as a form of SDC. This process entails substituting original data with synthetic versions, allowing for public disclosure without risking the privacy of the individuals who provided the data, as these synthetic versions lack enough detail to breach individual confidentiality.

It was only a decade later that the feasibility of procedures enabling the analysis of these synthetic data generated by multiple imputation became available. The works of Reiter (2002, 2003), Raghunathan et al. (2003), and Reiter (2005a,b) were instrumental in this development, driven by a Bayesian approach, allowing for the study of the parameter or vector of parameters involved.

However, in specific situations, where there is a significant risk of revealing a respondent's identity, it may become impractical to publish multiple iterations of the original data. Under these circumstances, the strategy shifts to the release of a single synthetic version, relying thus on data produced through single imputation (Abowd et al., 2006; Kinney et al., 2011, 2014; Bowen et al., 2020; Alam et al., 2020).

Evidently, the conventional methods for combining inferences are not suitable when just one synthetic dataset is released, leading to the question of whether valid inferences can still be drawn in these scenarios.

In the following sections, attention will be directed towards an extensive review of the literature concerning the generation and analysis of single imputation synthetic datasets. This exploration will cover a range of methods, highlight the key challenges faced, and discuss the significant advancements. The aim is to offer a thorough insight into how the field currently stands and to shed light on prospective developments in utilizing single imputation synthetic data as a tool for preserving privacy when releasing data.

[1] Regulation (European Commission) Number 223/2009, 2009.

5.2 Different Types of Synthetic Data

As mentioned before, we want to focus on the generation of synthetic data as a way of protecting the confidentiality and thus we must start to specify the different categories of synthetic data available.

5.2.1 Fully Synthetic Data

Introduced by Rubin (1993), the approach releases, instead of the original dataset, one or more synthetic datasets obtained by generating the overall data, that is, generating values for all variables, from a model that is assumed to fit the original data. If missing data is present in the original data, this process may even fill these gaps. Subsequently, only the synthetic versions of these datasets are released to the public. These publicly available datasets will not contain any of the original data. When circumstances dictate that only one synthetic dataset can be released, we will be in the single imputation scenario.

5.2.2 Partially Synthetic Data

Little (1993) suggested to partially synthesize the data. This method refers to situation where synthetic versions are synthesized only for those individual records deemed sensitive, that is, which release could compromise their confidentiality. The remaining values are left unchanged, ensuring protection without compromising the overall quality of the publicly disclosed data (Reiter, 2005c). Partially synthetic generated or imputed data may thus involve (1) the synthetic generation of some of the unit or observational records, for all the variables, or only for the sensitive variables, or (2) otherwise involve the synthetic generation of the values for all units (or only sensitive units) for given sensitive variables. Clearly, releasing some of the original data values to the public presents a greater risk of disclosure. However, as a trade-off, the dataset made available will more closely mirror the properties of the original dataset, especially when compared to a scenario where the available data is fully synthetic. Indeed, if the number of key variables or key values in the original data, which are supposed to remain unaltered, is reduced to zero, then the scenario of partially synthetic data effectively becomes equivalent to the case of fully synthetic data. However, the use of the partially synthetic data approach to create a fully imputed synthetic dataset will lead to a different inferential analysis approach.

5.3 Methods of Creating Synthetic Data

This survey presents an overview of the diverse techniques employed to generate synthetic data. We may say there are two main classes of methods for synthetic data generation, model-based and machine learning based.

5.3.1 *Model-Based*

The model-based approach to synthetic data generation heavily depends on assuming a specific model for the data and then synthesizing this data by sampling (randomly generating values) from the specified model. Model-based techniques include the use of various parametric models like logit models, exponential models, pareto, univariate and multivariate normal models, multiple and multivariate regression models, and sequential conditional regression models. Some references, to name just a few, include works by Kennickell (1997), Abowd et al. (2006), Reiter (2005a), An and Little (2007), Klein and Sinha (2015c,b,d,a), Moura et al. (2017, 2021), and Mishra and Barui (2022).

It is crucial to note that these model-based methods can be categorized into two distinct approaches when synthesizing multiple key variables.

One approach, known as *joint modeling*, involves specifying the joint distribution of the data. This method focuses on clearly defining a joint model that captures the existing relations among all variables, both those that are to be synthesized and those that may remain unchanged (Drechsler & Haensch, 2023).

The other approach is *sequential modeling*, where each sensitive variable is synthesized in sequence using models that are conditioned on the original variables that do not need to be replaced and on any of the remaining variables that need to be synthesized. To better understand the process of sequential modeling let us consider that we have a set of p variables from which $Y_1, \ldots, Y_{p_1}$ are considered to be sensitive, and which thus need to be replaced by the synthetic versions, and variables $X_{p_1+1}, \ldots, X_p$, which are the non-sensitive variables. One approach, is the sequential regression multivariate imputation (SRMI) where a sequence of conditional distributions $f(Y_j|Y_1, \ldots, Y_{j-1}, Y_{j+1}, \ldots, Y_{p_1}, X_{p_1+1}, \ldots, X_p)$ is used to generate synthetic versions of the original Y_j, for $j = 1, \ldots, p_1$ (Raghunathan et al., 2001). However, this process may be very complex and may require a large amount of time. Thus, a simplified algorithm commonly used considers the distributions conditioned only on the non-sensitive variables, $X_{p_1+1}, \ldots, X_p$ and the variables that have already been synthesized. That is, generates values sequentially by using the distributions $f(Y_1|X_{p_1+1}, \ldots, X_p)$, $f(Y_2|Y_1, X_{p_1+1}, \ldots, X_p)$, and so on, until we generate the values for the last variable from $f(Y_{p_1}|Y_1, \ldots, Y_{p_1-1}, X_{p_1+1}, \ldots, X_p)$ (Drechsler, 2011). Due to its simpler application, this is the algorithm that is often used in practice.

Rubin (1987) suggested that for an approximately unbiased estimation of variance in multiple imputation of missing data, the replacement data should be drawn from posterior predictive distributions. This idea formed the basis for generating synthetic data through *Posterior Predictive Sampling* (PPS). However, Reiter and Kinney (2012) established that this is not strictly necessary. They showed that one can simply use sample estimates, such as posterior modes or maximum likelihood estimates, and plug them in the model, drawing then the synthetic data from this model. This method of generation is commonly referred to as the *Plug-in Sampling* (PS) method.

In the PPS method, the synthetic data is essentially seen as predictions coming from the posterior distribution of the model, whose parameters are estimated from the confidential dataset. On the other hand, in the PS method, synthetic data is created directly from the parametric model by plugging in the sample estimates, which are also based on the confidential data. This approach differs from PPS in that it uses same sample estimates to generate the different imputations of synthetic data rather than using several posterior distributions to generate those imputations.

In the model-based approach to synthetic data generation, it is generally advised to use a large set of predictor variables for generating the synthetic variables. This way, when someone wants to analyze the data, using a simpler model than the one used to synthesized the data, it is still ensured a good degree of congruence between the two models.

5.3.2 Machine Learning Based

Contrary to the notion of using a parametric model to generate the data, one can consider adopting a non-parametric approach. This alternative methodology does not rely on predefined parametric models but instead uses methods that are more flexible and less dependent on assumptions about the data's underlying distribution. Reiter (2005c) proposed the use of Classification and Regression Trees (CART) for synthetic data generation. When there are multiple key variables to be replaced with synthetic values, a sequential regression approach, as mentioned earlier, is recommended. This method offers more flexibility compared to the model-based approach, allowing for the accommodation of non-linear relationships between variables and non-standard variable distributions. However, due to the sequential nature of generation in this method, the first variables are generated independently of those generated later. As a result, if there are any relationships between these early and later variables, such interdependencies may not be fully captured in the synthetic data creation process. We may see extensions of the very popular CART approach in the use of random forests (Caiola & Reiter, 2010), support vector machines (Drechsler, 2010) and even into the use of deep learning methods such as Generative adversarial networks (GANs) (Park et al., 2018; Xu et al., 2019; Koivu et al., 2020), and Variational Autoencoders (VAEs) (Srivastava et al., 2017; Vardhan & Kok, 2020; Ma et al., 2020).

5.4 Inferential Analysis of Single Imputation Synthetic Datasets

5.4.1 *Origins of Inferential Analysis for Synthetic Datasets*

Before delving into the inference analysis of the single imputation case, it is essential to first discuss the origins of inference analysis in this context. The concept of synthetic data, originating from the work of Rubin (1987), naturally led to the development of inferential procedures initially based on scenarios involving both fully and partially generated data, related to the multiple imputation framework. The first comprehensive rules for analyzing multiply imputed full synthetic data were derived by Raghunathan et al. (2003). In a similar way, methodologies for partially synthetic data were introduced by Reiter (2003). Further advancements were made by Reiter (2005c), who expanded these concepts to include multivariate techniques, thereby extending the application of Wald tests and Likelihood ratio tests.

To gain a deeper insight into the essence of these rules, consider an analyst who is focused on a specific parameter Q from the true model. When M synthetic datasets are released, the procedure entails gathering the point estimate q_j $(j = 1, \ldots, M)$ and its corresponding variance estimate u_j, from every synthetic dataset as though it were the original data. These rules, commonly designated as combination rules, as formulated by the previously mentioned authors, involve calculating the variance of the point estimate by combining the within-variance of the point estimates q_j with the average of the variance estimates u_j. Consequently, the analyst can use the mean of q_j as an unbiased estimator for Q and then use either a Student's t-distribution or an F-distribution, to infer about Q. For a more comprehensive understanding, readers are directed to the work of Reiter and Raghunathan (2007).

As the reader might have surmised, these methodologies and associated distributions need that the number of synthetic versions released to the public exceed 1. This requirement is critical, as a single dataset or a very small number of synthetic datasets would render the distributions used for inference impractical. In such scenarios, one would face the predicament of not being able to calculate the degrees of freedom, or, in cases where the number is very small, extremely limited degrees of freedom in the distributions.

As highlighted in the Introduction, certain scenarios (Abowd et al., 2006; Kinney et al., 2011, 2014; Bowen et al., 2020; Alam et al., 2020) need the release of a single synthetic dataset to mitigate the risk of data disclosure.

5.4.2 *The Single Imputation Scenario*

The inability to effectively handle single imputation requirements or a limited number of releases of synthetic versions of data was what prompted Klein and Sinha (2015d) to address this specific challenge. Beginning in 2013, these authors embarked on compiling a collection of papers focused on this issue. In their work, Klein and Sinha (2015c), they considered both Posterior Predictive Sampling (PPS) and Plug-in Sampling (PS) generation methods. They developed exact inference procedures for analyzing data, based on sufficient statistics, when it is presumed that the original data follows an exponential model. These exact procedures are applicable not only to the multiple imputation scenario but also to single imputation cases. For both PPS and PS cases unbiased estimators of the true parameters can be derived.

In the same year, the same authors, published three additional works (Klein & Sinha, 2015b,d,a) focusing on single imputation synthetic data for confidentiality purposes. These papers expanded the results of exact inference procedures for both PPS and PS methods, considering scenarios where the original data is assumed to follow univariate or multivariate normal, and for cases of partially generated data through multiple linear regression (MultiLR) models. In these works, the authors successfully derive the exact distributions for key sample measures in the synthetic data: the mean, variance, and regression coefficients.

All these studies are based on the assumption of normality in the data, but they differ in the type of synthetic data used. For the univariate and multivariate normal cases, it is assumed that the synthetic data will be fully synthesized. In contrast, for the MultiLR model, the released data is partially synthesized; here, the response variables are considered sensitive, and the predictor variables remain unchanged.

In the latter scenario, those authors explored cases where the imputer and/or analyst either overfits or underfits the regression model. They also examined cases where the analyst's regression model is the regression of the unsynthesized variable on the synthesized variable, which is the opposite of the generation case. These explorations were conducted under the PS method.

In the following year, Moura (2016) delved into the multivariate linear regression model (MLR), resulting in two more publications (Moura et al., 2017, 2021). In Moura et al. (2017), attention was given to the PPS method, whereas Moura et al. (2021) focused on the PS method. A major challenge tackled in these studies was the measurement of the size of confidence sets based on the test statistics employed in drawing inference about the matrix of regression coefficients. To tackle this problem, they introduced a new measure for the multivariate scenario, termed "*radius*", which may be seen as the distance between the centre and the edge of the confidence set. This measure, based on the quantiles of the distribution of the test statistic used for inference on the matrix of regression coefficients and on the determinant of the estimator of the variance-covariance matrix of the variables based on the synthetic data provides an useful approach to evaluating confidence in multivariate contexts, in situations where the confidence sets may have an infinite volume.

A critical aspect of Moura et al. (2021) was their approach to scenarios involving the release of synthetic data that was either non-normal or involved non-continuous response variables. Despite these challenges, the inferential procedures that were based on normality assumptions continued to perform effectively. The authors proposed adaptations to the PS method to suit these more complex data scenarios.

This showed the method's adaptability and efficacy in handling a diverse array of data types, highlighting the wide applicability of this synthetic data generation methods.

It's important to recognize that these studies did not just develop exact inference procedures for multiple imputation synthetic data generated by PPS and PS methods, but also extended these procedures to single imputation approaches. Additionally, they adapted Reiter (2005b) vector methodology, known for its combination rules, to a matrix parameter, as it is relevant in the context of the MLR. This adaptation is a significant advancement, as it extends the applicability of Reiter's methodology to more complex statistical models.

In 2022, Guin et al. (2022) addressed the inferential analysis for the MultiLR model under PPS and PS, but from a Bayesian inference perspective. They employed a non-informative diffuse prior, in contrast to the frequentist approach commonly seen in the literature. This study extended the work of Klein and Sinha (2015d,a) into the Bayesian realm.

The paper by Guin et al. (2022) explored two distinct methodologies for generating data. The first method focuses exclusively on using the sensitive data to estimate the model parameters. The rationale behind this approach is to ensure that in the released dataset, the synthetic portion, which represents the sensitive data, remains independent of the non-synthetic and non-sensitive part. This separation is key for maintaining the integrity of the non-sensitive data while protecting the confidentiality of the sensitive information. The second method involves utilizing the entire original dataset for parameter estimation. This comprehensive approach can potentially offer a more holistic view of the data's underlying structure, albeit with increased difficulty in maintaining data privacy.

Interestingly, in terms of estimating the regression coefficients, both approaches in this paper mirror the frequentist inference method. The estimator for the vector of regression coefficients, in the MultiLR model, is computed in the same way as the usual unbiased estimator used with the original data. This parallelism simplifies the computation of estimates when only one dataset is available.

In 2023, the same authors turned their focus to the Multivariate Normal Model, examining both PPS and PS scenarios through the lens of Bayesian inference (Guin et al., 2023). Once again, opting for a non-informative prior, they obtained Bayes estimators for the mean, the covariance matrix, and the generalized covariance. Notably, the Bayes estimator for the population mean mirrors precisely the estimator derived from traditional frequentist inference methods. Furthermore, they outlined methods for constructing credibility sets for both the generalized variance and the population mean, providing valuable insights into the convergence of Bayesian and frequentist methodologies in this context.

Klein et al. (2021) addressed a gap in previous research concerning inference about the variance structure under the Multivariate Normal assumption, particularly in the context of fully synthetic data generation. Their work significantly expanded the scope of inferential procedures to include more complex statistical assessments, especially when dealing with PS synthetic data.

Specifically, they developed inference procedures for several key statistical measures: the generalized variance, the sphericity test, and tests for the independence of two subsets of variables and for the regression of one subset on the other. This extension allows for a more comprehensive analysis of synthetic data, encompassing a broader range of inferential needs.

A notable feature of the proposed test statistics is their similarity to the conventional statistics commonly found in existing literature. This familiarity facilitates analysts' quick identification and computation of the desired test statistic values. The only challenge for analysts lies in utilizing the 'R' package 'PSinference' (Moura et al., 2023), which provides access to the Monte Carlo simulated distributions of these test statistics. This package represents a valuable tool, enabling analysts to accurately apply these advanced statistical procedures to synthetic data.

Under the same assumption of normal distribution, Basak and Sinha (2024) tackled the univariate case for fully synthetic data generated by both PS and PPS. Their work focused on inferential procedures for testing the equality of means across several populations, akin to ANOVA (Analysis of Variance). They provided insights into the distribution properties not just under the null hypothesis but also under the alternative hypothesis.

An important aspect of their study was examining the robustness of these proposed tests when the assumption of normality in the original data was violated. The results were quite noteworthy. When the original data is generated from distributions like Double Exponential, Student's t, Exponential, or Lognormal, the Type 1 error rates are significantly impacted in the traditional ANOVA F-test. However, this is not the case with their proposed tests for PS and PPS synthetic data. This finding indicates that the tests developed for synthetic data are more robust than the traditional tests, offering greater reliability when the underlying assumptions about the data distribution are not met.

Venturing beyond the normality assumption, Mishra and Barui (2022) explored the Pareto Model (specifically the two-parameter model) in the context of synthetic data generation. Their study encompassed both types of data synthesis: PS and PPS. In the case of PS, the authors developed inferential procedures for all possible scenarios: whether one of the parameters is known and the other unknown, or both parameters are unknown. This comprehensive approach in the PS framework allows for a broad application of the Pareto Model in various data contexts. On the other hand, for PPS, the focus was narrower. The inferential procedures were presented only for the shape parameter, under the assumption that the other parameter (scale parameter) is known.

While much of the work in synthetic data generation focuses on methods like Plug-in Sampling (PS) or Posterior Predictive Sampling (PPS), it's important to acknowledge alternative approaches. Klein and Datta (2018) explored a method based on the principle of sufficiency, particularly under the assumption of a normal MultiLR model. In this method, partial synthetic data is generated by sampling from the conditional distribution of the data, given the sufficient statistics which are free from unknown parameters.

This approach is particularly advantageous for several reasons. Firstly, when the regression model is correctly specified, the values of the sufficient statistics can be directly calculated from the synthetic version of the data. This allows the analyst to use standard regression methodologies. Furthermore, if the imputer overfits the regression model, valid inferences can still be obtained by the analyst, even when using an underfitted model. However, the same is not true if the imputer underfits the model and the analyst overfits model.

It's also crucial to note that this type of synthetic data generation and the subsequent release of synthetic data is contingent upon the institution holding the original data agreeing to release the sufficient statistics of those data, since with access to these sufficient statistics and the specified model, additional versions of the original data can be generated.

Finally, we should highlight the work of Raab et al. (2018), which introduced a novel variance estimator suitable for use with large synthetic datasets. This estimator is applicable regardless of whether the datasets are generated through non-parametric or parametric methods, and whether they are derived from the posterior predictive distribution or not. A key feature of this new variance estimator is that it does not account for variance between multiple synthetic datasets. As a result, it can be used in the inferential analysis of datasets that have undergone single imputation, using procedures somewhat akin to those found in Raghunathan et al. (2003).

Despite the promising nature of this development, there remains a keen interest in seeing its application in real-world scenarios, particularly through simulations and practical applications on actual datasets. Specifically, it would be informative to observe the performance of this variance estimate in cases where the number of synthetic versions is reduced to one, a scenario that was not examined in their work. Such investigations could provide valuable insights into the effectiveness and reliability of this variance estimator in different contexts of synthetic data analysis.

Table 5.1 presents a summary of the references in this section pertaining to singly imputation analysis, where the works are ordered chronologically. We may note that all works except the one by Raab et al. (2018) address exact inferential procedures for the analysis of synthetic data.

It is important to emphasize the significance of exact inferential methods, which are not only applicable in both single and multiple imputation scenarios but also serve as a vital tool for assessing the quality and effectiveness of asymptotic procedures in these contexts.

These exact methods provide a robust framework for researchers and practitioners, ensuring the reliability and accuracy of statistical inferences drawn from synthetic data, thereby playing a crucial role in advancing the methodologies and applications of SDC. As such, these exact inferential methods are an essential component to consider in the ongoing evolution and refinement of SDC techniques.

Table 5.1 Summary of the work done for the single imputation case

Author(s)	Year	Model	Generation method	Type of synthetic generation
Klein and Sinha (2015c)	2015	Exponential	PPS/PS	Fully/Partially
Klein and Sinha (2015b)	2015	Univariate normal	PPS/PS	Fully
Klein and Sinha (2015d)	2015	Multivariate normal and MultiLR	PPS	Fully/Partially
Klein and Sinha (2015a)	2015	Multivariate normal and MultiLR	PS	Fully/Partially
Moura et al. (2017)	2017	MLR	PPS	Partially
Klein and Datta (2018)	2018	MultiLR	Sufficiency principle	Partially
Raab et al. (2018)	2018	Generic	PPS/PS	Fully/Partially
Moura et al. (2021)	2021	MLR	PS	Partially
Klein et al. (2021)	2021	Multivariate normal (Tests for the covariance structure)	PS	Fully
Guin et al. (2022)	2023	MultiLR (Bayesian inference)	PPS/PS	Partially
Mishra and Barui (2022)	2022	Pareto	PPS/PS	Fully
Guin et al. (2023)	2023	Multivariate normal (Bayesian inference)	PPS/PS	Fully
Basak and Sinha (2024)	2024	ANOVA	PPS/PS	Fully

5.5 Multivariate Normal Model: A Theoretical Overview

Given the range of theoretical results surrounding single imputation, our discussion will focus specifically on the Multivariate Normal Model. This model is selected for its comprehensive suite of inferential procedures and its broad applicability. Our focus will be on the PS generation method due to its straightforward implementation and the fact that a single synthetic dataset may suffice for drawing inferences. This approach not only simplifies the process but also ensures an adequate level of privacy protection, making it a highly suitable choice for our analysis.

To start, let us delve into the generation of synthetic data using the PS method, within the context of the Multivariate Normal Model. For that, we begin to consider the original data as $\mathbf{X} = (\mathbf{x}_1, \ldots, \mathbf{x}_n)$, where

$$\mathbf{x}_i \overset{i.i.d.}{\sim} N_p(\boldsymbol{\mu}, \boldsymbol{\Sigma}), i = 1, \ldots, n \ , \tag{5.1}$$

for $\mathbf{x}_i = (x_{1i}, \ldots, x_{pi})'$, where n is the sample size and p the number of variables involved. The initial phase is estimating the mean vector $\boldsymbol{\mu}$ and the variance-covariance matrix $\boldsymbol{\Sigma}$, from the original dataset, respectively by the sample mean, $\bar{\mathbf{x}} = \frac{1}{n}\sum_{i=1}^{n}\mathbf{x}_i$, and the sample variance-covariance matrix

$$\hat{\boldsymbol{\Sigma}} = \frac{\mathbf{S}}{n-1} = \frac{1}{n-1}\sum_{i=1}^{n}(\mathbf{x}_i - \bar{\mathbf{x}})(\mathbf{x}_i - \bar{\mathbf{x}})'. \tag{5.2}$$

The second step, is to plug-in the estimates in to the model and generate the synthetic versions of $\mathbf{x}_i$, $\mathbf{v}_i$, by randomly drawing

$$\mathbf{v}_i \overset{i.i.d.}{\sim} N_p\left(\bar{\mathbf{x}}, \hat{\boldsymbol{\Sigma}}\right), i = 1, \ldots, n .$$

Thus, we end up with one fully synthetic dataset, $\mathbf{V} = (\mathbf{v}_1, \ldots, \mathbf{v}_n)$. If someone wishes to reproduce more than one fully synthetic dataset, it has only to multiply create $\mathbf{V}_1, \ldots, \mathbf{V}_M$ $(M > 1)$, independently from each other. As one may see, the process is very straightforward, since we draw from a multivariate Normal model where the sample estimates are plugged-in.

The idea of releasing a single synthetic version of the original dataset and not the sample mean and sample variance is due to the fact that institutions are usually not comfortable in doing so, since that way someone may produce multiple synthetic versions of the original dataset and come up closer to the original dataset, possibly compromising the confidentiality of the respondents (Reiter et al., 2014). As such, in this section we will only address the case of singly imputed datasets.

5.5.1 *Synthetic Mean and Variance-Covariance Matrix*

Now that the synthetic version of the dataset is available, obtaining estimators for the population mean and covariance matrix becomes a crucial next step. Fortunately, this process is straightforward.

Using the synthetic version of the data, we can calculate the synthetic sample mean as

$$\bar{\mathbf{v}} = \frac{1}{n}\sum_{i=1}^{n}\mathbf{v}_i,$$

and the synthetic variance-covariance matrix as

$$\hat{\boldsymbol{\Sigma}}_{PS} = \frac{\mathbf{S}_{PS}}{n-1} = \frac{1}{n-1}\sum_{i=1}^{n}(\mathbf{v}_i - \bar{\mathbf{v}})(\mathbf{v}_i - \bar{\mathbf{v}})'.$$

Some key properties of $\bar{\mathbf{v}}$ and $\hat{\boldsymbol{\Sigma}}_{PS}$ are:

1. They are jointly sufficient statistics for $(\boldsymbol{\mu}, \boldsymbol{\Sigma})$, ensuring efficient use of data for estimation (Klein & Sinha, 2015b).
2. $\bar{\mathbf{v}}$ is the maximum likelihood estimator of $\boldsymbol{\mu}$, providing an unbiased estimate with a variance of $\frac{2\boldsymbol{\Sigma}}{n}$ (Klein & Sinha, 2015b).
3. $\hat{\boldsymbol{\Sigma}}_{PS}$ offers an unbiased estimation of $\boldsymbol{\Sigma}$, enhancing the reliability of the synthetic data (Klein & Sinha, 2015b).
4. With a non-informative joint prior, $\pi(\boldsymbol{\mu}, \boldsymbol{\Sigma}) \propto |\boldsymbol{\Sigma}|^{-\gamma/2}$, the Bayes estimator for $\boldsymbol{\mu}$ aligns with $\bar{\mathbf{v}}$, and the Bayes estimator for $\boldsymbol{\Sigma}$ is determined as $\frac{(n-1)^2\hat{\boldsymbol{\Sigma}}_{PS}}{(n-2p+\gamma-3)^2}$, with $n > \max\{p, 2p-\gamma+3\}$ (Guin et al., 2023).

These properties offer a framework for analyzing synthetic data, since they ensure that the estimates of the mean and covariance matrix are both statistically sound and practical for further analysis.

5.5.2 *Confidence and Credibility Sets for $\boldsymbol{\mu}$*

A fundamental step in conducting inference on a dataset involves making inferences about the mean. The synergistic contributions of Klein et al. (2021) and Guin et al. (2023) methodologies facilitates the creation of both confidence and credibility sets for $\boldsymbol{\mu}$. This dual approach encompasses the principles of frequentist and Bayesian inference, offering two ways to draw inference about the mean in the context of normal distribution.

5.5.2.1 Confidence Ellipsoid for $\boldsymbol{\mu}$

To infer about $\boldsymbol{\mu}$ using synthetic data, consider the random variable

$$T^2 = \frac{n}{n-1}(\bar{\mathbf{v}} - \boldsymbol{\mu})' \hat{\boldsymbol{\Sigma}}_{PS}^{-1} (\bar{\mathbf{v}} - \boldsymbol{\mu}), \tag{5.3}$$

as defined by Klein and Sinha (2015b). Its distribution can be simulated through a series of steps to generate a random value of T^2:

- Generate a $p \times p$ random matrix $\boldsymbol{\Sigma}^*$ from a Wishart distribution, with parameter $\mathbf{I}_p$ and $n-1$ degrees of freedom
- Compute $\lambda_1, \ldots, \lambda_p$, as the solution of $\left|(n-1)\mathbf{I}_p + (1-\lambda)\boldsymbol{\Sigma}^*\right| = 0$
- Generate χ^2_{1j}, $j = 1, \ldots, p$, independently, from a χ^2 distribution with 1 degree of freedom
- Compute $T_2 = \sum_{j=1}^{p} \lambda_j \chi^2_{1j}$

- Generate T_1, independently of T_2, from a inverse χ^2 distribution with $n - p$ degrees of freedom
- Compute $T^2 = T_1 \times T_2$.

To construct a $(1-\alpha)$ confidence ellipsoid for $\boldsymbol{\mu}$, calculate the $(1-\alpha)$ percentile, $c_{n,p;\alpha}$ from the empirically simulated values from the distribution of T^2. This ellipsoid is given by

$$\Delta(\boldsymbol{\mu}) = \{\boldsymbol{\mu} : T^2 \leq c_{n,p,\alpha}\}$$

with a volume of

$$V_{\mu}(\mathbf{V}) = \frac{(n-1)^{p/2}\pi^{p/2}}{n^{p/2}\Gamma(p/2+1)}(c_{n,p;\alpha})^{p/2}|\hat{\boldsymbol{\Sigma}}_{PS}|^{1/2},$$

and corresponding expected volume

$$E[V_{\mu}(\mathbf{V})] = \frac{\pi^{p/2}}{n^{p/2}\Gamma(p/2+1)}(c_{n,p;\alpha})^{p/2}\prod_{j=1}^{p}\left[\frac{2^{1/2}\Gamma\left(\frac{n-j+1}{2}\right)}{\Gamma\left(\frac{n-j}{2}\right)}\right]|\boldsymbol{\Sigma}|^{1/2}.$$

5.5.2.2 Credibility Set for $\boldsymbol{\mu}$

From a Bayesian perspective, the random variable defined in (5.3) is also considered, but taking into account the posterior distribution of the synthetic data. Guin et al. (2023), proposed the following steps to repeatedly generate a value of its distribution:

- Generate a $p \times p$ random matrix $\mathbf{B}$ from an inverse Wishart distribution, with parameter $\mathbf{I}_p$ and $n - p + \gamma - 2$. Here one must establish one value for γ, abiding by the restriction in Sect. 5.1. In their simulations study, Guin et al. (2023) suggest setting the value of γ at 10 for the PS scenario
- Generate a $p \times p$ random matrix $\mathbf{A}^*$ from an inverse Wishart distribution, with parameter $(n-1)\mathbf{B}$ and $n - p + \gamma - 2$ degrees of freedom
- Compute $\mathbf{A} = \mathbf{A}^* + \mathbf{B}$
- Compute $\lambda_1, \ldots, \lambda_p$, as the solution of $|\mathbf{A} - \lambda \mathbf{I}_p| = 0$
- Generate χ^2_{1j}, $j = 1, \ldots, p$, independently, from a χ^2 distribution with 1 degree of freedom
- Compute $T^2 = \sum_{j=1}^{p} \lambda_j \chi^2_{1j}$.

To construct a $(1-\alpha)$ credibility ellipsoid for $\boldsymbol{\mu}$, obtain the $(1-\alpha)$ percentile, $k_{n,p,\gamma;\alpha}$, from the empirically simulated distribution of T^2. This ellipsoid is given by

$$\Delta_{Bayes}(\boldsymbol{\mu}) = \{\boldsymbol{\mu} : T^2 \leq k_{n,p,\gamma;\alpha}\}$$

with volume

$$V_{Bayes,\boldsymbol{\mu}}(\mathbf{V}) = \frac{(n-1)^{p/2}\pi^{p/2}}{n^{p/2}\Gamma(p/2+1)}(k_{n,p,\gamma;\alpha})^{p/2}|\hat{\boldsymbol{\Sigma}}_{PS}|^{1/2}$$

and corresponding expected volume

$$E[V_{Bayes,\boldsymbol{\mu}}(\mathbf{V})] = \frac{\pi^{p/2}}{n^{p/2}\Gamma(p/2+1)}(k_{n,p,\gamma;\alpha})^{p/2}\prod_{j=1}^{p}\left[\frac{2^{1/2}\Gamma\left(\frac{n-j+1}{2}\right)}{\Gamma\left(\frac{n-j}{2}\right)}\right]|\boldsymbol{\Sigma}|^{1/2}.$$

It's noteworthy that the expected volumes for the confidence and credibility sets, derived from the frequentist and Bayesian approaches respectively, exhibit a high degree of similarity. This resemblance arises primarily due to the analogous nature of their construction, with the only distinction being the specific cut-off points chosen from the distributions of T^2 for each approach.

5.5.3 *Variance-Covariance Structure*

Based on the result found in Klein and Sinha (2015b), conditionally on $\mathbf{S}$, defined in (5.2),

$$\mathbf{S}_{PS} \sim W_p\left(n-1, \frac{\mathbf{S}}{n-1}\right),$$

Klein et al. (2021) presented four inferential procedures regarding the structure of the variance-covariance matrix, which are presented in the following subsections.

5.5.3.1 Generalized Variance

We have seen, previously, that both the expected volumes for the confidence set and credibility set for $\boldsymbol{\mu}$ are a function of $|\boldsymbol{\Sigma}|$, the generalized variance. The generalized variance is an important measure, used not only to determine the volume of those sets but also to measure the spread of the dataset.

Consider the random variable

$$T_1^{PS} = (n-1)^p\frac{|\mathbf{S}_{PS}|}{|\boldsymbol{\Sigma}|}.$$

The distribution of T_1^{PS} can be obtained using Monte Carlo simulation, following these steps, to generate one of its values:

- Generate p independent random variables A_j and, independently from these, p independent random variables B_j, for $j = 1, \ldots, p$, all from a chi-square distribution with $n - j$ degrees of freedom
- Calculate $T_1^{PS} = \left(\prod_{j=1}^{p} A_j\right)\left(\prod_{j=1}^{p} B_j\right)$.

Consequently, we may use the empirical distribution of T_1^{PS} to calculate the αth percentile, $t_{1;\alpha}^{PS}$, and use it to construct the $(1 - \alpha)$ level confidence interval for $|\mathbf{\Sigma}|$ which is given by

$$\left(\frac{(n-1)^p\,|\mathbf{S}_{PS}|}{t_{1;1-\alpha/2}^{PS}}, \frac{(n-1)^p\,|\mathbf{S}_{PS}|}{t_{1;\alpha/2}^{PS}}\right).$$

Remarkably, for those preferring the Bayesian methodology over the frequentist approach, the test statistic suggested by Guin et al. (2023), denoted as T_1^{PS}, remains the same. However, the distribution of T_1^{PS} differs due to a modification in the generation process of the independent B_j random variables. Specifically, when generating these variables, the degrees of freedom should be adjusted from $n - j$ to $n - p + \gamma - j$, reflecting the Bayesian framework's unique considerations, considering the restriction for γ in Sect. 5.1.

5.5.3.2 Sphericity Test

In multivariate statistics, the assumption of sphericity represents a critical condition for the validity of certain inferential tests, related to MANOVA, repeated measures ANOVA, and multivariate analysis of covariance (MANCOVA).

The sphericity test is designed to assess this assumption, providing a means to test the hypotheses

$$H_0 : \mathbf{\Sigma} = \sigma^2\mathbf{I}_p \;\; versus \;\; H_1 : \mathbf{\Sigma} \neq \sigma^2\mathbf{I}_p.$$

Defining

$$T_2^{PS} = \frac{|\mathbf{S}_{PS}|^{1/p}}{tr\left(\mathbf{S}_{PS}\right)/p}, \tag{5.4}$$

Klein et al. (2021) demonstrated that its distribution, assuming that $\mathbf{\Sigma} = \sigma^2\mathbf{I}_p$, can be empirically obtained through Monte Carlo simulation by iteratively executing the steps:

- Generate $\mathbf{W}_1 \sim W_p\left(n - 1, \mathbf{I}_p/(n-1)\right)$ and $\mathbf{W}_2 \sim W_p\left(n-1, \mathbf{I}_p\right)$, independently

- Calculate $T_2^{PS} = \frac{|W_1 W_2|^{1/p}}{tr(W_1 W_2)/p}$.

With the empirical distribution we may conduct a sphericity test, that is, the hypotheses in (5.4), rejecting the null hypothesis, for a level of significance α, if the observed value of T_2^{PS} falls bellow $t_{2;\alpha}^{PS}$, the αth percentile of T_2^{PS}.

5.5.3.3 Test for the Independence of Two Subsets of Variables

Assessing the independence between two subsets of variables is crucial for uncovering complex relationships within a dataset. This analysis is key to understand the variables' structure, aiding in feature selection, dimensionality reduction, and other processes. The independence test aims to clarify whether variations in one variable set are statistically related to variations in another set or if the two sets behave independently.

Considering the vectors $\mathbf{x}_i$ in (5.1) split into two subvectors or sets of variables $\mathbf{x}_i = [\mathbf{x}_{1i}', \mathbf{x}_{2i}']'$ where the first set has size p_1, and the second set has size p_2, with $p = p_1+p_2$, to carry out this test, the matrices $\boldsymbol{\Sigma}$ and $\mathbf{S}_{PS}$ are partitioned as follows:

$$\boldsymbol{\Sigma} = \begin{bmatrix} \boldsymbol{\Sigma}_{11} & \boldsymbol{\Sigma}_{12} \\ \boldsymbol{\Sigma}_{21} & \boldsymbol{\Sigma}_{22} \end{bmatrix}, \quad \mathbf{S}_{PS} = \begin{bmatrix} \mathbf{S}_{PS,11} & \mathbf{S}_{PS,12} \\ \mathbf{S}_{PS,21} & \mathbf{S}_{PS,22} \end{bmatrix},$$

where both $\boldsymbol{\Sigma}_{11}$ and $\mathbf{S}_{PS,11}$ are square matrices of dimension p_1.

Once assumed the normality of the $\mathbf{x}_i$, to test the hypothesis of independence of the two subsets of variables is equivalent to test the hypotheses

$$H_0 : \boldsymbol{\Sigma}_{12} = \mathbf{0} \ \ versus \ \ H_1 : \boldsymbol{\Sigma}_{12} \neq \mathbf{0}. \tag{5.5}$$

Once defined the random variable

$$T_3^{PS} = \frac{|\mathbf{S}_{PS}|}{|\mathbf{S}_{PS,11}| \, |\mathbf{S}_{PS,22}|}$$

it is feasible to simulate its distribution, under H_0 in (5.5), by iteratively following the following steps (Klein et al., 2021):

- Generate $\boldsymbol{\Omega}_1$ from a Wishart distribution with parameters $n-1$ and $\mathbf{I}_p$
- Generate $\boldsymbol{\Omega}_2$ from a Wishart distribution with parameters $n-1$ and $\frac{\boldsymbol{\Omega}_1}{n-1}$
- Partition $\boldsymbol{\Omega}_2$ as

$$\boldsymbol{\Omega}_2 = \begin{bmatrix} \boldsymbol{\Omega}_{2,11} & \boldsymbol{\Omega}_{2,12} \\ \boldsymbol{\Omega}_{2,21} & \boldsymbol{\Omega}_{2,22} \end{bmatrix}$$

  where $\boldsymbol{\Omega}_{2,11}$ is a square matrix with dimension p_1
- Compute $T_3^{PS} = \frac{|\boldsymbol{\Omega}_2|}{|\boldsymbol{\Omega}_{2,11}||\boldsymbol{\Omega}_{2,22}|}$.

The null hypothesis in (5.5) should then be rejected, for a level of significance α, if the observed value of T_3^{PS} falls bellow $t_{3;\alpha}^{PS}$, the αth percentile of T_3^{PS}.

The strategic choice for $\mathbf{\Omega}_1$, whose Wishart distribution parameter matrix is an identity matrix, makes the derivation of the empirical values of T_3^{PS} easier, and also laid the groundwork for the next result (for details see Klein et al. 2021).

5.5.3.4 Test for the Regression of One Set of Variables on the Other

The regression of one set of variables on another emerges as a powerful tool, enabling researchers to explore and quantify the relationships between two distinct sets of variables. It enables the understanding of how variations in one group of variables can explain or predict variations in another group, uncovering the underlying dynamics and dependencies within the data. Tests for the matrix of regression coefficients of one set of variables on another set involves the test of the hypotheses

$$H_0 : \mathbf{\Delta} = \mathbf{\Delta}_0 \quad \text{versus} \quad H_1 : \mathbf{\Delta} \neq \mathbf{\Delta}_0, \tag{5.6}$$

where $\mathbf{\Delta} = \mathbf{\Sigma}_{12}\mathbf{\Sigma}_{22}^{-1}$, and where $\mathbf{\Delta}_0$ represents any specified $p_1 \times (p - p_1)$ matrix.

Consider the test statistic

$$T_4^{PS} = \frac{\left|\left(\mathbf{S}_{PS,12}\mathbf{S}_{PS,22}^{-1} - \mathbf{\Delta}_0\right)\mathbf{S}_{PS,22}\left(\mathbf{S}_{PS,12}\mathbf{S}_{PS,22}^{-1} - \mathbf{\Delta}_0\right)'\right|}{\left|\mathbf{S}_{PS,11} - \mathbf{S}_{PS,12}\mathbf{S}_{PS,22}^{-1}\mathbf{S}_{PS,21}\right|}.$$

To empirically derive the distribution of T_4^{PS}, one can utilize Monte Carlo simulation. This involves the following set of iterative steps:

- Generate $\mathbf{\Omega}_1$ from a Wishart distribution with parameters $n-1$ and $\mathbf{I}_p$
- Generate $\mathbf{\Omega}_2$ from a Wishart distribution with parameters $n-1$ and $\frac{\mathbf{\Omega}_1}{n-1}$
- Partition $\mathbf{\Omega}_2$ as

$$\mathbf{\Omega}_2 = \begin{bmatrix} \mathbf{\Omega}_{2,11} & \mathbf{\Omega}_{2,12} \\ \mathbf{\Omega}_{2,21} & \mathbf{\Omega}_{2,22} \end{bmatrix}$$

where $\mathbf{\Omega}_{2,11}$ is a square matrix with dimension p_1

- Compute $T_4^{PS} = \frac{\left|\mathbf{\Omega}_{2,12}\mathbf{\Omega}_{2,22}^{-1}\mathbf{\Omega}_{2,21}\right|}{\left|\mathbf{\Omega}_{2,11} - \mathbf{\Omega}_{2,12}\mathbf{\Omega}_{2,22}^{-1}\mathbf{\Omega}_{2,21}\right|}$.

For a given significance level α, we should reject the null hypothesis in (5.6), if the observed value of T_4^{PS} exceeds the $(1-\alpha)$th percentile, $t_{4,1-\alpha}^{PS}$, of the T_4^{PS} distribution.

5.5.4 As a Summary

As shown, in the context of the fully synthetic PS method under the assumption of a normal distribution, exact inferential procedures for the mean, generalized variance, and tests for the covariance structure are readily available. Although comprehensive validation across all scenarios is pending, Moura et al. (2021) have explored the implications of assuming non-normal and even non-continuous distributions for the data, particularly for response variables, within the context of partially synthetic datasets. Their results indicate that the normality assumption holds up well, even when not fully met, showing its practical effectiveness. This paper posits that such effectiveness is likely to extend to fully synthetic datasets, highlighting the adaptability and utility of these inferential approaches even when faced with deviations from normality.

Moreover, because these procedures are exact and adaptable to multiple imputation contexts for synthetic data, they establish a benchmark for evaluating asymptotic inferential techniques tailored to both single (Raab et al., 2018) and multiple imputation frameworks (Raghunathan et al., 2003). Contrasting confidence sets generated from synthetic data using these exact procedures against those derived from asymptotic approaches offers a way to gauge the effectiveness and precision of asymptotic methods. Traditionally, the evaluation of inferential methods' quality involved comparisons with results from the original dataset. However, the exact nature of the procedures discussed in this manuscript provides an opportunity to compare expected values of sample estimators obtained from these exact procedures against the findings from asymptotic methods.

5.6 Conclusion and Last Remarks

This manuscript has traveled across the comprehensive landscape of generating and analyzing single imputation synthetic datasets within the domain of Statistical Disclosure Control (SDC). Through a detailed analysis of the literature on the development of inferential procedures for single imputation, this work gives a description of the evolution and the current state of methodologies for creating and utilizing synthetic datasets to uphold individual privacy while enabling statistical analysis.

The exact inferential procedures for analyzing synthetic data, as showcased by the contributions of several authors, show the potential of synthetic datasets in maintaining statistical integrity, while protecting the privacy. These procedures not only provide a basis for making robust inferences from synthetic data but also offer a benchmark for assessing the quality of asymptotic inferential methods in both single and multiple imputation scenarios. The adaptability of these exact methods to the Bayesian framework further extends their applicability, offering a versatile toolset for researchers and practitioners in SDC.

One of the key insights from this review is the effectiveness of the normality assumption in synthetic data analysis. Despite potential deviations from normality in real-world datasets, the procedures developed under this assumption can still show remarkable robustness, suggesting that their utility extends beyond strictly normal conditions. Broader studies of these methodologies, in the presence of non-normal or non-continuous data types, would be desirable and worthy.

Furthermore, the exploration into the variance-covariance structure and the development of tests for sphericity, independence, and regression analysis within the synthetic data context have enriched the toolkit available for SDC. These advances not only facilitate a deeper understanding of the data underlying structure but also enable analyses that were previously challenging in the realm of synthetic data.

In conclusion, the work presented in this manuscript aims at being a valid contribution to the field of SDC by providing a detailed overview of the generation and analysis of single imputation synthetic datasets. As SDC continues to adapt to the challenges posed by modern data environments, the insights and approaches detailed herein may play a role in shaping the future of privacy-preserving statistical analysis.

Acknowledgments The research conducted by Ricardo Moura and Carlos Agra Coelho was funded by national funds through the *Fundação para a Ciência e a Tecnologia* (FCT), I.P., Center for Mathematics and Applications (NOVA Math) under the scope of the projects UIDB/00297/2020 (https://doi.org/10.54499/UIDB/00297/2020) and UIDP/00297/2020 (https://doi.org/10.54499/UIDP/00297/2020). Bimal Sinha is thankful to Tommy Wright at the Center for Statistical Research and Methodology (CSRM) of the US Census Bureau for his kind support and encouragement.

References

Abowd, J. M., Stinson, M. H., & Benedetto, G. (2006). Final report to the social security administration on the SIPP/SSA/IRS public use file project. https://ecommons.cornell.edu/items/d7cf980e-5ec3-4322-aea2-a58b344c3ac6

Abowd, J., Benedetto, G., Garfinkel, S., Dahl, S., Dajani, A., Graham, M., Hawes, M., Karwa, V., Kifer, D., Kim, H., Leclerc, P., Machanavajjhala, A., Reiter, J., Rodrigues, R., Schmutte, I., Sexton, W., Singer, P., & Vilhuber, L. (2020). The modernization of statistical disclosure limitation at the U.S. Census Bureau. Working Paper Number CED-WP-2020-009.

Alam, M. J., Dostie, B., Drechsler, J., & Vilhuber, L. (2020). Applying data synthesis for longitudinal business data across three countries. *Statistics in Transition New Series, 21*(4), 212–236.

An, D., & Little, R. J. A. (2007). Multiple imputation: An alternative to top coding for statistical disclosure control. *Journal of the Royal Statistical Society, Series A, 170*, 923–940.

Basak, B. and Sinha, B. (2024). Analysis of one-way anova model using synthetic data. *Sankhya B*, pages 1–27.

Bowen, C. M., Bryant, V., Burman, L., Khitatrakun, S., McClelland, R., Stallworth, P., Ueyama, K., & Williams, A. R. (2020). A synthetic supplemental public use file of low-income information return data: Methodology, utility, and privacy implications. In *Privacy in statistical databases: UNESCO chair in data privacy, international conference, PSD 2020, Tarragona, Spain, September 23–25, 2020, Proceedings* (pp. 257–270). Springer International Publishing.

Caiola, G., & Reiter, J. P. (2010). Random forests for generating partially synthetic, categorical data. *Transactions on Data Privacy*, *3*(1), 27–42.

Drechsler, J. (2010). Using support vector machines for generating synthetic datasets. In *Privacy in statistical databases: UNESCO chair in data privacy, international conference, PSD 2010, Corfu, Greece, September 22–24, 2010. Proceedings* (pp. 148–161). Springer Berlin Heidelberg.

Drechsler, J. (2011). *Synthetic datasets for statistical disclosure control: Theory and implementation* (Vol. 201). Springer Science & Business Media.

Drechsler, J., & Haensch, A. C. (2023). 30 years of synthetic data. Preprint, arXiv:2304.02107.

Guin, A., Roy, A., & Sinha, B. (2022). Bayesian analysis of multiply imputed synthetic data under the multiple linear regression model. *International Journal of Statistical Sciences*, *22*(2), 25–38.

Guin, A., Roy, A., & Sinha, B. (2023). Bayesian analysis of singly imputed synthetic data under the multivariate normal model. *International Journal of Statistical Sciences*, *23*(2), 1–18.

Hundepool, A., Domingo-Ferrer, J., Franconi, L., Giessing, S., Lenz, R., Longhurst, J., Nordholt, E. S., Seri, G., & Wolf, P. (2010). Handbook on statistical disclosure control. *ESSnet on Statistical Disclosure Control*. https://cros.ec.europa.eu/system/files/2023-12/SDC_Handbook.pdf

Kennickell, A. B. (1997). Multiple imputation and disclosure protection: The case of the 1995 survey of consumer finances. *Record Linkage Techniques*, *1997*, 248–267.

Kinney, S. K., Reiter, J. P., Reznek, A. P., Miranda, J., Jarmin, R. S., & Abowd, J. M. (2011). Towards unrestricted public use business microdata: The synthetic longitudinal business database. *International Statistical Review*, *79*(3), 362–384.

Kinney, S. K., Reiter, J. P., & Miranda, J. (2014). *Improving the synthetic longitudinal business database*. US Census Bureau Center for Economic Studies.

Klein, M., & Sinha, B. (2015a). Inference for singly imputed synthetic data based on posterior predictive sampling under multivariate normal and multiple linear regression models. *Sankhya B*, *77*, 293–311.

Klein, M., & Sinha, B. (2015b). Likelihood based finite sample inference for singly imputed synthetic data under the multivariate normal and multiple linear regression models. *Journal of Privacy and Confidentiality*, *7*(1), 43–98.

Klein, M., & Sinha, B. (2015c). Likelihood-based finite sample inference for synthetic data based on exponential model. *Thailand Statistician*, *13*(1), 33–47.

Klein, M., & Sinha, B. (2015d). Likelihood-based inference for singly and multiply imputed synthetic data under a normal model. *Statistics & Probability Letters*, *105*, 168–175.

Klein, M. D., & and Datta, G. S. (2018). Statistical disclosure control via sufficiency under the multiple linear regression model. *Journal of Statistical Theory and Practice*, *12*, 100–110.

Klein, M., Moura, R., & Sinha, B. (2021). Multivariate normal inference based on singly imputed synthetic data under plug-in sampling. *Sankhya B*, *83*, 273–287.

Koivu, A., Sairanen, M., Airola, A., & Pahikkala, T. (2020). Synthetic minority oversampling of vital statistics data with generative adversarial networks. *Journal of the American Medical Informatics Association*, *27*(11), 1667–1674.

Little, R. J. A. (1993). Statistical analysis of masked data. *Journal of Official Statistics*, *9*, 407–426.

Ma, C., Tschiatschek, S., Turner, R., Hernández-Lobato, J. M., & Zhang, C. (2020). Vaem: A deep generative model for heterogeneous mixed type data. *Advances in Neural Information Processing Systems*, *33*, 11237–11247.

Mishra, N., & Barui, S. (2022). Likelihood-based finite sample inference for synthetic data from pareto model. *REVSTAT-Statistical Journal*, *20*(5), 655–676.

Moura, R. P. (2016). *Likelihood-based inference for multivariate regression models using synthetic data*. Ph.D. Thesis, Universidade NOVA de Lisboa, Portugal.

Moura, R., Klein, M., Coelho, C. A., & Sinha, B. (2017). Inference for multivariate regression model based on synthetic data generated under fixed-posterior predictive sampling. *REVSTAT: Statistical Journal*, *15*(2), 155–186.

Moura, R., Klein, M., Zylstra, J., Coelho, C. A., & Sinha, B. (2021). Inference for multivariate regression model based on synthetic data generated using plug-in sampling. *Journal of the American Statistical Association, 116*(534), 720–733.

Moura, R., Norouzirad, M., & Mazarei, D. (2023). Psinference: Inference for released plug-in sampling single synthetic dataset. *Computer software*. https://github.com/ricardomourarpm/PSinference.

Park, N., Mohammadi, M., Gorde, K., Jajodia, S., Park, H., & Kim, Y. (2018). Data synthesis based on generative adversarial networks. *Proceedings of the VLDB Endowment, 11*(10), 1071–1083.

Raab, G. M., Nowok, B., & Dibben, C. (2018). Practical data synthesis for large samples. *Journal of Privacy and Confidentiality, 7*(3), 67–97.

Raghunathan, T. E., Lepkowski, J. M., Van Hoewyk, J., Solenberger, P., et al. (2001). A multivariate technique for multiply imputing missing values using a sequence of regression models. *Survey Methodology, 27*(1), 85–96.

Raghunathan, T. E., Reiter, J. P., & Rubin, D. B. (2003). Multiple imputation for statistical disclosure limitation. *Journal of Official Statistics, 19*, 1–16.

Reiter, J. P. (2002). Satisfying disclosure restrictions with synthetic data sets. *Journal of Official Statistics, 18*(4), 531.

Reiter, J. P. (2003). Inference for partially synthetic, public use microdata sets. *Survey Methodology, 29*, 181–188.

Reiter, J. P. (2005a). Releasing multiply imputed, synthetic public use microdata: An illustration and empirical study. *Journal of the Royal Statistical Society: Series A (Statistics in Society), 168*(1), 185–205.

Reiter, J. P. (2005b). Significance tests for multi-component estimands from multiply imputed, synthetic microdata. *Journal of Statistical Planning and Inference, 131*(2), 365–377.

Reiter, J. P. (2005c). Using CART to generate partially synthetic public use microdata. *Journal of Official Statistics, 21*(3), 441–462.

Reiter, J. P., & Kinney, S. K. (2012). Inferentially valid, partially synthetic data: Generating from posterior predictive distributions not necessary. *Journal of Official Statistics, 28*(4), 583.

Reiter, J. P., & Raghunathan, T. E. (2007). The multiple adaptations of multiple imputation. *Journal of the American Statistical Association*, 1462–1471.

Reiter, J. P., Wang, Q., & Zhang, B. (2014). Bayesian estimation of disclosure risks for multiply imputed, synthetic data. *Journal of Privacy and Confidentiality, 6*(1).

Rubin, D. B. (1987). *Multiple imputation for nonresponse in surveys*. Wiley.

Rubin, D. B. (1993). Discussion: Statistical disclosure limitation. *Journal of Official Statistics, 9*, 461–468.

Srivastava, A., Valkov, L., Russell, C., Gutmann, M. U., & Sutton, C. (2017). Veegan: Reducing mode collapse in gans using implicit variational learning. *Advances in Neural Information Processing Systems, 30.*

Vardhan, L. V. H., & Kok, S. (2020). Generating privacy-preserving synthetic tabular data using oblivious variational autoencoders. In *Proceedings of the workshop on economics of privacy and data labor at the 37th international conference on machine learning (ICML).*

Xu, L., Skoularidou, M., Cuesta-Infante, A., & Veeramachaneni, K. (2019). Modeling tabular data using conditional GAN. In *Advances in neural information processing systems* (Vol. 32). Curran Associates, Inc.

Part II
Modeling with Spherically Symmetric, Heavy-Tailed, Skewed, Circular-Linear, Mixture Distributions and Neural Networks

Chapter 6
Some Empirical Findings on Neural Network-Based Forecasting When Subjected to Autoregressive Resampling

J. T. Ferreira, D. Wrbka, and A. van der Merwe

6.1 Introduction

6.1.1 Background

Time series analysis is a cornerstone in the statistical discipline when data exhibits some temporal dependence. Uncertainty continues to plague modern time series analysis, particularly in the realm of forecasting. Often, models fitted to historical data produce generally reliable forecasts; however, many of these approaches may result in some inaccuracies. Parameter estimation underpinning these techniques mostly rely on Maximum Likelihood Estimation (MLE); deviation from and critical violations of the underlying assumptions relevant to MLE estimation may lead to inaccurate estimates and increased uncertainty. This is due to temporal data exhibiting structural dependence in the form of autocorrelation. It is clear that the time domain includes a vast and appropriate array of alternative forecasting approaches such as those utilized in non-linear, non-parametric and heteroscedastic models; examples include the usual autoregressive integrated moving average (ARIMA) approach (Cryer & Chan, 2008), as well as several instances within the machine learning (ML) environment (Lim & Zohren, 2021).

Woodward et al. (2022) explains that the main reason for fitting a time series model is to forecast future behaviour when presented with historical realizations,

J. T. Ferreira (✉)
Department of Statistics, University of Pretoria, Pretoria, South Africa

National Institute for Theoretical and Computational Sciences (NITheCS), South Africa
e-mail: johan.ferreira@up.ac.za

D. Wrbka · A. van der Merwe
Department of Statistics, University of Pretoria, Pretoria, South Africa
e-mail: ane.vandermerwe@up.ac.za

C. A. Coelho, D.-G. Chen (eds.), *Statistical Modeling and Applications*, Emerging Topics in Statistics and Biostatistics, https://doi.org/10.1007/978-3-031-69622-0_6

and makes a crucial assumption regarding forecasted values in that future behaviour is directly explained by its relationship to its past behaviour. Forecasting based on ARIMA analysis suffers a drawback in that the forecasted values will almost surely tend to the mean of the series. This phenomenon is directly affected by the loss of structural dependence in the long-run. Singh (1981) confirmed the superiority of Efron's independent and identically distributed (IID) bootstrap (abbreviated to bs) method in the case of IID data, but this method subsequently proves inferior when applied to dependent data. Results illustrated that the conditional distribution of the normalized sample mean under Efron's method did indeed converge to a normal distribution but that the variance was incorrect, ultimately leading to inconsistency of the bs estimator. Thus, it remains essential to consult and employ alternate resampling strategies most suited to data exhibiting temporal dependence. More recent literature on forecasting approaches in conjunction with the block bootstrap approach includes Shah et al. (2022), Li (2021), and Sroka (2022).

ML is a powerful tool for obtaining forecasts in the time domain, including methods such as random forests, support vector machines, linear discriminant analysis, and artificial neural networks (ANN). These methods seek to offer accuracy and computational performance gains over the classic ARIMA based forecasting strategy. Specifically regarding ANN, literature provides a myriad of neural networking strategies in the time domain. Firat et al. (2010) performed a comparative analysis of various ANN by investigating the best fit forecasting structure of water consumption over time. They illustrate adequate forecasting accuracy via the Normalized Root Mean Squared Error and the Average Absolute Relative Error of the Feed Forward Neural Network (FNN), even whilst being outperformed by its competitors. The FNN was established as an accurate and reliable forecasting strategy in the time domain. Medeiros et al. (2006) made use of a single hidden layer FNN to model temporal data and reported on the improved forecasting accuracy to that of its non-linear competitors. Multilayer perceptron neural networks characterised by backpropagation are some of the most widely considered and applied ML strategies in challenging the drawbacks of more conventional forecasting methods in the time domain. Various works Specht et al. (1991), Aliev et al. (2008), Hwang & Little (1996), Hüsken & Stagge (2003), López et al. (1993), Zhang & Qi (2005), Cigizoglu (2005), and Chen & Leung (2004) illustrate positive gains in accuracy and increased computational performance in the realm of forecasting when a neural network approach is applied to the time domain.

Literature provides many variations of the neural network. In particular, the Generalized Regression Neural Network (GRNN) will serve to be a suitable ML candidate for the purpose of realising performance gains of forecasts in the time domain. Li et al. (2008) outlines the main advantages of this method to be improved forecasting performance by taking into account the statistical volatility exhibited in the time domain alongside its applications in linear and non-linear modelling. The GRNN is an extension of the Radial Basis Function neural network (RFB) via efficient single-pass learning. This network comprises RFB neurons in its hidden layer alongside a multivariate Gaussian neuron in its center. The output of the GRNN provides a measure of minimum distance of the input layer to the center.

The objective of this architecture is to obtain a weighted average of the training set in its central neuron. These weights are determined by the smoothing parameter and decay the further they diverge from the training set. The smoothing parameter will therefore play a pivotal role when subjected to autoregressive (AR) resampling. If this parameter is too large, the forecast will tend to the mean due to being assigned similar weights. Therefore, a small smoothing parameter is necessary in order to obtain weights that will add significant value in the forecasting strategy. The resampling strategies under consideration will assist in investigating improved forecasting.

In this spirit, this chapter seeks to empirically discuss the following question: How is the neural network-based forecast of a stationary autoregressive process affected, when subject to different bs methods? This chapter will be focusing predominantly on the forecasting of an AR(1) process, when being subjected to the moving block bootstrap (MBB) as well as the stationary block bootstrap (SBB). Various accuracy measures will be implemented to compare the performances under these resampling schemes.

6.1.2 The Autoregressive Model of Order 1

The value of an AR model is emphasised by Maleki et al. (2020) and Christmas and Everson (2010) as a flexible tool for modelling dependent stationary data; and is the focal candidate in this chapter. As defined by Cryer and Chan (2008), AR processes are regressions upon themselves at previous points in time. Maleki et al. (2020) extend this definition by stating that the AR model determines the probabilistic behaviour of current values Y_t for $t = 1, 2, \ldots, n$ based on a linear combination of previously observed values $(Y_{t-1}, Y_{t-2}, \ldots, Y_{t-p})$. The drawbacks of forecasting will seek to be exposed in the case of the selected forecasting strategy when subjected to resampling. The AR(1) is defined by

$$Y_t = \theta + \phi_1 Y_{t-1} + a_t \tag{6.1}$$

where ϕ_1 is the unknown AR parameter, with the process mean given by $\mu = \theta(1 - \phi_1)^{-1}$. The variance of Y_t is given as by

$$\text{var}(Y_t) = \gamma_0 = \sigma_Y^2 = \frac{\sigma_a^2}{1 - \phi_1^2}$$

where σ_Y^2 is a finite constant since σ_a^2 is assumed to be finite and $|\phi_1| < 1$. The autocovariance between Y_t and Y_{t-1} is given by $\text{Cov}(Y_t, Y_{t-1}) = \gamma_1 = \phi_1 \sigma_Y^2$, and the autocorrelation at any lag $k = 0, 1, 2, \ldots$ is given by

$$\rho_k = \phi^k. \tag{6.2}$$

The h-step forecast of this process can be written (in the usual ARIMA case) as

$$\hat{Y}_{t+h} = \mu + \phi^h (Y_t - \mu).$$

For full details on this process, the reader is referred to Cryer and Chan (2008).

6.1.3 Outline of Chapter

In Sect. 6.2, a broader empirical contextualisation is discussed with computational examples together with the description of considered accuracy measures. Section 6.3 contains further computational comparisons, with specific focus on the MBB and SBB methods. Section 6.4 concludes the chapter with a brief discussion and indications for further work.

6.2 Empirical Contextualisation of the bs

6.2.1 Issues of bs *When Dealing with Autoregressive Data*

Here, we simulate $n = 203$ stationary AR(1) values such that $Y_t = (Y_1, \cdots, Y_{203})$ with $\phi_1 = 0.6$, $\mu = 0$ and $\sigma_1 = 1$. A training sample of size $n = 200$ was selected for the purpose of measuring the performance of forecasting 1 step-ahead. A testing sample of size $n = 3$ was taken in order to compare the forecasting performances of the GRNN and ARIMA based methods, when subjected to the IID bs method. Figure 6.1 below illustrates this performance comparison.

The GRNN method slightly outperforms the ARIMA based method in forecasting 1 step-ahead. Unfortunately, even the GRNN strategy struggles to retain lagged-covariance in the long-run. In addition to this, the IID bs neglects to account for any type of temporal dependency that is inherent in (this case) an AR(1) process (see Eq. (6.2)). Figures 6.2a and b illustrate significance of the current series mean and variance under their respective sampling distributions. Upon further investigation, Fig. 6.2b will prove to be quite misleading. The inclusion of Fig. 6.3 serves to expose the possible reason for this attraction to the mean. This is done by exploring whether the autocorrelations under the IID bs actually add significant value to the forecasting strategy or not.

Since 0 lies within the confidence interval of the autocorrelation sampling distribution, no significant value is added to the forecasting strategy, resulting in the continued attraction to the mean. Inconsistency of the sampling distribution variance is confirmed through further simulation study and is appropriately illustrated in Fig. 6.4. In each case here, $n = 300$ bootstrapped variances are calculated from an AR(1) process.

It is clear to see that as the IID bs series tends to non-stationarity through an iteratively increasing value of ϕ, the variance is incorrect. This is a crucial

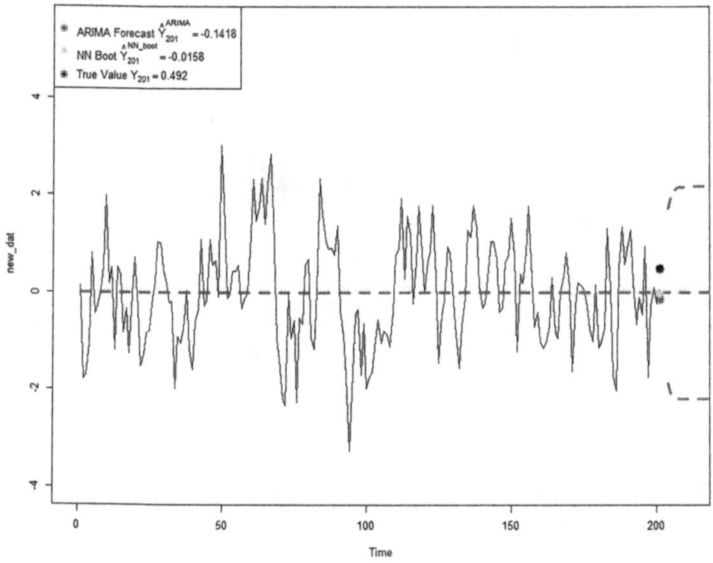

Fig. 6.1 Comparison of ARIMA based and GRNN $h = 1$ step-ahead forecasting when subjected to the IID bs

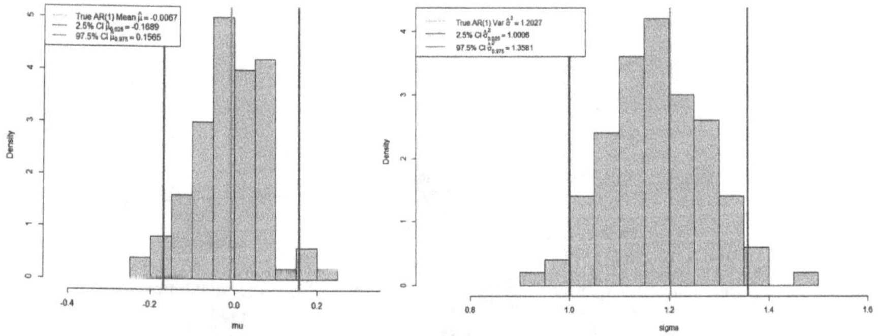

(a) (b)

Fig. 6.2 Testing for significance of the series mean and variance under their respective sampling distributions under IID bs resampling. (**a**) IID bs sampling distribution $\hat{\mu}_{NNboot}$. (**b**) IID bs sampling distribution $\hat{\sigma}^2_{NNboot}$

violation of variance assumptions under normality. Singh (1981) discovered similar behaviour and concluded that Efron's method failed when presented with dependent data. The IID bs method ignores sequential dependence and as a result, fails to account for lag-covariance terms in the asymptotic variance. It is clear that without an appropriate resampling strategy, forecasting would prove to be quite challenging and ultimately misleading. Resampling techniques in the time domain have advantageously evolved over time to being viewed through two contrasting lenses; bs of residuals as a model-based method for the sake of estimation and bo

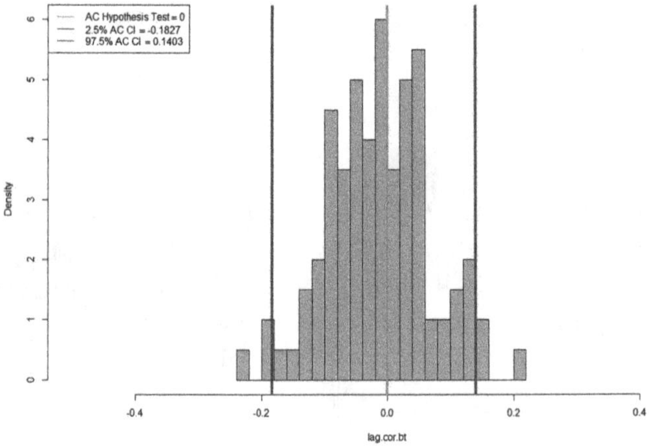

Fig. 6.3 Testing for significance of the autocorrelations in the long-run forecasting strategy under the IID bs

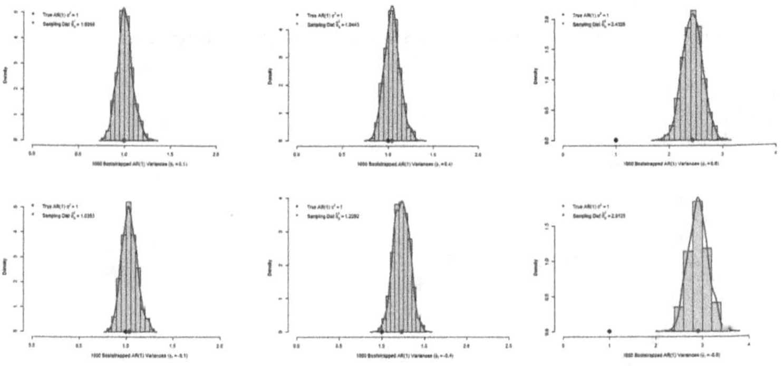

Fig. 6.4 Sampling distribution of σ^2 alongside a showcase of σ^2 inconsistency, for increasing values of ϕ

of blocks in time itself. Here, block bs will be of interest and will be expanded upon further in Sect. 6.2.2. As previously illustrated, an alternative forecasting strategy to ARIMA based analysis lies in the realm of ML. The neural network will continue to fall under consideration as an appropriate forecasting strategy when subjected to AR resampling.

6.2.2 Resampling Methods for Autoregressive Data

The idea of the block bs was initially proposed by Carlstein (1986) and later expanded upon by Kunsch (1989) and Liu (1992) as a resampling scheme applicable to dependent data, without the need for any parametric model assumptions. Chernick and LaBudde (2014) illustrated detailed applications of block bs to AR modelling as a method of producing forecasts and prediction intervals of greater accuracy whilst maximizing computational efficiency. Additionally, they outlined the key idea in the development of block bs for stationary time series. When observations are characterized by a high level of separation, in time, they are considered to be nearly uncorrelated. Therefore, separated blocks can be treated as exchangeable. Exchangeability of blocks allows for all possible permutations to share similar distribution characteristics, and in turn, allows for the data to be broken up into a number of disjoint blocks of a specific length. They continued to emphasise the importance of selecting a long enough block length such that the correlation structure within the blocks is preserved. Consecutive blocks of a specified length are then resampled with replacement and pasted together in order to retain the original length of the series. Lahiri (2003) documented functionally appropriate resampling methods for dependent data including the stationary, circular, tapered, dependent wild, sieve and Bayesian block bs techniques. However, focus will primarily be placed on AR resampling via the MBB and SBB methods alongside a comparative analysis on their respective computational aspects.

The creation of the MBB is attributed to the non-overlapping block bs (NBB), originally proposed by Carlstein (1986), through development of the blocking rule. This rule allows for non-overlapping segments of the data to define the blocks. A later breakthrough by Kunsch (1989) and Liu (1992) led to the formulation of the proposed MBB method. This method is particularly appropriate to dependent data without the need for any parametric model assumptions, and aims to resample blocks of sequential observations at a specified time—leading to preservation of structural dependence within each block. An advantage of this method lies in its ability to effectively capture certain underlying process characteristics determined by shared structural dependence of observations at shorter lags. It is important to note that Chernick and LaBudde (2014) and Lahiri (2003) discussed three challenges shared by these methods:

1. pasting resampled blocks together does not perfectly resemble the correlation structure in the data;
2. structural dependency within sampled blocks can either be weakened or strengthened depending on the distance between adjacent blocks; and
3. inconsistency when the length of the pasted blocks is equal to the length of the series.

The second challenge pervades when this method is subjected to computational comparison; and subsequently prompts possible future investigation into methods tailored at estimating optimal block length. The SBB proposed by Politis and Romano (1994) was specifically designed to alleviate the shortcomings of the MBB method and will thus take center stage as the direct competitor.

The SBB method is unique in that the length of each block is not fixed, but is randomly sampled from a geometric distribution and arranged sequentially. Lahiri (2003) explored the main advantage of this method in that, conditional on the original data, the bs observations retain stationarity. They prove consistency of the variance estimator and explore the possibility of successfully applying this method to possibly non-stationary data. However, this method is not without its own disadvantages. Chernick and LaBudde (2014) would discover that even though the data has a finite length, the block lengths are unbounded and could thus result in infinite resampling. Potential solutions to this problem are proposed by firstly taking the average block size instead of it being selected randomly; and secondly the suggestion for truncating the bs sample at the endpoints, prompting the need for normalization, ultimately resulting in the sum of probabilities to 1.

6.2.3 Considered Accuracy Measures

Comparative studies rely on various accuracy measures in order to determine the most accurate forecasting methods. This chapter considers often encountered measures for accuracy. Firstly, we consider the mean absolute error (MAE), given by

$$\text{MAE} = \frac{1}{h}\sum_{t=1}^{h} |\epsilon_t|,$$

where the forecasting error ϵ_t, is defined as $(Y_t - \hat{Y}_t)$, with $\hat{Y}_t$ being the forecasted series value and h, the size of the training set under rolling origin validation (see subsequent section) alongside the selected h-step ahead forecasting strategy. This method is well known in literature for being the primary measure of accuracy and performance for forecasting models. A drawback of this method is that it may produce biased results when presented with extreme outliers.

Next, we consider the root mean square error (RMSE). This method was proposed in order to deal with the drawback of MSE of being vulnerable to outliers due to its weighting strategy. The RMSE is defined as

$$\text{RMSE} = \sqrt{\frac{1}{h}\sum_{t=1}^{h} \epsilon_t^2},$$

and slightly addresses the drawback of MSE by scaling the data Chen et al. (2017). This method is still sensitive to outliers. Percentage-based methods are also of interest. To ensure scale-independence, methods of percentage errors are utilised. First we begin by exploring the mean absolute percentage error (MAPE) as a suitable forecasting performance metric. This method is defined as

$$\text{MAPE} = \frac{1}{h}\sum_{t=1}^{h} \frac{|\epsilon_t|}{|Y_t|},$$

and is considered to be one of the major accuracy measures present in the literature. However, should the target forecasted value be close to 0, then the percentage errors could become excessively inflated Chen et al. (2017). This is due to asymmetry in the forecasting range. Lastly, we define symmetric mean absolute percentage error (SMAPE) as

$$\text{SMAPE} = \frac{1}{h}\sum_{t=1}^{h} \frac{2 \times |\epsilon_t|}{|Y_t| + |\hat{Y}_t|}.$$

In order to deal with the asymmetry characteristic of the MAPE method, SMAPE is introduced as a symmetric alternative alongside having advantages in terms of not being excessively large or infinite and it is not so sensitive to outliers.

6.3 Computational Comparison and Discussion

The following computational comparison will involve the application of the GRNN forecasting strategy to both cases of a simulated process being resampled. The purpose of this simulation study will be to observe and report on the performance of and the direct effect on the proposed forecasting strategy when subjected to either of the resampling contenders.

6.3.1 The Moving Block bs (MBB)

The MBB was initially explored by Kunsch (1989) and Liu (1992) as an evolved alternative of the NBB. The problem with the NBB is that it does not capture adequate variation for the purpose of proper inferences. The main difference is that for the MBB, the selected blocks are allowed to overlap. For this method, a block length is chosen such that $l = \frac{n}{k}$ where n is the length of the series and k is the effective number of blocks to randomly resample. It is important to note that the selected block length needs to be large enough such that observations are adequately spread apart as to ensure near independence. Vogel and Shallcross (1996) explain that throughout the process of resampling blocks of length l, correlation between observations within each block is retained whilst attempting to have consecutive blocks remain as uncorrelated as possible. This resampling and the subsequent creation of the blocks of size l_n gives $\mathsf{B}_{i,l_n} = (Y_i, \ldots, Y_{i-l_n+1})$.

This process proceeds by randomly drawing K indices where $(I_j)_{1 \leq k \leq K}$ uniformly on $\{1, \ldots, n - l_n + 1\}$. The final step involves defining the bs set to be $\mathsf{D}_n^{\star} = (\mathsf{B}_{I1}, \ldots, \mathsf{B}_{IK})$ once a block is set to be associated with a randomly generated index, $(\mathsf{B}_{Ik})_{1 \leq k \leq K}$. Algorithm 6.2 below outlines the entire algorithmic process surrounding the computational comparison of the MBB.

Algorithm 6.1 Simulation of AR(1) data when subjected to moving block bootstrapping

Step 1: Simulate 203 stationary AR(1) observations with the following equation:

$$Y_t = 0.6Y_{t-1} + \epsilon_t$$

where $\epsilon_t \sim \mathcal{N}(0, 1)$ (Gaussian) white noise innovations.
Step 2: Select an appropriate block length $(l_n) = 10$ such that $l_n \in [1, n]$, where $l \rightarrow \infty$ as $n \rightarrow \infty$.

Step 3: Select l_n to be the length of consecutive blocks to be picked.
Step 4: Define the observed sample as follows:

$$\mathcal{D}_n = ((Y_1), \ldots, (Y_n)).$$

Step 5: Transmute the sample $\mathcal{Y}_t$ into a block of size $k \times l$.
Step 6: Randomly sample moving blocks of data from the sample, with replacement such that $\mathcal{B}_1^*, \ldots, \mathcal{B}_k^*$ is the randomly drawn sample of blocks. The sampling method is performed as follows:

$$\mathcal{B}_{i,l_n} = ((Y_i) : (Y_{i+l_n-1})) \quad i \in \{1, \ldots, n - l + 1\}.$$

Step 7: Reconstruct and define the bootstrap set as follows:

$$\mathcal{D}_n^\star = (B_{l1}, \ldots, B_{lk}).$$

Step 8: Calculate the statistics of interest.

In order to investigate the computational aspects of this method, we begin by simulating $n = 203$ stationary AR(1) realizations such that $Y_t = (Y_1, \cdots, Y_{203})$ with $\phi_1 = 0.6$, $\mu = 0$ and $\sigma_1 = 1$. A training sample of size $n = 200$ is selected for the purpose of forecasting 1:3 steps-ahead via the GRNN strategy. A testing sample of size $n = 3$ was taken in order to compare the forecasting performance to the ARIMA based method when subjected to the AR MBB resampling method. Ultimately, these results are captured for the purpose of providing a comprehensive comparison of the effect that this resampling method has on the GRNN forecasting strategy to that of its afore-mentioned competitor. Figure 6.5 below illustrates the performance comparison of a 1:3 step-ahead forecast to the forecast of an ARIMA based analysis.

It is clear that the GRNN strategy, when subjected to AR MBB resampling, outperforms the ARIMA based strategy in forecasting 1:3 step-ahead. It is also clear to see that this method accurately forecasts the true value, justifying its ability to perform exceptionally well in the time domain. Various performance metrics for this forecasting strategy were obtained via rolling origin validation and subsequently represented in Table 6.1 below.

The above table begins by stating the h-step ahead forecasting strategy, specifically from $h = 1$ to $h = 3$. Next, the true value is given in order to have

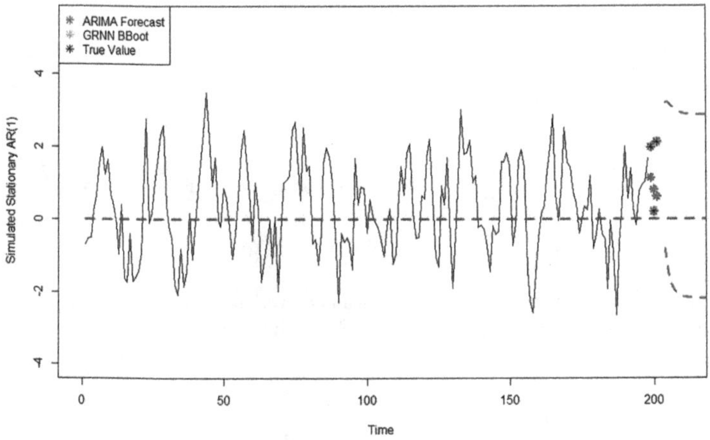

Fig. 6.5 Performance comparison 1: ARIMA based 1:3 step-ahead forecast and GRNN 1:3 step-ahead forecast when subjected to AR MBB resampling

Table 6.1 Performance Metrics 1: ARIMA based 1:3 step-ahead forecasts and GRNN 1:3 step-ahead forecasts when subjected to AR MBB resampling

h-step strategy	True value	ARIMA forecast	GRNN forecast	RMSE	MAE	MAPE	SMAPE
$h = 1$	1.98	1.20	1.98	0.11	0.11	5.36	5.50
$h = 2$	0.20	1.17	0.19	0.10	0.10	27.83	32.32
$h = 3$	2.13	1.14	2.14	1.44	1.44	55.74	77.28

a comparative reference point of what should be expected from the various implemented methods. The next two columns illustrate the ARIMA based 1:3 step-ahead forecasts and GRNN 1:5 step-ahead forecasts when subjected to AR MBB resampling. It is clear to see from comparing these values that the GRNN forecasting strategy outperforms the ARIMA forecasting strategy at every h-step strategy. Finally, the values of various performance metrics are given in order to illustrate the excellent performance of the GRNN forecasting strategy, when subjected to AR MBB resampling, over its ARIMA counterpart. However, in order to fully understand the differences between these two forecasting methods, we have to investigate further. Figure 6.6a and b illustrate significance of the current series mean and variance under their respective bs sampling distributions.

These figures illustrate significance of the mean and variance when the series is subjected to AR MBB resampling. Figure 6.7 illustrate the core difference between both cases. Here we test whether the autocorrelations add significant long-run value to the GRNN forecasting strategy or not, when subjected to AR MBB resampling.

From Fig. 6.7 it is clear that since the value of 0 lies outside the confidence interval, the autocorrelations add significant value in long-run GRNN forecasting

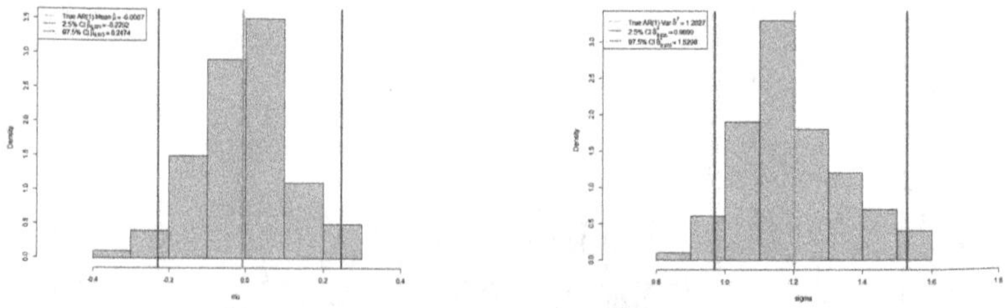

(a) (b)

Fig. 6.6 Testing for significance of the series mean and variance under their respective sampling distributions under AR MBB resampling. (**a**) MBB sampling distribution of $\hat{\mu}_{NNBboot}$. (**b**) MBB sampling distribution of $\hat{\sigma}^2_{NNBboot}$

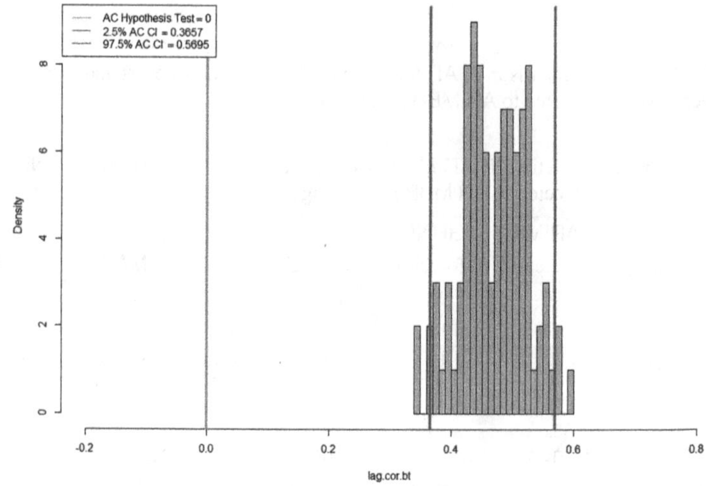

Fig. 6.7 Testing for significance of the autocorrelations in long-run GRNN forecasting, when subjected to AR MBB resampling

strategy, when subjected to AR MBB resampling. The ARIMA based strategy, when subjected to IID bs, fails in this regard and as a result, will always rely on the autocorrelated strength provided by the mean of the series when forecasting. It can therefore be concluded that the GRNN forecasting strategy, when subjected to AR MBB resampling, heavily outperforms the ARIMA based strategy, when subjected to IID bs.

6.3.2 *The Stationary Block* bs *(SBB)*

Radovanov and Marcikić (2014) argues the SBB to be a superior resampling technique in the time domain due to its ability to be generally applied to weakly dependent stationary data in the time domain. The MBB suffers from the fact that resampled blocks are considered to be non-stationary. The main objective of the SBB is to not only retain structural dependence within resampled blocks but to ensure stationarity during this process. Radovanov and Marcikić (2014) make a crucial comparison to the MBB in that the resampling process allows for the lengths of blocks to vary. Due of this, the length of each block is actually approximated by the geometric distribution. This resampling and the subsequent creation of the blocks of size l_n gives $\mathsf{B}_{il} = (Y_i, \ldots, Y_{i+l_n-1})$.

The transition function of the circular block bootstrap (see Lahiri 2003) is implemented in order to ensure equal probability in the selection of observations, forming yet another crucial difference to that of the MBB. The probability mass function of the geometric distribution is given by $P(S_i = m) = (1-p)^{m-1}p$ where S_i is independent of Y_t, $p \in \{0, 1\}$ where $p \to 0$ as the geometric mean, $np \to \infty$ as $n \to \infty$. The resampled series is finally generated via the sequential random selection of block lengths. Algorithm 6.2 below outlines the algorithmic process surrounding the computational comparison of the SBB.

Algorithm 6.2 Simulation of AR(1) data when subjected to stationary block bootstrapping

---

Step 1: Simulate 203 stationary AR(1) observations with the following equation:

$$Y_t = 0.6Y_{t-1} + \epsilon_t$$

where $\epsilon_t \sim \mathcal{N}(0, 1)$ (Gaussian) white noise innovations.
Step 2: Appropriate block lengths (l) are approximated by the geometric distribution.
Step 3: The observed sample is given as follows:

$$\mathcal{D}_n = ((Y_1), \ldots, (Y_n)).$$

Step 4: Transmute the sample $\mathcal{Y}_t$ into blocks of size l that are allowed to vary.
Step 5: The process of ensuring Stationary blocks is achieved in the following manner:

$$\mathcal{B}_{il} = (Y_i, \ldots, Y_{i+l_n-1}).$$

Step 6: Randomly draw non-overlapping blocks of data from the sample, with replacement such that $\mathcal{B}_1^*, \ldots, \mathcal{B}_k^*$ is the randomly drawn sample of blocks, denoted by $\mathcal{D}_n^*$.
Step 7: Reconstruct the resampled time series and calculate statistics of interest.

In order to investigate the computational aspects of this method, we simulate $n = 203$ stationary AR(1) realizations such that $Y_t = (Y_1, \cdots, Y_{203})$ with $\phi_1 = 0.6$, $\mu = 0$ and $\sigma_1 = 1$. A training sample of size $n = 200$ is selected for the purpose of forecasting 1:3 steps-ahead via the GRNN strategy. A testing sample of size n = 3 was taken in order to compare the forecasting performance to the ARIMA based method when subjected to the AR SBB resampling method. Ultimately, these results are captured for the purpose of providing a comprehensive comparison of the effect that this resampling method has on the GRNN forecasting strategy to that of its before-mentioned competitor. Figure 6.8 below illustrates the performance comparison of a 1:3 step-ahead forecast to the forecast of an ARIMA based analysis.

It is clear that the GRNN strategy, when subjected to AR SBB resampling, outperforms the ARIMA based strategy in forecasting 1:3 steps-ahead. It is also clear to see that this method accurately forecasts the true value, justifying its ability to perform exceptionally well in the time domain. Various performance metrics for this forecasting strategy were obtained vie rolling origin validation and subsequently represented in Table 6.2 below.

The above table begins by stating the h-step ahead forecasting strategy, specifically from $h = 1$ to $h = 3$. Next, the true value is given in order to have a comparative reference point of what should be expected from the various implemented methods. The next two columns illustrate the ARIMA based 1:3 step-ahead forecasts and GRNN 1:3 step-ahead forecasts when subjected to AR SBB resampling. It is clear to see from comparing these values that the GRNN

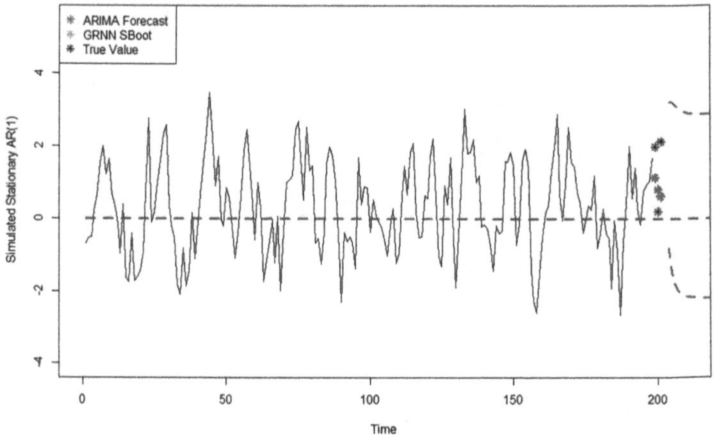

Fig. 6.8 Performance comparison 2: ARIMA based 1:3 step-ahead forecast and GRNN 1:3 step-ahead forecast when subjected to AR SBB resampling

Table 6.2 Performance Metrics 2: ARIMA based 1:3 step-ahead forecasts and GRNN 1:3 step-ahead forecasts when subjected to AR SBB resampling

h-step strategy	True value	ARIMA forecast	GRNN forecast	RMSE	MAE	MAPE	SMAPE
$h = 1$	1.98	1.20	1.98	0	0	0	0
$h = 2$	0.20	0.81	0.19	2.34	2.34	92.46	171.95
$h = 3$	2.13	0.61	2.12	0.92	0.92	53.70	73.42

forecasting strategy outperforms the ARIMA forecasting strategy at every h-step strategy. Finally, the values of various performance metrics are given in order to illustrate the excellent performance of the GRNN forecasting strategy, when subjected to AR SBB resampling, over its ARIMA counterpart. It is therefore concluded by comparison to the MBB method, that the GRNN strategy, when subjected to AR SBB resampling, performs slightly better in terms of capturing the true value. However, in order to fully understand the differences between these two forecasting methods, we have investigate further yet again. Figure 6.9a and b below serve to illustrate significance of the current series mean and variance under their respective bs sampling distributions.

From Fig. 6.10 it is clear that since the value of 0 lies outside the confidence intervals of the autocorrelation sampling distribution, they add significant value to

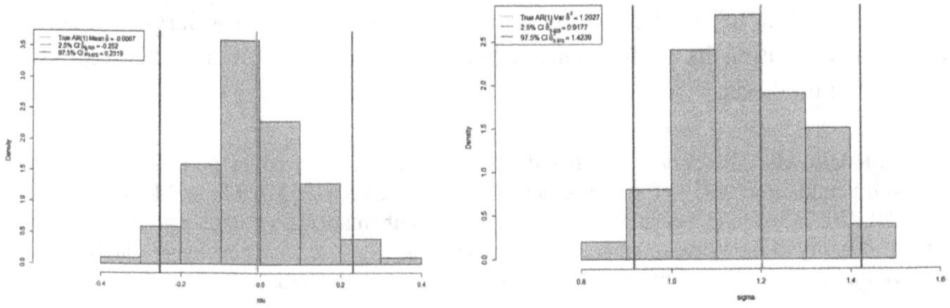

(a) (b)

Fig. 6.9 Testing for significance of the series mean and variance under their respective sampling distributions under AR SBB resampling. (**a**) SBB sampling distribution of $\hat{\mu}_{NNBboot}$. (**b**) SBB sampling distribution of $\hat{\sigma}^2_{NNBboot}$

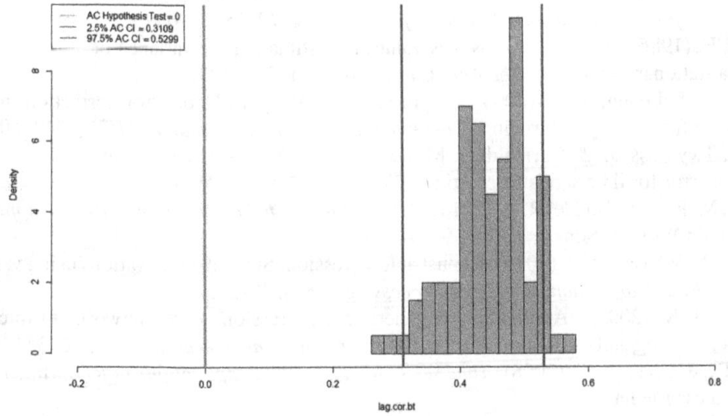

Fig. 6.10 Testing for significance of the autocorrelations in long-run GRNN forecasting, when subjected to AR SBB resampling

the forecasting strategy, hence the improved forecasting accuracy in the long-run under the SBB when compared to the ARIMA method.

6.4 Conclusion

This chapter investigates the challenge of forecasting stationary autoregressive data by using a neural network approach, when data has been subjected to different resampling strategies relevant to autocorrelated data. A computational investigation illustrates the impact and relevance of using a neural network based approach, and highlights the drawbacks when relying on the usual bs approach; here, when data are not IID. Further work includes the consideration and potential implementation of other ML forecasting techniques within this temporal domain of data over time, the investigation of the behaviour of different approaches for larger values of h, and investigating its performance when subjected to alternative autoregressive resampling methods.

Acknowledgments This work was based upon research supported in part by the National Research Foundation (NRF) of South Africa (SA), grant RA201125576565, nr 145681 and grant MND210720624363; and the Centre of Excellence in Mathematical and Statistical Sciences based at the University of the Witwatersrand (SA). The opinions expressed and conclusions arrived at are those of the authors and are not necessarily to be attributed to the NRF.

References

Aliev, R. A., Fazlollahi, B., Aliev, R. R., & Guirimov, B. (2008). Linguistic time series forecasting using fuzzy recurrent neural network. *Soft Computing*, *12*, 183–190.

Carlstein, E. (1986). The use of subseries values for estimating the variance of a general statistic from a stationary sequence. *The Annals of Statistics*, 1171–1179.

Chen, A.-S., & Leung, M. T. (2004). Regression neural network for error correction in foreign exchange forecasting and trading. *Computers & Operations Research*, *31*(7), 1049–1068.

Chen, C., Twycross, J., & Garibaldi, J. M. (2017). A new accuracy measure based on bounded relative error for time series forecasting. *PloS One*, *12*(3), e0174202.

Chernick, M. R., & LaBudde, R. A. (2014). *An introduction to bootstrap methods with applications to R*. John Wiley & Sons.

Christmas, J., & Everson, R. (2010). Robust autoregression: Student-*t* innovations using variational Bayes. *IEEE Transactions on Signal Processing*, *59*(1), 48–57.

Cigizoglu, H. K. (2005). Application of generalized regression neural networks to intermittent flow forecasting and estimation. *Journal of Hydrologic Engineering*, *10*(4), 336–341.

Cryer, J. D., & Chan, K.-S. (2008). *Time series analysis: With applications in R*. Springer Science & Business Media.

Firat, M., Turan, M. E., & Yurdusev, M. A. (2010). Comparative analysis of neural network techniques for predicting water consumption time series. *Journal of Hydrology*, *384*(1–2), 46–51.

Hüsken, M., & Stagge, P. (2003). Recurrent neural networks for time series classification. *Neurocomputing*, *50*, 223–235.

Hwang, J.-N., & Little, E. (1996). Real time recurrent neural networks for time series prediction and confidence estimation. In *Proceedings of international conference on neural networks (ICNN'96)* (Vol. 4, pp. 1889–1894). IEEE.

Kunsch, H. R. (1989). The jackknife and the bootstrap for general stationary observations. *The Annals of Statistics*, 1217–1241.

Lahiri, S. N. (2003). *Resampling methods for dependent data*. Springer Science & Business Media.

Li, J. (2021). Block bootstrap prediction intervals for parsimonious first-order vector autoregression. *Journal of Forecasting*, *40*(3), 512–527.

Li, W., Luo, Y., Zhu, Q., Liu, J., & Le, J. (2008). Applications of AR*-GRNN model for financial time series forecasting. *Neural Computing and Applications*, *17*, 441–448.

Lim, B., & Zohren, S. (2021). Time-series forecasting with deep learning: A survey. *Philosophical Transactions of the Royal Society A*, *379*(2194), 20200209.

Liu, R.Y. & Singh, K. (1992). Moving blocks jackknife and bootstrap capture weak dependence. In *Exploring the limits of bootstrap*, eds. LePage and Billard, Vol. 270, John Wiley & Sons.

López, V., Huerta, R., & Dorronsoro, J. R. (1993). Recurrent and feedforward polynomial modeling of coupled time series. *Neural Computation*, *5*(5), 795–811.

Maleki, M., Mahmoudi, M., Wraith, D., & Pho, K.-H. (2020). Time series modelling to forecast the confirmed and recovered cases of Covid-19. *Travel Medicine and Infectious Disease*, *37*, 101742.

Medeiros, M. C., Teräsvirta, T., & Rech, G. (2006). Building neural network models for time series: A statistical approach. *Journal of Forecasting*, *25*(1), 49–75.

Politis, D. N., & Romano, J. P. (1994). The stationary bootstrap. *Journal of the American Statistical Association*, *89*(428), 1303–1313.

Radovanov, B., & Marcikić, A. (2014). A comparison of four different block bootstrap methods. *Croatian Operational Research Review*, 189–202.

Shah, I., Iftikhar, H., & Ali, S. (2022). Modeling and forecasting electricity demand and prices: A comparison of alternative approaches. *Journal of Mathematics*, *2022*(1), 3581037.

Singh, K. (1981). On the asymptotic accuracy of Efron's bootstrap. *The Annals of Statistics*, 1187–1195.

Specht, D. F., et al. (1991). A general regression neural network. *IEEE Transactions on Neural Networks*, *2*(6), 568–576.

Sroka, Ł. (2022). Applying block bootstrap methods in silver prices forecasting. *Econometrics*, *26*(2), 15–29.

Vogel, R. M., & Shallcross, A. L. (1996). The moving blocks bootstrap versus parametric time series models. *Water Resources Research*, *32*(6), 1875–1882.

Woodward, W. A., Sadler, B. P., & Robertson, S. (2022). *Time series for data science: Analysis and forecasting*. CRC Press.

Zhang, G. P., & Qi, M. (2005). Neural network forecasting for seasonal and trend time series. *European Journal of Operational Research*, *160*(2), 501–514.

Chapter 7
Enriched Lognormal Models for Income Data

S. Makgai, J. T. Ferreira, J. Pillay, A. Bekker, and F. Marques

7.1 Introduction

7.1.1 Income Modelling

An important step in measuring income inequality is the estimation of the income distribution (Mazza & Punzo, 2019; Warchold, Pradhan, and Kropp, 2021). Finding appropriate distributions for modelling income data sets has been the subject of numerous studies (Kleiber & Kotz, 2003), and has been central in economic debates for policy makers to understand and find effective ways of balancing unequal financial scales (Barroso et al., 2020; Silber, 2023). Innovative and improved modelling of income data may help in a better understanding of a nation's economic growth and to ensure social protection for the poor and vulnerable, as set by the first sustainable development goal (Warchold et al., 2021).

Schneider and Scharfenaker (2020) reasons for the revival of a probabilistic approach to study the distributional properties of income to be able to infer the generating mechanism (see Dagum, 2008). It is also argued that one should "let the

S. Makgai (✉) · J. T. Ferreira · J. Pillay
Department of Statistics, University of Pretoria, Pretoria, South Africa
e-mail: seite.makgai@up.ac.za; johan.ferreira@up.ac.za

A. Bekker
Department of Statistics, University of Pretoria, Pretoria, South Africa

Centre for Environmental Studies, Department of Geography, Geoinformatics and Meteorology, University of Pretoria, Pretoria, South Africa
e-mail: andriette.bekker@up.ac.za

F. Marques
Departamento de Matemática, Faculdade de Ciências e Tecnologia da Universidade Nova de Lisboa, Caparica, Portugal
e-mail: fjm@fct.unl.pt

C. A. Coelho, D.-G. Chen (eds.), *Statistical Modeling and Applications*, Emerging Topics in Statistics and Biostatistics, https://doi.org/10.1007/978-3-031-69622-0_7

data speak for itself" and rather apply a non-parametric approach than to impose a restricted parametric model. However, the latter approach can lead to unreliable inference in the presence of heavy tails (Lubrano & Ndoye, 2016). This emphasizes the continued importance of considering parametric methods. An abundant number of parametric models have been suggested in literature (see Kleiber and Kotz, 2003), however the lognormal distribution for income modelling is echoed by several works (see Akhundjanov and Toda, 2020, and Mitzenmacher, 2004). Choosing the lognormal is justified because of its well known footprint in income data modelling; furthermore, it is due to its convenient characterization of being defined on a strictly positive support, unimodal property, and being positively skewed (see Hlasny, 2021 and references therein).

Navigated by the need and inherent difficulties that the modern-day statistician experiences when modelling data sets with vastly different underlying characteristics, we propose an approach to model an income data set via a data-enriched approach, which incorporates knowledge from a regression function (deduced from the observed data) within a lognormal distribution. Similar investigations include that of Wesolowski (1995) and Roux and Bekker (1989). Thus, this model is proposed from an interpretable perspective other than just creating parameter rich models (which may lack interpretative implementation (Ley, 2015; Ley et al., 2021; Jones, 2015). Inspired by the non-Gaussian nature of income data and a conditional distribution with a stochastic parameter, this work focuses on developing a data-enriched lognormal distribution to model income data sets. For this setup, the regression coefficient indicates the dependence structure and behaves as an inequality parameter.

7.1.2 Data Description

Consider an income data set detailing public employee compensation values across Atlanta city in the United States[1] in the year 2015. This data comprises of 8246 observed names of employees, ages, genders, ethnicities, job titles, organisations, and annual salaries. The monthly salary (in USD) summary statistics are presented in Table 7.1 accompanied by a boxplot in Fig. 7.1. The disparities in income will be studied for different age groups. We aim to explore the relationship between salary and age.

From the boxplot in Fig. 7.1, one can observe that the salary data is heavily skewed to the right. Thus, it is reasonable to consider the natural log of the salaries versus the age as well. Figure 7.2 illustrates the histogram for the age variable (a) and its natural log transformation (b), motivating a consideration of a finite mixture lognormal model. A mode parameterized lognormal will also be considered for this data set for age groups (each of size 2). There exists economic motivations for

[1] https://data.world/brentbrewington/atlanta-city-employee-salaries.

Table 7.1 Income data set: Salary summary statistics

Minimum	25th Quantile	Median	Mean	75th quantile	Maximum
20,800	36,855	43,554	48,633	54,873	240,698

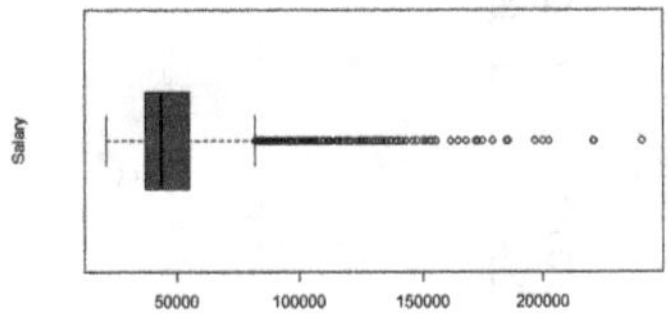

Fig. 7.1 Income data set: Salary boxplot

mixture models; Yakovenko and Silva (2005) proposed the two-class structure for the observed distribution of income data to reflect heterogeneity, where the grouping variable is unknown and accommodates segmentation in the income data set.

It is not reasonable to fit a curve on the raw data, as shown by Fig. 7.3, where the figure illustrates multiple observed log salaries per age. To overcome this challenge, the sample mean of log salary per age, as well as the mode of log salary per age group of size two will be considered as explanatory variables (see Fig. 7.4a and b).

The data set is truncated to observations whose ages range from 18 to 60, which is the age range representative of employees entering the work field and employees at the age generally considered to retire. From Figs. 7.2 and 7.4 we may emphasize that:

- Figure 7.2b illustrates that a finite mixture may be considered to model the logarithm of the ages,
- Figure 7.4b shows a dynamic change (transition) in the mode of the earnings, after the age of 40.

The rest of the chapter is organized as follows: Sect. 7.2 provides a general framework, including a finite mixture approach of the enriched model via a regression curve. In Sect. 7.3, an enriched model is explored via the mode, infused by the mode graphical display to modal income data. Numerical studies, in Sect. 7.4, illustrate the value added by the proposed methodology of a data-enriched lognormal model. Section 7.5 finalises the chapter with last thoughts.

7.2 Enriched Lognormal Model via Regression Curve

7.2.1 *General Framework*

The modelling approach in this chapter comprises three distinct building blocks:

- Firstly, the lognormal as the conditional distribution, with Y as the response variable that models income, and X as the random covariate, age, with the distribution of X referred to as the mixture distribution;

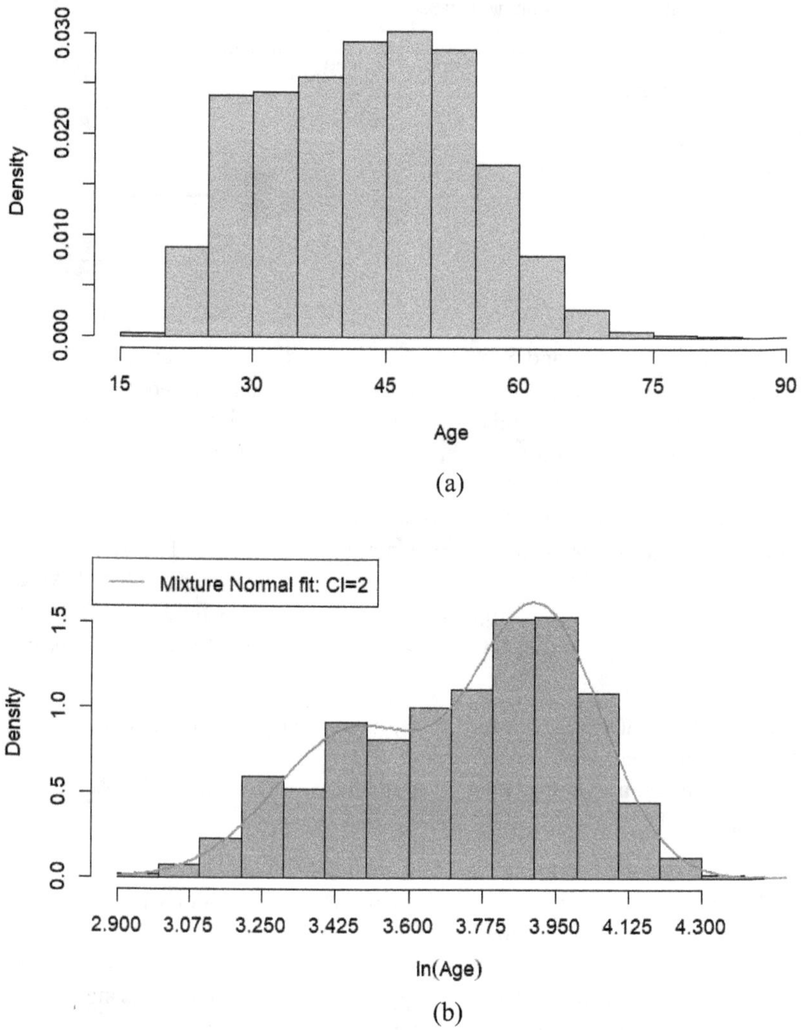

Fig. 7.2 Income data set: (**a**) Age and (**b**) the log age histograms

- Secondly, a functional relationship for the expectation of $Y|x$ is assumed by observing the regression curve; and
- Thirdly, the enriched marginal (compound; unconditional) distribution, is formed by superimposing the basic random component (mixture distribution) on the conditional distribution, to model Y unconditionally.

The joint probability density function (pdf) of the random variables Y and X can be factorized as

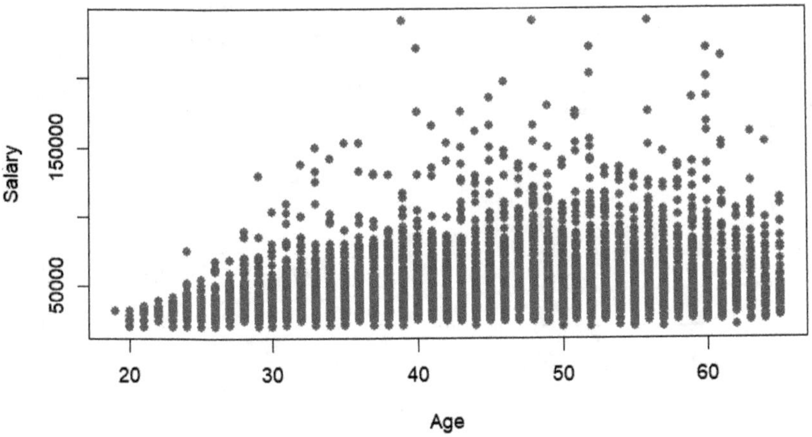

(a)

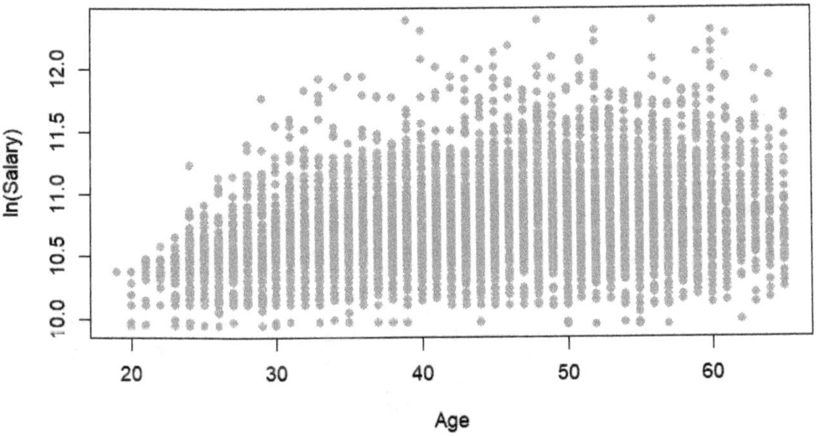

(b)

Fig. 7.3 Plot of salaries per age (**a**) and log salaries per age (**b**)

$$p(x, y; \boldsymbol{\xi}) = p\left(y|x; \xi_y\right) p(x; \xi_x) \tag{7.1}$$

where $\boldsymbol{\xi} = \left(\xi_x^{'}; \xi_y^{'}\right)$ is a vector of the unknown parameters in (7.1). The parametric link function between the response variable Y and the covariate X, is given as $E[Y|x;\ \boldsymbol{\beta}] = \mu_Y(x; \boldsymbol{\beta})$, where $\boldsymbol{\beta}$ denotes the vector of the regression parameters. From the basis of the income data set, let $Y|x \sim LOG(\mu_Y(x), \sigma)$ follow a lognormal distribution with pdf given by

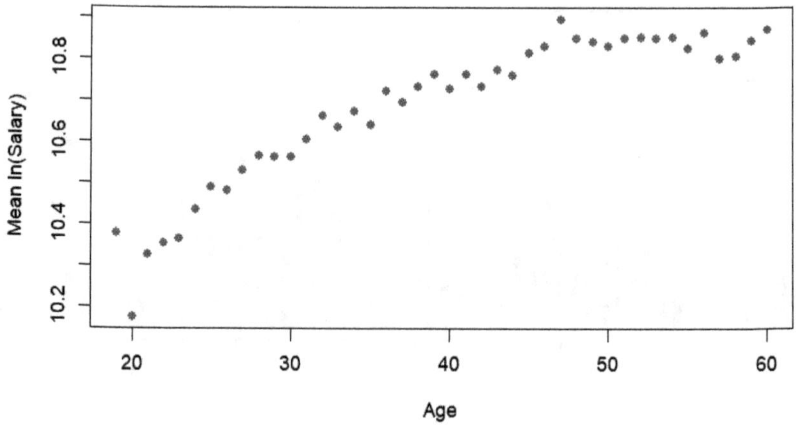

(a) Mean log salaries per age.

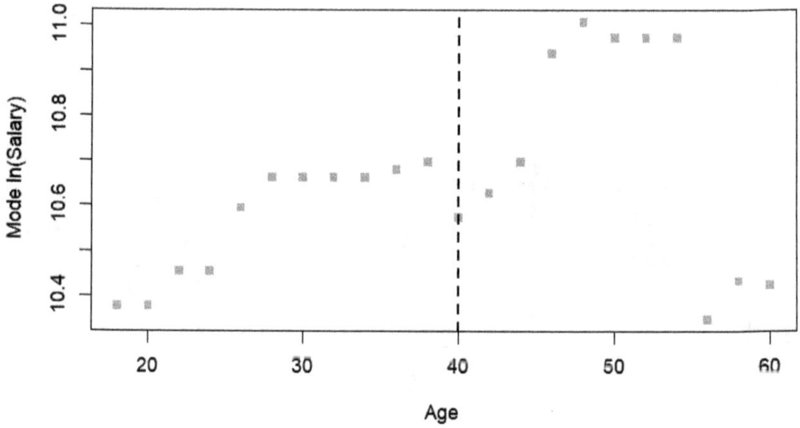

(b) Mode log salaries per age group of size two. The black dashed line serves as a benchmark in the change of trend from linear to sinusoidal.

Fig. 7.4 Mean of log salary per age (**a**) and mode of log salary per age group (**b**)

$$p\left(y|x;\,\mu_Y\left(x;\boldsymbol{\beta}\right),\sigma\right)=\frac{1}{\sqrt{2\pi}\sigma y}\exp\left(-\frac{1}{2\sigma^2}\left(\ln y-\mu_Y\left(x\right)\right)^2\right),\quad y>0, \tag{7.2}$$

where $\mu_Y(x) \in \mathbb{R}$ and $\sigma > 0$, and

$$E\left[Y\,|x;\,\boldsymbol{\beta}\right]=\exp\left(\mu_Y\left(x\right)+\frac{1}{2}\sigma^2\right),\quad \beta_0,\beta_1\in\mathbb{R}. \tag{7.3}$$

Historically, the bivariate (and multivariate) normal is mathematically an attractive tool for analyzing data, and has both the marginal $p\left(x;\xi_x\right)$- and conditional $p\left(Y|x;\xi_y\right)$ distributions as normal—often called pseudo-models. The k-dimensional Pseudo-F distribution was introduced by Filus et al. (2009), while Arnold and Manjunath (2021) considered the Pseudo-Poisson models. In this chapter, it is assumed that the conditional and marginal are of lognormal nature, thus constituting a pseudo-model. The following theorem develops the pdf of the enriched lognormal distribution.

Theorem 7.1 *Let* $Y|x \sim LOG(\mu_Y(x), \sigma)$ *and* $X \sim LOG(\mu, \zeta)$. *Assume that* $E\left[Y\,|x;\ \boldsymbol{\beta}\right] = \exp\left(\beta_1 \ln(x) + \beta_0\right)$, *with* β_0 *and* β_1 *as some constant values. Then* $Y \sim LOG\left(\beta_0 + \beta_1\mu - \frac{\sigma^2}{2}, \sqrt{\beta_1^2\zeta^2 + \sigma^2}\right)$.

Proof From $E\left[Y\,|x;\ \boldsymbol{\beta}\right] = \exp\left(\beta_1 \ln(x) + \beta_0\right)$ and (7.3) follows that $\mu_Y(x) = \beta_0 + \beta_1 \ln x - \frac{1}{2}\sigma^2$. Therefore the conditional distribution of $Y|x$ is

$$p\left(y|x\right) = \frac{1}{\sqrt{2\pi}\sigma y} \exp\left(-\frac{1}{2\sigma^2}\left(\ln y - \beta_0 - \beta_1 lnx + \frac{1}{2}\sigma^2\right)^2\right), \quad y > 0,$$

where $\beta_0, \beta_1 \in \mathbb{R}$ and $\sigma > 0$. We have that

$$p\left(x;\mu,\zeta\right) = \frac{1}{\sqrt{2\pi}\zeta x} \exp\left(-\frac{1}{2\zeta^2}\left(\ln x - \mu\right)^2\right), \quad x > 0, \tag{7.4}$$

where $\mu \in \mathbb{R}$ and $\zeta > 0$. Therefore the unconditional (marginal) distribution of Y is given by

$$p\left(y\right) = \frac{1}{2\pi\sigma\zeta y}\times$$

$$\int_0^\infty \exp\left(-\frac{1}{2}\left(\frac{1}{\zeta^2}\left(\ln x - \mu\right)^2 + \frac{1}{\sigma^2}\left(\ln y - \beta_0 - \beta_1 lnx + \frac{1}{2}\sigma^2\right)^2\right)\right) x^{-1}dx. \tag{7.5}$$

By completing the square of the exponent term in (7.5), it follows that:

$$p\left(y\right) = \frac{1}{2\pi\sigma\zeta y}\times \tag{7.6}$$

$$\int_0^\infty \exp\left(-\frac{\left(\beta_1^2\zeta^2 + \sigma^2\right)}{2\zeta^2\sigma^2}\left(\left(\ln x - \frac{\beta_1\zeta^2(\ln y - \beta_0 + \frac{1}{2}\sigma^2) + \sigma^2}{\beta_1^2\zeta^2 + \sigma^2}\right)^2\right)\right) x^{-1}dx$$

$$\times \exp\left(-\frac{\left(\left(\ln y-\beta_0+\frac{1}{2}\sigma^2\right)^2-2\beta_1\mu\left(\ln y-\beta_0+\frac{1}{2}\sigma^2\right)+\mu^2\beta_1^2\right)}{2\left(\beta_1^2\zeta^2+\sigma^2\right)^2}\right).$$

The integral in (7.6) can be identified as proportional to a lognormal distribution with mean $\frac{\beta_1\zeta^2(\ln y-\beta_0+\frac{1}{2}\sigma^2)+\sigma^2\mu}{\beta_1^2\zeta^2+\sigma^2}$ and variance $\frac{\zeta^2\sigma^2}{(\beta_1^2\zeta^2+\sigma^2)}$. Hence, the above simplifies to:

$$p(y)=\frac{1}{\sqrt{2\pi}\,y\left(\beta_1^2\zeta^2+\sigma^2\right)^{\frac{1}{2}}}$$
$$\times \exp\left(-\frac{1}{2\left(\beta_1^2\zeta^2+\sigma^2\right)}\left(\ln y-\beta_0-\beta_1\mu+\frac{1}{2}\sigma^2\right)^2\right), \tag{7.7}$$

where $\mu, \beta_1, \beta_0 \in \mathbb{R}$ and $\zeta, \sigma > 0$. □

An expression for the moments of Y follow immediately as

$$E(Y^h)=\exp\left\{h\left(\beta_0+\beta_1\mu-\frac{1}{2}\sigma^2\right)+\frac{h^2}{2}\left(\beta_1^2\zeta^2+\sigma^2\right)\right\}. \tag{7.8}$$

To determine the functional form of $\mu_Y(x;\boldsymbol{\beta})$, we use the graphical information obtained from the regression function, observed from the data (the functional form that the curve represents). As an intuitive case, consider $\mu_Y(x;\boldsymbol{\beta})=\beta_0+\beta_1 x$, $\boldsymbol{\beta}=(\beta_0,\beta_1)$, then $\exp\left(\mu_Y(x)+\frac{1}{2}\sigma^2\right)=\beta_0+\beta_1 x$ and $\mu_Y(x)$ can be written in terms of $\boldsymbol{\beta}$ and σ.

Assume that Y represents the income with pdf $p\left(Y|x;\xi_y\right)$, and X the level of income with pdf $p\left(x;\xi_x\right)$. In this case, $E\left[Y|x;\ \boldsymbol{\beta}\right]$ indicates the expected income for a specific level of a factor that has an impact on income. Figure 7.5 illustrates an example, for a given factor x, the expected income for a given career path is y' versus the expected income for other career path, y''. Subsequently, one can refer to the parameter β_1 as an inequality parameter when modelling income. It is assumed that for a given x, the group of individuals at this level is homogeneous with the understanding that a homogeneous group in the income context implies that they do not follow necessarily the same career but they are homogeneous with respect to income factors for example age, education and experience.

7.2.2 *Finite Mixture Approach*

A close inspection of the age profile displayed in Fig. 7.2, reflects the possibility of a two-class component for modelling. A two-class income distribution framework

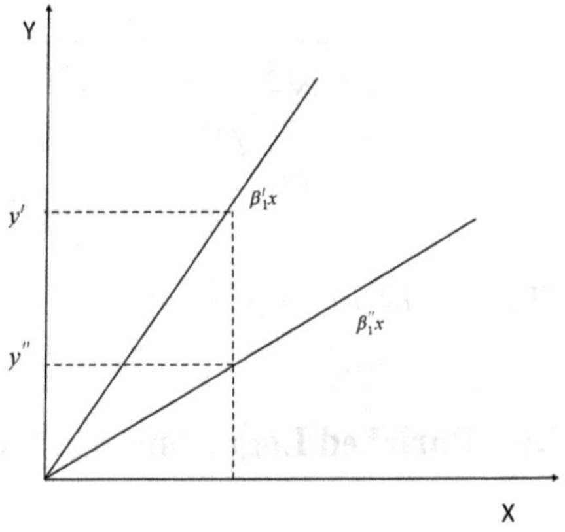

Fig. 7.5 Inequality parameter

was suggested by Yakovenko and Silva (2005), which paved the way for finite mixture modelling (see also Flachaire and Nunez, 2007). By assuming a finite mixture model for the variable X, leads to a finite mixture distribution for Y that supports the accepted labour market segmentation profile where the income of most populations exhibits strong heterogeneity. Following the assumptions from Theorem 1, the component unimodal pdfs should all belong to the same parametric family. The finite mixture $p(x)$ then assumes the form

$$p\left(x;\boldsymbol{\phi},\boldsymbol{\mu},\boldsymbol{\zeta}\right)=\sum_{m=1}^{k}\phi_m f\left(x;\mu_m,\zeta_m\right),\qquad x>0. \tag{7.9}$$

In (7.9):

- $f(x;\mu_m,\zeta_m)$ is the component with pdf in (7.4) with parameters μ_m and ζ_m;
- $\boldsymbol{\phi}=(\phi_1,\ldots,\phi_k)$ is the vector of mixture weights where $\phi_m\in(0,1]$, $\sum_{m=1}^{k}\phi_m=1$;
- $\boldsymbol{\mu}=(\mu_1,\ldots,\mu_k)$ is the vector of component modes μ_m;
- $\boldsymbol{\zeta}=(\zeta_1,\ldots\zeta_k)$ is the component parameters ζ_m.

We assume a two-component finite mixture model ($k=2$). Therefore the unconditional distribution of Y has the following pdf

$$p\left(y\right)=\frac{\phi_1}{\sqrt{2\pi}\,y\left(\beta_1^2\zeta_1^2+\sigma^2\right)^{\frac{1}{2}}}\times\exp\left(-\frac{1}{2\left(\beta_1^2\zeta_1^2+\sigma^2\right)}\left(\ln y-\beta_0-\beta_1\mu_1+\frac{1}{2}\sigma^2\right)^2\right)$$

$$+\frac{(1-\phi_1)}{\sqrt{2\pi}y\left(\beta_1^2\zeta_2^2+\sigma^2\right)^{\frac{1}{2}}}$$
$$\times\exp\left(-\frac{1}{2\left(\beta_1^2\zeta_2^2+\sigma^2\right)}\left(\ln y-\beta_0-\beta_1\mu_2+\frac{1}{2}\sigma^2\right)^2\right), \quad (7.10)$$

for $\mu_1, \mu_2, \beta_0, \beta_1 \in \mathbb{R}$ and $\zeta, \sigma > 0$.

7.3 Enriched Lognormal via the Mode

Modal regression presents an appropriate alternative, where the regression function of variable Y conditioned on the value of X is regarded as the mode of the conditional density when analysing highly asymmetric data sets, with the assumption that the conditional density of $Y|X$ is unimodal (Sager & Thisted, 1982; Yao et al., 2012). The mode is easy to locate graphically and it is of vital importance to identify where the majority of the data is in large data sets.

Most often, real life data sets exhibit asymmetric behaviour and symmetric distributions are not worthy contenders for fitting these type of data sets. The limitations of the mean and median as opposed to the mode are illustrated by Chacón (2020), since the mean might be positioned far from the mode. The mode is known to be less affected by larger or smaller values (Carreira-Perpinan, 2000; Nolan, 1998). Further motivation for the mode reparameterization of a distribution, is that a finite mixture of mode parameterized distributions may fit into the framework of finite mixtures model-based clustering. The latter is commended for their ability to model multi-modal distributions where each mode (bump) characterizes a certain entity. These benefits have contributed to the current interest in conducting research on mode statistics (see for example Arias-Castro and Qiao, 2023, Dalenius, 1965, Cheng, 1995, Lee, 1989, Kemp and Silva, 2012, Sando and Hino, 2020, and Yao et al., 2012).

Several authors built flexible models by focusing on mode parameterized versions of existing candidates (see Punzo et al., 2018, Punzo, 2019, Bagnato and Punzo, 2013, and Chen, 2000). In this way, such re-parameterized models allow for interpretable parameters by accounting for location and variability. The pdf of a random variable with a lognormal distribution is given by

$$p\left(y;\mu,\sigma\right)=\frac{1}{\sqrt{2\pi}\sigma y}\exp\left(-\frac{1}{2\sigma^2}\left(\ln y-\mu\right)^2\right), \quad y>0, \quad (7.11)$$

where $\mu \in \mathbb{R}$ and $\sigma > 0$. The mode parameterized lognormal distribution, proposed by Mazza and Punzo (2019), has the following pdf

$$p_{mode}\,(y;\theta,\nu)=\frac{1}{\sqrt{2\pi\nu y}}\exp\left(-\frac{1}{2\nu}\,(\ln y-\ln\theta-\nu)^2\right),\quad y>0,\tag{7.12}$$

where $\theta>0$ and $\nu>0$. In this parameterization, θ represents the mode of Y, and ν is a parameter not equal to the variance, but is related to the variability of Y. The pdfs in (7.11) and (7.12) are linked as follows

$$\begin{cases}\mu=\ln\theta+\nu,\\ \sigma^2=\nu\end{cases}\Longrightarrow\begin{cases}\theta=\exp(\mu-\sigma^2)\\ \nu=\sigma^2\end{cases}.$$

The emphasis of Sect. 7.3 is to extend the idea of developing an enriched lognormal model with insight from the behaviour of the mode. The approach mirrors Roux and Bekker (1989) and Wesolowski (1995), where the information revealed by the regression curve is incorporated in the conditional distribution of the response variable.

The data-enriched approach via the mode is outlined as follows: From (7.12), we assume that the conditional distribution is given by

$$p_{mode}\,(y|x;\theta\,(x)\,,\nu)=\frac{1}{\sqrt{2\pi\nu y}}\exp\left(-\frac{1}{2\nu}\,(\ln y-\ln\theta\,(x)-\nu)^2\right),\quad y>0,\tag{7.13}$$

with the focus on the $mode(Y|x;\boldsymbol{a})\equiv\theta\,(x)$, where $\boldsymbol{a}$ denotes the mode parameters. $\theta\,(x)$ is determined from the data.

Theorem 7.2 *Let* $Y|x\sim modeLOG(\theta\,(x)\,,\nu)$, $X\sim modeLOG(m,\varsigma)$. *Assume that* $mode(Y;\boldsymbol{a})=ax^b$, *with* $a,\ b\in\mathbb{R}$ *constants. Then* Y *has a lognormal distribution with parameters* $\xi=(a,b,\sigma,m,\varsigma)$.

Proof From (7.13)

$$p_{mode}\,(x;m,\varsigma)=\frac{1}{\sqrt{2\pi\varsigma x}}\exp\left(-\frac{1}{2\varsigma}\,(\ln x-\ln m-\varsigma)^2\right)\tag{7.14}$$

follows that

$$p\,(y;\xi)=\frac{1}{2\pi\sqrt{\nu\varsigma y}}\times\tag{7.15}$$

$$\int_0^\infty\exp\left(-\frac{1}{2}\left(\frac{1}{\varsigma}\,(\ln x-\ln m-\varsigma)^2+\frac{1}{\nu}\,(\ln y-\ln a-b\ln x-\nu)^2\right)\right)x^{-1}dx.$$

By completing of the square of the exponent term in (7.15), it follows that

$$p\left(y;\xi\right)=\frac{1}{2\pi\sqrt{\nu\varsigma y}}\times \tag{7.16}$$

$$\int_0^{\infty}\exp\left(-\frac{\left(b^2\varsigma+\nu\right)}{2\nu\varsigma}\left(\ln x-\frac{\left(\ln y-\ln a-\nu\right)b\varsigma+\nu\left(\ln m+\varsigma\right)}{b^2\varsigma^2+\nu}\right)^2\right)x^{-1}dx\ \times$$

$$\exp\left(\frac{2\left(\ln y-\ln a-\nu\right)b\left(\ln m+\varsigma\right)-\left(\ln y-\ln a-\nu\right)^2-b^2\left(\ln m+\varsigma\right)^2}{2\left(b^2\varsigma+\nu\right)}\right).$$

The integral in (7.16) can be identified as proportional to a lognormal distribution. Subsequently, the result follows as

$$\begin{aligned}p\left(y;\xi\right)=&\frac{1}{2\pi\sqrt{\left(b^2\varsigma+\nu\right)y}}\\&\times\exp\left(-\frac{1}{2\left(b^2\varsigma+\nu\right)}\left(\ln y-\left(\ln a+\nu+b\left(\ln m+\varsigma\right)\right)\right)^2\right).\end{aligned} \tag{7.17}$$

□

The pdf of the enriched marginal (compound; unconditional) distribution, formed by superimposing the mixing distribution, $p_{mode}\left(x;\xi_x\right)$, on the conditional distribution to model the income data, is given by

$$p_{mode}\left(y;\xi\right)=\int_{\mathbb{R}+}p_{mode}\left(Y|x;\xi_y\right)p_{mode}\left(x;\xi_x\right)dx.$$

To accommodate the observations from graph age and graph mode (see Figs. 7.2 and 7.3), an enrichment approach will be followed via the mode being of stochastic nature.

7.4 Numerical Studies

This section evaluates the performance of the two types of enriched models derived from (7.2)–(7.17) and comparing them to that of the usual lognormal distribution given by (7.11). Estimating the location and scale parameters in (7.11) is done using its maximum likelihood estimators (MLE). That is, for μ_o, σ_o^2 the parameters of a lognormal distribution, their respective MLEs are

$$\hat{\mu_o} = \frac{1}{n}\sum_{i=1}^{n} ln(y_i), \text{ and} \tag{7.18}$$

$$\hat{\sigma_o^2} = \frac{1}{n}\sum_{i=1}^{n} \left(ln(y_i) - \hat{\mu_o}\right)^2. \tag{7.19}$$

7.4.1 *Simulation Study 1*

Consider a simulated random sample of size 300 of a random variable X with distribution $LOG(3.77, 0.113)$ and set $y = e^{\beta_0} x^{\beta_1} E$ such that $\beta_0 = 0, \beta_1 = 2$, and the error random variable $E \sim LOG(9.46, 0.153)$. This is visually illustrated in Fig. 7.6.

The usual lognormal (7.2), as well as the enriched lognormal pdfs (7.7) are fitted to the sample y, where their performances are analysed using AIC and BIC (Akaike, 1981; Schwarz, 1978) as goodness-of-fit measures. The loglikelihood is maximized with respect to $(\mu, \sigma, \zeta, \beta_0, \beta_1)$ using the `optim` function ('L-BFGS-B' method in R). The lower bounds and the starting values of each parameter are set as $(-\infty, 0, 0.01, -\infty, -\infty)$ and $(0, 0, 0.1, 0, 0.1)$, respectively. The above-mentioned steps are ran repeatedly to derive the sampling distributions of the estimates and the goodness-of-fit values. The results are displayed in Fig. 7.7 as boxplots.

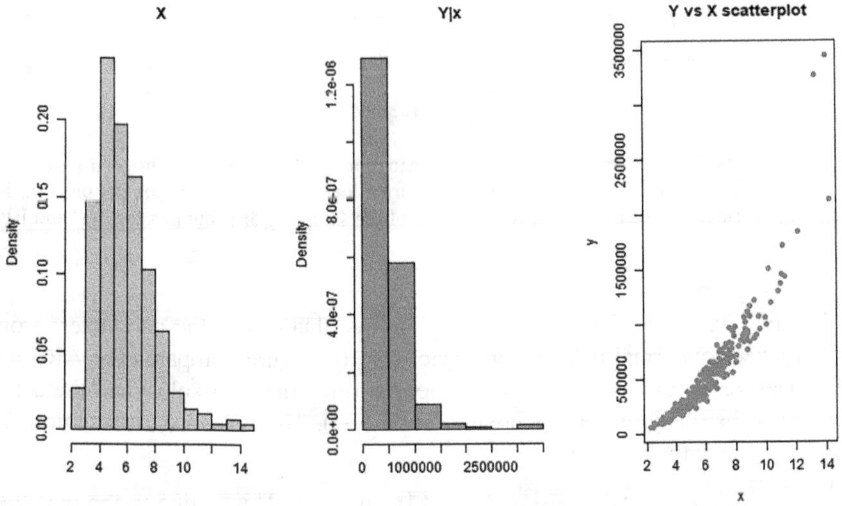

Fig. 7.6 Histogram of simulated samples: X (left), $Y|x$ (middle), and a scatterplot demonstrating correlation between Y and X (right)

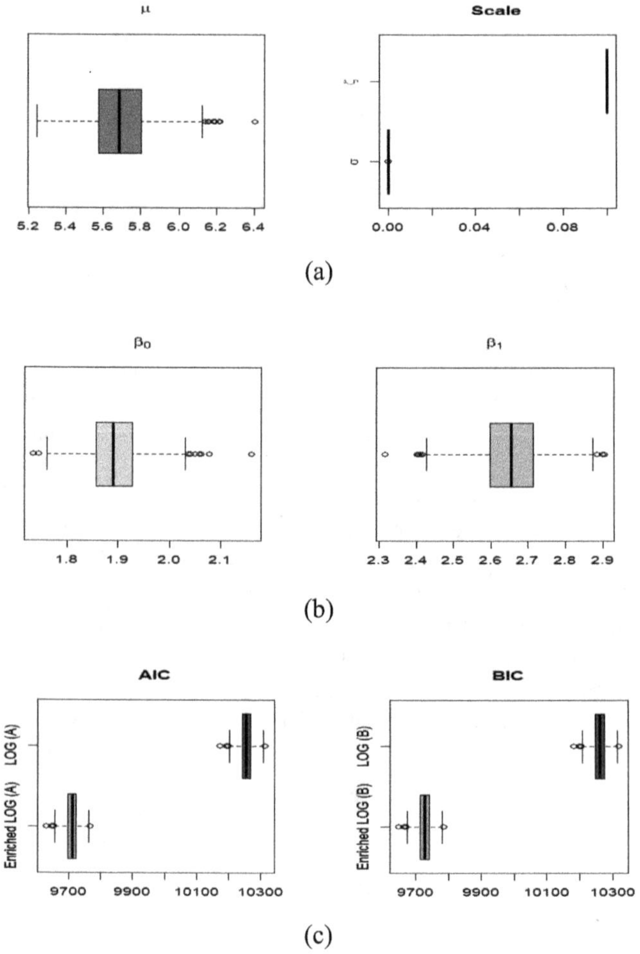

Fig. 7.7 Comparison between the lognormal distribution (7.2) and the data-enriched lognormal distribution (7.7): (**a**) Boxplots of the sampling distributions for μ, ζ, and σ. (**b**) Boxplots of the sampling distributions for β_0 and β_1. (**c**) Boxplots of the sampling distributions for AIC and BIC measures

Figure 7.7a and b illustrates the sampling distributions of the parameters from the enriched lognormal distribution. Panel c of the figure compares the AIC and BIC measures of the fitted enriched lognormal (the orange boxplot) and the usual lognormal distributions (purple boxplot at the top). The sampling distributions in Fig. 7.7 indicate the following:

- σ approaches zero when using the `optim` function. This is due to the bias that $\frac{1}{2}\sigma^2$ introduces into the enriched lognormal pdf at (7.7). Thus, maximising the likelihood results in the variance getting close to zero.

- The location of β_0 is not close to zero, but it is common for regression intercepts to be particularly sensitive to the bias.
- The spread of the AIC and BIC measures of the enriched lognormal pdf (7.7) are smaller than that of the usual lognormal pdf (7.2). Smaller measures are indicative of a better fit. Despite the bias present in the regression estimates, the enriched lognormal pdf outperforms the usual lognormal pdf (7.2).

7.4.2 Simulation Study 2

In Sect. 7.4.1 we illustrate the performance and parameter recovery of the enriched lognormal pdf by simulating data originating from said enriched lognormal pdf given by (7.7). In this section we aim to do the same, but by using the enriched lognormal pdf via the mode given by (7.17). That is, consider a simulated random sample, of size 300, for random variable X, with distribution $modeLOG(55, 0.1)$ and set $y = e^{\beta_0} x^{\beta_1} E$ such that $\beta_0 = 0, \beta_1 = 2$, and the error random variable $E \sim LOG(9.46, 0.153)$. The results of this simulation process are illustrated in Fig. 7.8.

In this simulation study, a two step approach is followed for estimation: β_0, β_1 are estimated numerically, separate from m, ζ, and ν. Thereafter, the usual lognormal (7.2) and the enriched lognormal pdf (7.17) are fitted to the sample y, where their performances are compared using AIC and BIC as goodness-of-fit measures. The loglikelihood is maximized with respect to (m, ζ, ν) using the `optim` function ('L-

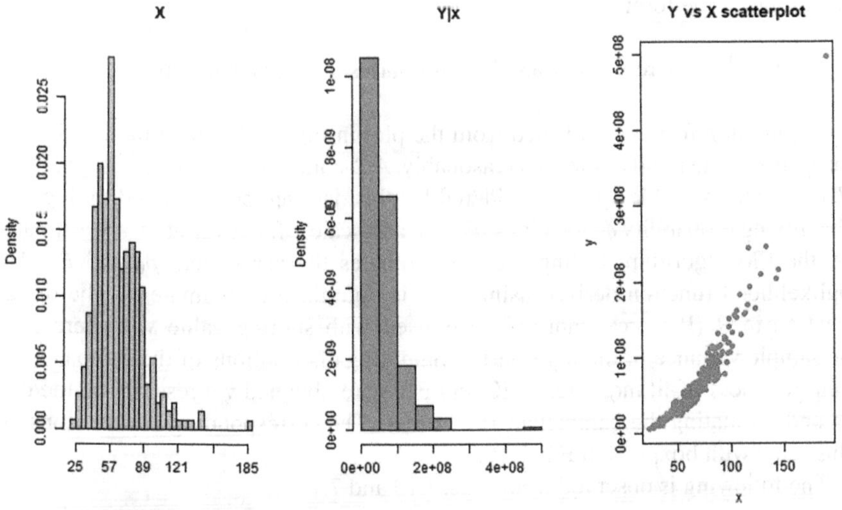

Fig. 7.8 Histogram of simulated samples: X (left), $Y|x$ (middle), and a scatterplot demonstrating correlation between Y and X (right)

BFGS-B' method in R R Core Team (2024)). The lower bounds of the method's search algorithm and the starting values of each parameter are set as $(0, 0, 0.01)$ and $(0, 0, 0.1)$, respectively. The above-mentioned steps are ran repeatedly to derive sampling distributions of the estimates and the goodness-of-fit measures, with the results displayed in Fig. 7.9.

Figure 7.9a and b illustrates the boxplots of the sampling distributions of the parameters of the enriched mode lognormal distribution. Figure 7.9c compares the AIC and BIC measures of the fitted enriched mode lognormal distribution (the orange boxplot) and the usual lognormal distributions (purple boxplot at the top). These sampling distributions indicate the following:

- ζ approaches zero when using the `optim` function, which is due to the potential bias introduced by $b \times \zeta$ in the pdf (7.17). Thus, maximising the likelihood results in the variance component getting closer to zero.
- The location of β_0 does not approach zero.
- The spread of the AIC and BIC measures of the enriched mode lognormal pdf (7.17) are smaller than that of the usual lognormal pdf (7.2). Smaller measures are indicative of a better fit. Despite the bias present in the regression estimates, the enriched mode lognormal model outperforms the usual lognormal model.

7.4.3 *Real Data Fitting*

Consider the income data set discussed in the introduction. The proposed data-enriched lognormal (7.7) is fitted to the income data set, where the chosen regression function is of the form:

$$y = e^{\beta_0} x^{\beta_1}, \text{ where } y \text{ and } x \text{ are the log salaries and ages, respectively.} \tag{7.20}$$

The regression form is motivated from the plot in Fig. 7.12, where the fitted curve encapsulates the trend of the data reasonably. A 2-component mixture of lognormals (7.10) for the variable age is considered for this data set, as illustrated by Fig. 7.2. The mixing probability ϕ, locations μ_1, μ_2, and scale ζ for covariate x are estimated via the EM algorithm. Fitting a curve estimates the parameters β_0 and β_1. The loglikelihood function derived using (7.7) is optimized to estimate σ. The `optim` function in R (R Core Team, 2024) is used with starting value s_y, where s_y is the sample variance of the log salaries. Sampling distributions of the estimates and their goodness-of-fit measures (AIC and BIC) are obtained via resampling the data set and estimating the parameters 1000 times. The corresponding distributions are illustrated with boxplots in Fig. 7.10:

The following is observed from Figs. 7.10 and 7.11:

- The mixing probability in Fig. 7.10 does not include zero in its spread, indicating that opting for a mixture of lognormal distributions for age is appropriate. This is

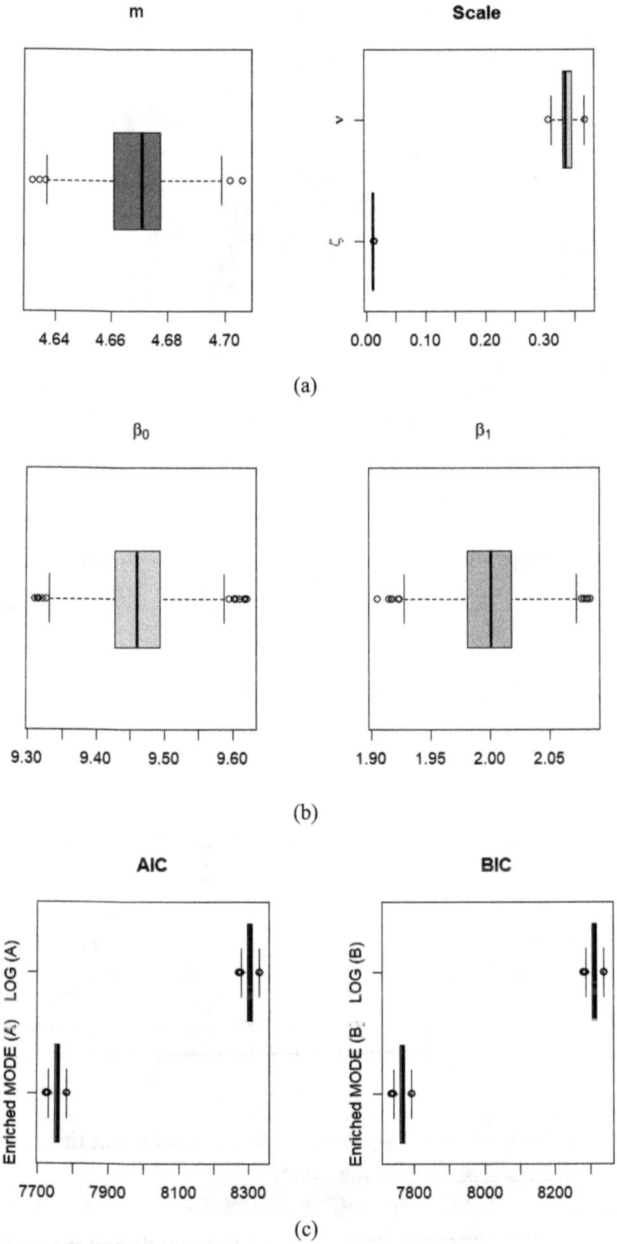

Fig. 7.9 Comparison between the lognormal distribution (7.2) and the data-enriched mode lognormal distribution (7.17): (**a**) Boxplots of the sampling distributions for m, ζ, and ν. (**b**) Boxplots of the sampling distributions for β_0 and β_1. (**c**) Boxplots of the sampling distributions for AIC and BIC measures

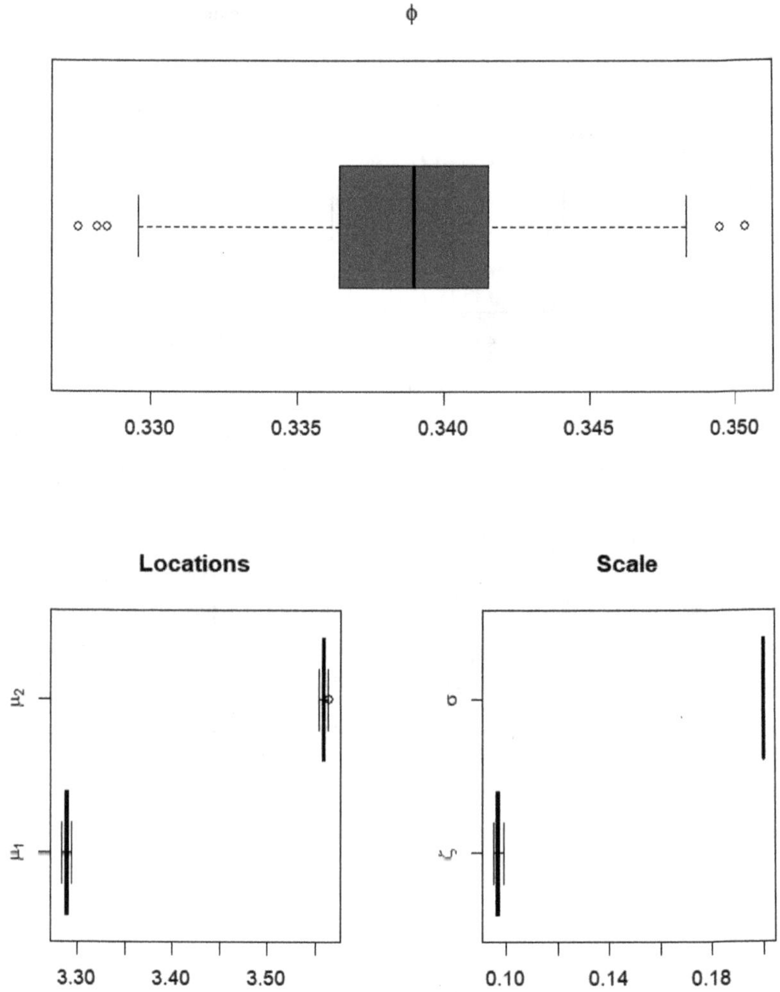

Fig. 7.10 Boxplots of the sampling distributions of the estimated mixing proportion, mean and scale parameters for the finite mixture of the data-enriched model (7.10)

confirmed when looking at the spread of μ_1, μ_2. Notice that they do not overlap, despite their distances being less than one.

- By the spread of the AIC and BIC measures for the enriched finite mixture model and the usual model in Fig. 7.11, the enriched model measures are lower when compared to the usual model. That is, the mixture enrichment to the model allowed it to fit better to the data, outperforming the usual lognormal model's fit.

As illustration for the mode enriched lognormal distribution to a real data application, we consider again the income data set as described previously. However, by

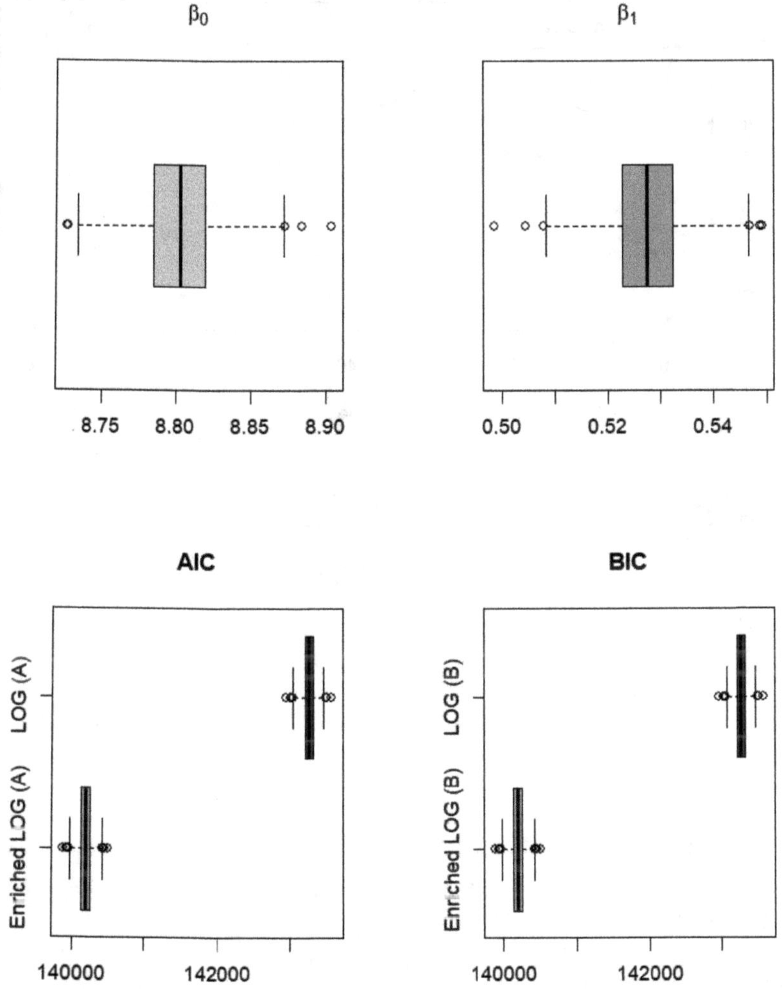

Fig. 7.11 Boxplots of the sampling distributions of the estimated regression coefficients and goodness-of-fit measures for the finite mixture of data-enriched model (7.10)

investigation (see Fig. 7.4b), to obtain meaningful results, we truncate the covariate (Age) between 20 and 40. The mode per age is plotted in Fig. 7.13:

We fit the pdf given by (7.7); particularly a curve to the mode per age, leaving only m, ζ, ν to be estimated using the `optim` function in R (R Core Team, 2024). Sampling distributions for $m, \zeta, \nu, \beta_0, \beta_1$ are obtained via bootstrapping 100 times, see Figs. 7.12 and 7.13.

To evaluate its performance, the AIC and BIC values are computed. The following is observed from Figs. 7.14 and 7.15:

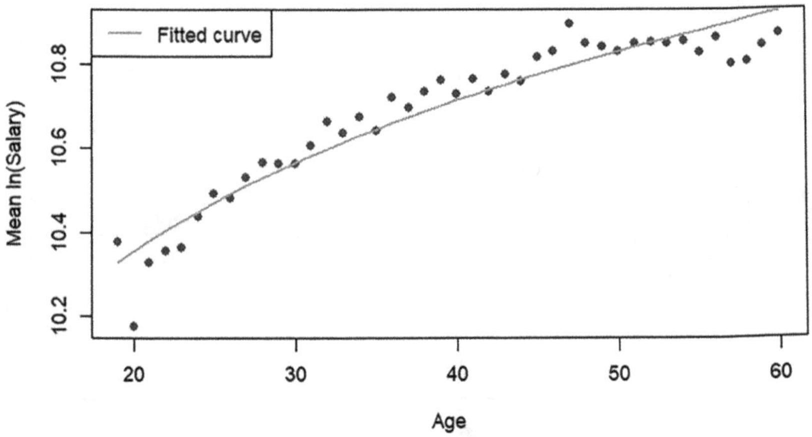

Fig. 7.12 Mean log salary per age

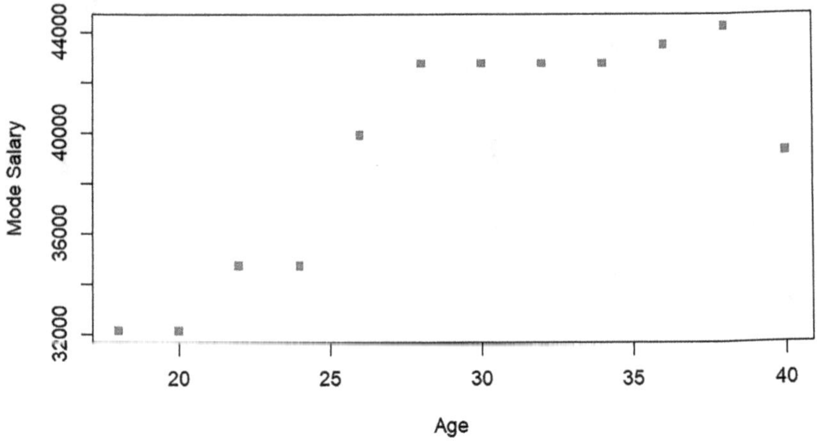

Fig. 7.13 Mode salary per age

- The mode m covers the age group of the young working class, which is in line with our truncation approach.
- By the spread of the AIC and BIC measures for the enriched mode pdf (7.17) and the usual lognormal model (7.2), the enriched model measures are lower when compared to the usual model. That is, the finite mixture enrichment to the model allowed it to fit better to the data, outperforming the usual lognormal model's fit.

Note, the truncated data set was used for modelling by the enriched mode lognormal. This is just an illustration of the approach for the practitioner.

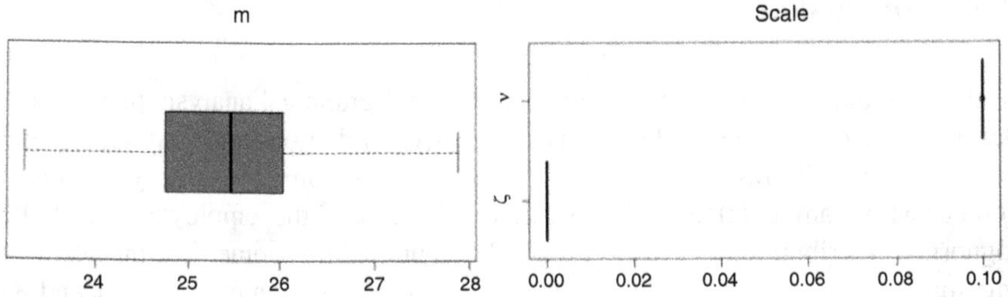

Fig. 7.14 Boxplots of the sampling distributions of estimated scale parameters

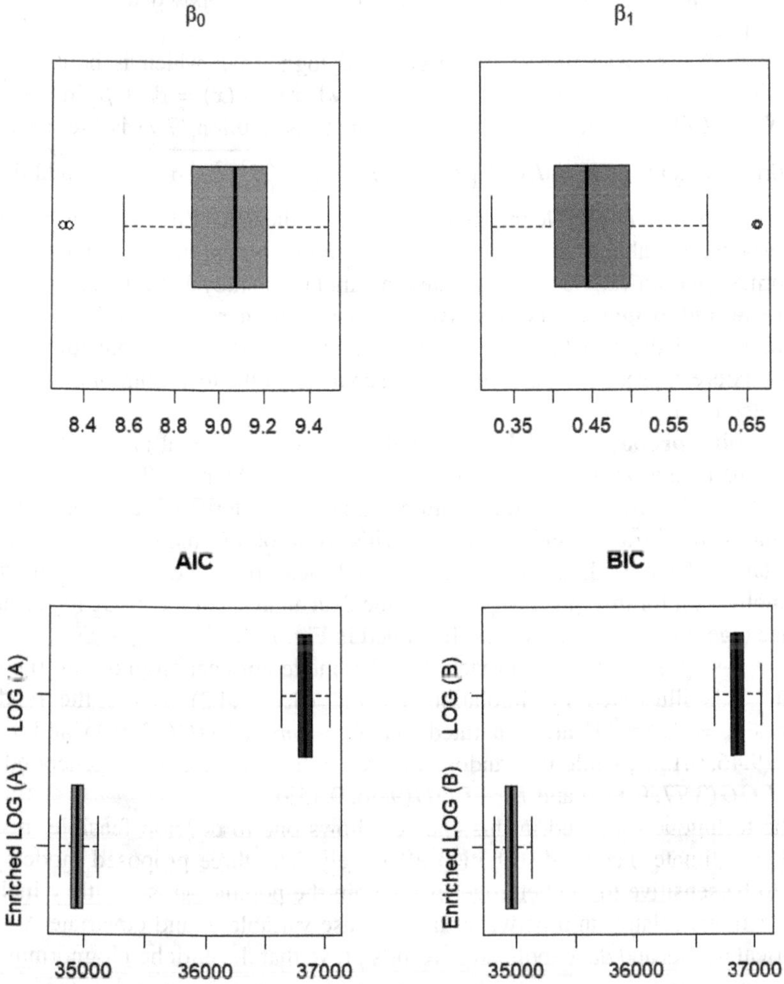

Fig. 7.15 Boxplots of the parameter estimates of data-enriched mode models (7.17), β_0, β_1 and goodness-of-fit measures of a lognormal (7.2) and a data-enriched mode lognormal distribution (7.17)

7.5 Conclusion

In this chapter, we demonstrate the role that initial graphical analysis plays in the development of flexible models. This perspective is demonstrated via an income data set, where the response variable Y denotes the income (i.e. yearly employee compensation amounts) and the covariate X denotes the employee ages. The lognormal distribution, as a well known distribution for income, is considered in this study, as the suitable conditional and marginal distribution for $Y|X = x$ and X, respectively. The aim of this study, is to develop enriched lognormal distributions for the response variable Y by incorporating the functional relationship between income, Y, and the employee ages, X. Three models are proposed and developed in this chapter.

The **first proposed model** is the enriched lognormal, which is built via the conditional mean of $Y|x \sim LOG(\mu_Y(x), \sigma)$, where $\mu_Y(x) = \beta_0 + \beta_1 \ln x - \frac{1}{2}\sigma^2$ and $X \sim LOG(\mu, \zeta)$. The resulting marginal distribution (7.7) is the enriched lognormal denoted as $Y \sim LOG\left(\beta_0 + \beta_1\mu - \frac{\sigma^2}{2}, \sqrt{\beta_1^2\zeta^2 + \sigma^2}\right)$. The additional parameters of the enriched lognormal distribution indicates the dependency between the response variable Y and covariate X. A simulation approach for an enriched lognormal random variate y is illustrated in simulation study 1 (Sect. 7.4.1).

The **second proposed model** is a two-component finite mixture model developed via the pdfs of the enriched lognormal distribution (7.10). The motivation of the finite mixture approach stems from the data analysis of the logarithm of the ages, as illustrated in Fig. 7.2b.

The **third proposed model** is the mode-enriched lognormal (7.17), developed via a re-parameterization of the enriched lognormal distribution. The mode parameterized enriched lognormal distribution, allows for interpretable parameters by accounting for location and variability. This view of the mode emphasizes the importance of the mode as another considered measure of location in regression. The motivation for this model stems from the data analysis of the mode log salaries per age (age groups of size two) as illustrated in Fig. 7.4b.

A two-step simulation approach for the mode-enriched lognormal random variate y is illustrated in simulation study 2 (Sect. 7.4.2), where the random variates $y = e^{\beta_0} x^{\beta_1} E$ are generated via $X \sim modeLOG(55, 0.1)$ and $E \sim LOG(9.46, 0.153)$, while the random variates y in simulation 1 are generated via $X \sim LOG(3.77, 0.113)$ and $E \sim LOG(9.46, 0.153)$.

The technique proposed in this chapter allows one to develop feasible models that can estimate income distribution effectively. The three proposed models are built to be sensitive to the heterogeneity within the population, since they include the functional relationship between the response variable Y and covariate X. The empirical results and data application results prove that the enriched lognormal distributions, when compared to the lognormal distribution, are contenders for income modelling. For future exploration, an enriched contaminated mode lognormal model and finite mixtures of the enriched mode lognormal model can be considered in the development of flexible and feasible income data distributions.

Acknowledgments We thank the editors as well as the two anonymous reviewers for their constructive feedback and allowing us to present an improved manuscript. This work was based upon research supported in part by the National Research Foundation (NRF) of South Africa (SA), grant RA201125576565 & RA211204653274, nr 145681 & 151035; NRF ref. SRUG2204203865, and the Centre of Excellence in Mathematical and Statistical Sciences, based at the University of the Witwatersrand (SA). This work was also funded in part by national funds through the FCT—Fundação para a Ciência e a Tecnologia, I.P., under the scope of the projects UIDB/00297/2020 (https://doi.org/10.54499/UIDB/00297/2020) and UIDP/00297/2020 (https://doi.org/10.54499/UIDP/00297/2020) (Center for Mathematics and Applications). The opinions expressed and conclusions arrived at are those of the authors and are not necessarily to be attributed to the NRF.

References

Akaike, H. (1981). Likelihood of a model and information criteria. *Journal of Econometrics, 16*(1), 3–14.

Akhundjanov, S. B., & Toda, A. A. (2020). Is gibrat's "economic inequality" lognormal? *Empirical Economics, 59*, 2071–2091.

Arias-Castro, E., & Qiao, W. (2023). A unifying view of modal clustering. *Information and Inference: A Journal of the IMA, 12*(2), 897–920.

Arnold, B. C., & Manjunath, B. (2021). Statistical inference for distributions with one poisson conditional. *Journal of Applied Statistics, 48*(13–15), 2306–2325.

Bagnato, L., & Punzo, A. (2013). Finite mixtures of unimodal beta and gamma densities and the k-bumps algorithm. *Computational Statistics, 28*(4), 1571–1597.

Barroso, F. J. C., Garcia-Perez, C., & Prieto-Alaiz, M. (2020). Modelling income distribution using the log student'st distribution: New evidence for european union countries. *Economic Modelling, 89*, 512–522.

Carreira-Perpinan, M. A. (2000). Mode-finding for mixtures of gaussian distributions. *IEEE Transactions on Pattern Analysis and Machine Intelligence, 22*(11), 1318–1323.

Chacón, J. E. (2020). The modal age of statistics. *International Statistical Review, 88*(1), 122–141.

Chen, S. X. (2000). Probability density function estimation using gamma kernels. *Annals of the Institute of Statistical Mathematics, 52*(3), 471–480.

Cheng, Y. (1995). Mean shift, mode seeking, and clustering. *IEEE Transactions on Pattern Analysis and Machine Intelligence, 17*(8), 790–799.

Dagum, C. (2008). A new model of personal income distribution: specification and estimation. In: Chotikapanich, D. (Ed.), Modeling income distributions and Lorenz curves. Economic studies in equality, social exclusion and well-being (Vol. 5). New York: Springer. https://doi.org/10.1007/978-0-387-72796-7_1

Dalenius, T. (1965). The mode–a neglected statistical parameter. *Journal of the Royal Statistical Society Series A (General)*, 110–117.

Filus, J. K., Filus, L. Z., & Arnold, B. C. (2009). Families of multivariate distributions involving "triangular" transformations. *Communications in Statistics-Theory and Methods, 39*(1), 107–116.

Flachaire, E., & Nunez, O. (2007). Estimation of the income distribution and detection of subpopulations: An explanatory model. *Computational Statistics & Data Analysis, 51*(7), 3368–3380.

Hlasny, V. (2021). Parametric representation of the top of income distributions: Options, historical evidence, and model selection. *Journal of Economic Surveys, 35*(4), 1217–1256.

Jones, M. (2015). On families of distributions with shape parameters. *International Statistical Review, 83*(2), 175–192.

Kemp, G. C., & Silva, J. S. (2012). Regression towards the mode. *Journal of Econometrics, 170*(1), 92–101.

Kleiber, C., & Kotz, S. (2003). *Statistical size distributions in economics and actuarial sciences.* John Wiley & Sons.

Lee, M. J. (1989). Mode regression. *Journal of Econometrics, 42*(3), 337–349.

Ley, C. (2015). Flexible modelling in statistics: past, present and future. *Journal de la Société Française de Statistique, 156*(1), 76–96.

Ley, C., Babić, S., & Craens, D. (2021). Flexible models for complex data with applications. *Annual Review of Statistics and Its Application, 8*, 369–391.

Lubrano, M., & Ndoye, A. A. J. (2016). Income inequality decomposition using a finite mixture of log-normal distributions: A bayesian approach. *Computational Statistics & Data Analysis, 100*, 830–846.

Mazza, A., & Punzo, A. (2019). Modeling household income with contaminated unimodal distributions. In *New statistical developments in data science: SIS 2017, Florence, Italy, June 28–30* (pp. 373–391). Springer.

Mitzenmacher, M. (2004). A brief history of generative models for power law and lognormal distributions. *Internet Mathematics, 1*(2), 226–251.

Nolan, J. P. (1998). Parameterizations and modes of stable distributions. *Statistics & Probability Letters, 38*(2), 187–195.

Punzo, A. (2019). A new look at the inverse gaussian distribution with applications to insurance and economic data. *Journal of Applied Statistics, 46*(7), 1260–1287.

Punzo, A., Bagnato, L., & Maruotti, A. (2018). Compound unimodal distributions for insurance losses. *Insurance: Mathematics and Economics, 81*, 95–107.

R Core Team (2024). *R: A language and environment for statistical computing*. R Foundation for Statistical Computing. https://www.R-project.org/.

Roux, J., & Bekker, A. (1989). On the unconditional lognormal distribution. *Suid-Afrikaanse Tydskrif vir Natuurwetenskap en Tegnologie, 8*(4), 144–148.

Sager, T. W., & Thisted, R. A. (1982). Maximum likelihood estimation of isotonic modal regression. *The Annals of Statistics*, 690–707.

Sando, K., & Hino, H. (2020). Modal principal component analysis. *Neural Computation, 32*(10), 1901–1935.

Schneider, M. P., & Scharfenaker, E. (2020). Mixing it up: The case for finite mixture models to study the distribution of income. *The European Physical Journal Special Topics, 229*, 1685–1704.

Schwarz, G. (1978). Estimating the dimension of a model. *Annals of Statistics, 6*(2), 461–464.

Silber, J. (2023). *Research handbook on measuring poverty and deprivation.* Edward Elgar Publishing.

Warchold, A., Pradhan, P., & Kropp, J. P. (2021). Variations in sustainable development goal interactions: Population, regional, and income disaggregation. *Sustainable Development, 29*(2), 285–299.

Wesolowski, J. (1995). Bivariate distributions via a pareto conditional distribution and a regression function. *Annals of the Institute of Statistical Mathematics, 47*, 177–183.

Yakovenko, V. M., & Silva, A. C. (2005). Two-class structure of income distribution in the usa: Exponential bulk and power-law tail. *Econophysics of Wealth Distributions: Econophys-Kolkata I*, 15–23.

Yao, W., Lindsay, B. G., & Li, R. (2012). Local modal regression. *Journal of Nonparametric Statistics, 24*(3), 647–663.

Chapter 8
A New Approach to Estimate Semi-Parametric Gaussian Mixtures of Regressions with Varying Mixing Proportions

Sphiwe B. Skhosana, Salomon M. Millard, and Frans H. J. Kanfer

8.1 Introduction

Mixture models have become a useful tool for studying any variable, say y, that takes its values from a population that is made up of several sub-populations. Moreover, the identity of the latter is unknown. For the theory and applications of mixture models, see Titterington et al. (1985) and McLachlan and Peel (2000) and more recently Fruhwirth-Schnatter et al. (2019). Suppose the variable y depends on a set of D variables $\mathbf{x} = (x_1, x_2, \ldots, x_D)$ and we are interested in studying this dependence. For this purpose, mixture models have been extended to a class of models referred to as mixtures of regression models. The latter model is useful for studying the relationship between a response variable y and a set of covariates $\mathbf{x}$.

Mixtures of regression models were first introduced by Quandt (1972) as switching regression models. Consider a population with two components. Associated with each component is a regression model of y on $\mathbf{x}$. It is assumed that "*nature chooses between regimes* [regression models] *with probabilities* π *and* $1 - \pi$". Mixture of regression models have received widespread adoption in areas such as economics (Quandt & Ramsey, 1978), marketing (DeSarbo & Cron, 1988), machine learning (Jacobs et al., 1991), environmental economics (Hurn et al., 2003), medicine (Schlattmann, 2009), among many other fields. See Chapter 8 of Frühwirth-Schnatter (2006) for more details on the theory of mixtures of regression models.

S. B. Skhosana (✉) · S. M. Millard · F. H. J. Kanfer
Department of Statistics, University of Pretoria, Pretoria, South Africa
e-mail: spiwe.skhosana@up.ac.za; sollie.millard@up.ac.za; frans.kanfer@up.ac.za

C. A. Coelho, D.-G. Chen (eds.), *Statistical Modeling and Applications*, Emerging Topics in Statistics and Biostatistics, https://doi.org/10.1007/978-3-031-69622-0_8

The form of the mixture of regression model introduced by Quandt (1972) is given as

$$f(y|\mathbf{X}=\mathbf{x})=\sum_{k=1}^{K}\pi_k\mathcal{N}\{y|m_k(\mathbf{x}),\sigma_k^2\} \tag{8.1}$$

where $m_k(\mathbf{x})$ and σ_k^2 are the regression function and variance of the kth component regression model, respectively. The component regression functions (CRFs) have the form

$$m_k(\mathbf{x})=\mathbf{x}^\top\boldsymbol{\beta}_k \tag{8.2}$$

where $\mathbf{x}=(1,\mathbf{x})^\top$ and $\boldsymbol{\beta}_k=(\beta_0,\beta_1,\ldots,\beta_K)$ are the regression coefficients of the kth component. The π_k, positive and summing to unity, are known as the mixing proportions or probabilities. As a probability, π_k specifies the probability that a given observation belongs to the kth sub-population. On the other hand, as a proportion, it specifies the relative size of the kth sub-population. Assuming that K is known, where K is the number of components, model (8.1) is a finite Gaussian mixture of linear regressions (GMLRs). $\mathcal{N}\{y|\mu,\sigma^2\}$ denotes that y has a Gaussian distribution with mean μ and variance σ^2. This implies that each component regression model is assumed to have a Gaussian error distribution.

Model (8.1) is not suitable for application in areas where the probability of belonging to a given sub-population depends on the state or value of some covariate. That is, the covariates $\mathbf{x}$ may contain some information about the mixing probabilities π_k, for $k=1,2,\ldots,K$. This is often the case in practice, as most, if not all, probabilities are conditional on some prior information (see Lindley 2000). In this case, the mixing proportion is a covariate-varying function and the appropriate model should have the following general form

$$f(y|\mathbf{X}=\mathbf{x},\mathbf{T}=\mathbf{t})=\sum_{k=1}^{K}\pi_k(\mathbf{t})\mathcal{N}\{y|\mathbf{x}^\top\boldsymbol{\beta}_k,\sigma_k^2\}, \tag{8.3}$$

where $\mathbf{t}\in\mathbb{R}^{D_2}$ can be equal to $\mathbf{x}\in\mathbb{R}^{D_1}$ or different. In model (8.3), the mixing proportions $\pi_k(\mathbf{t})$ are assumed to be smooth unknown functions of the covariates $\mathbf{t}$ (hence, non-parametric).

Model (8.3) was first introduced and studied by Young and Hunter (2010). Due to the curse-of-dimensionality, the model is not useful in practice for large D_2. Theoretically, the multivariate non-parametric estimator has a slow convergence rate. In practice, as pointed out by the authors, the performance of the estimators may deteriorate as D_2 increases. To circumvent the curse-of-dimensionality, Xiang and Yao (2020) recently proposed to make use of a single index $\mathbf{t}^\top\boldsymbol{\alpha}\in\mathbb{R}$ to non-parametrically model the mixing proportions as $\pi_k(\mathbf{t}^\top\boldsymbol{\alpha})$, where $\boldsymbol{\alpha}$ is the single index parameter vector. An alternative way to overcome the curse-of-dimensionality is to use a parametric function for each $\pi_k(\mathbf{t})$, for $k = 1, 2, \ldots, K$. If we use a logistic function, then model (8.3) is the mixture of experts (ME) introduced by Jacobs et al. (1991). If we use a neural network to model the mixing proportion functions $\pi_k(\mathbf{t})$, we get the neural network embedded model of Xue and Yao (2022).

The advantage of model (8.3) is that it combines the flexibility of a non-parametric model and the simplicity, in particular interpretability, of a parametric model. However, in practice, due to the presence of both parametric and non-parametric terms, model (8.3) poses an estimation and computational challenge. First, traditional non-parametric estimation uses data in a neighbourhood of a local point whereas parametric estimation uses all the observed data. How can we construct an estimation procedure that is appropriate for estimating both the parametric and non-parametric terms? Second, a likelihood approach for estimating the non-parametric functions $\pi_k(\cdot)$ requires that we maximize a set of local likelihood functions. Using the Expectation-Maximization (EM) algorithm (Dempster et al., 1977) to separately maximize each local-likelihood function may lead to label switching (see Huang and Yao 2012).

In this Chapter, we propose a unified approach to address all of these challenges. The proposed approach has two stages. In the first stage, we propose a model-based approach to address the label-switching problem. Briefly, note that, locally, model (8.3) is a GMLRs (8.1). Thus, each local-likelihood function is a likelihood function of a GMLRs. Next, we rewrite the model (8.3) as a mixture of GMLRs. Estimating the latter is equivalent to simultaneously maximizing the local-likelihood functions. Estimation of the mixture of GMLRs is carried out using the classical EM algorithm. In the second stage, we propose one-step backfitting estimates of the parametric and non-parametric terms. To simplify the presentation, we consider the case when $D_2 = 1$, model (8.3) reduces to the form

$$f(y|\mathbf{X} = \mathbf{x}, T = t) = \sum_{k=1}^{K} \pi_k(t) \mathcal{N}\{y|\mathbf{x}^\top \boldsymbol{\beta}_k, \sigma_k^2\}. \tag{8.4}$$

The mixing proportion functions are now univariate functions of the covariate t and there is no curse of dimensionality problem. However, note that the proposed method can be extended to the case of a multivariate $\mathbf{t}$. The rest of the document is organized as follows: In order to demonstrate the label-switching problem, Sect. 8.2 presents the traditional local likelihood approach used to estimate model (8.4). Section 8.3 presents the proposed estimation procedure and the fitting algorithm to implement the procedure. Sections 8.4 and 8.5 presents a simulation study and an application on an environmental dataset to demonstrate the performance of the proposed methods, respectively. Section 8.6 concludes the paper after and then provides direction for future research.

8.2 Estimation

Consider a random sample $\{(t_i, \mathbf{x}_i, y_i) : i = 1, 2, \dots, n\}$ of size n obtained from model (8.4). The corresponding log-likelihood function is given as

$$\ell(\boldsymbol{\theta}) = \sum_{i=1}^{n} \log\left[\sum_{k=1}^{n} \pi_k(t_i)\mathcal{N}\{y_i|\mathbf{x}_i^{\mathsf{T}}\boldsymbol{\beta}_k, \sigma_k^2\}\right] \tag{8.5}$$

where $\boldsymbol{\theta} = (\boldsymbol{\pi}, \boldsymbol{\beta}, \boldsymbol{\sigma}^2) = (\boldsymbol{\pi}_1, \boldsymbol{\pi}_2, \dots, \boldsymbol{\pi}_K; \boldsymbol{\beta}_1, \boldsymbol{\beta}_2, \dots, \boldsymbol{\beta}_K; \sigma_1^2, \sigma_2^2, \dots, \sigma_K^2)$, with $\boldsymbol{\pi}_k = (\pi_k(t_1), \pi_k(t_2), \dots, \pi_k(t_n))$, for $k = 1, 2, \dots, K$, is the vector of all the model parameters.

In order to estimate model (8.4), we must estimate $\boldsymbol{\theta}$, this is done using a likelihood approach. Direct maximization of the log-likelihood function (8.5) with respect to $\boldsymbol{\theta}$ poses a challenge due to the presence of both a parametric term $(\boldsymbol{\beta}, \boldsymbol{\sigma}^2)$ (global parameters, henceforth) and a non-parametric term $\boldsymbol{\pi}$. Given the non-parametric term, it is straightforward to maximize (8.5) with respect to the global parameters. However, maximizing (8.5) with respect to $\boldsymbol{\pi}$ without any constraints or restrictions on the mixing proportion functions $\boldsymbol{\pi}$ would result in estimates that are practically useless due to overfitting (Tibshirani & Hastie, 1987). To overcome this problem, we make use of the local-likelihood estimation (LLE) (Tibshirani & Hastie, 1987). LLE is an extension of local-polynomial kernel estimation (see Fan and Gijbels 1996) to likelihood-based problems (see Tibshirani and Hastie 1987 for more details).

Let $\mathcal{U} = \{u_1, u_2, \dots, u_N\}$ be a set of N local grid points on the domain of the covariate t. Assume that at each $u \in \mathcal{U}$, each mixing proportion function $\pi_k(t)$, for $k = 1, 2, \dots, K$, has a $(p+1)$th derivative. By Taylor expansion, a pth degree polynomial function can be used to locally approximate each mixing proportion function $\pi_k(t)$, for $k = 1, 2, \dots, K$, in the neighbourhood of u, as

$$\begin{aligned}\pi_k(t) &\approx \pi_k(u) + \pi_k^{(1)}(u)[t-u] + \dots + \frac{\pi_k^{(p)}(u)}{p!}[t-u]^p \\ &= \sum_{j=0}^{p} \pi_{kj}(u)[t-u]^j\end{aligned} \tag{8.6}$$

where $\pi_k^{(r)}(u)$ denotes the rth derivative of $\pi_k(u)$ at local point u and $\pi_{kj}(u) = \frac{\pi_k^{(j)}(u)}{j!}$ for $k = 1, 2, \dots, K$.

Let $\boldsymbol{\pi}(u) = (\boldsymbol{\pi}_1(u), \boldsymbol{\pi}_2(u), \dots, \boldsymbol{\pi}_K(u))$, with $\boldsymbol{\pi}_k(u) = (\pi_{k0}(u), \pi_{k1}(u), \dots, \pi_{kp}(u))$, be the vector of all local parameters at grid point u. The estimate of $\boldsymbol{\pi}_k(u)$, denoted $\hat{\boldsymbol{\pi}}_k(u)$, for $k = 1, 2, \dots, K$ and $u \in \mathcal{U}$, is obtained by maximizing the following weighted (local) log-likelihood function

$$\ell[\boldsymbol{\pi}(u)] = \sum_{i=1}^{n} \log\left[\sum_{k=1}^{K}\sum_{j=0}^{p} \pi_{kj}(u)[t_i-u]^j\mathcal{N}\{y_i|\mathbf{x}_i^{\mathsf{T}}\boldsymbol{\beta}_k, \sigma_k^2\}\right]K_h(t_i-u) \tag{8.7}$$

where $K_h(t) = K(t/h)/h$ and $K(t)$ is a kernel function used to assign weights to the data points in the neighbourhood of a given local grid point u and $h > 0$ is the bandwidth used to specify the size of the neighbourhood.

From (8.6), we obtain the estimator of $\pi_k(u)$, denoted by $\hat{\pi}_k(u)$, for $k = 1, 2, \ldots, K$, as

$$\hat{\pi}_k(u) = \hat{\pi}_{k0}(u). \tag{8.8}$$

$\hat{\pi}_{k0}(u)$ can be referred to as a local-polynomial likelihood (LPL) estimator. To estimate $\pi_k(u)$, for $k = 1, 2, \ldots, K$ and $u \in \mathcal{U}$, we repeat the above maximization for all $u \in \mathcal{U}$. To obtain the estimated mixing proportion functions $\hat{\pi}_k(x_i)$, for $i = 1, 2, \ldots, n$ and $k = 1, 2, \ldots, K$, we can use interpolation over $\hat{\pi}_k(u_t)$, for $t = 1, 2, \ldots, N$ and $k = 1, 2, \ldots, K$.

8.2.1 *The Naïve Fitting Algorithm and the Label-Switching Problem*

To maximize the likelihood function for any mixture model, the standard algorithm is the EM algorithm. Recall that we have both global and local parameters. Estimation of the latter uses only the data in a neighborhood of some local point whereas efficient estimation of the former requires the use of all the data. Thus, to satisfy these competing interests, the estimation procedure must be implemented in two stages. One approach is to first locally estimate $\boldsymbol{\pi}$, $\boldsymbol{\beta}$ and $\boldsymbol{\sigma}^2$. Secondly, given $\hat{\pi}_k(x_i)$, for $i = 1, 2, \ldots, n$ and $k = 1, 2, \ldots, K$, from the first-stage, estimate $\boldsymbol{\beta}$ and $\boldsymbol{\sigma}^2$ by maximizing

$$\ell(\boldsymbol{\beta}, \boldsymbol{\sigma}^2) = \sum_{i=1}^{n} \log\left[\sum_{k=1}^{n} \hat{\pi}_k(t_i) \mathcal{N}\{y_i | \mathbf{x}_i^\top \boldsymbol{\beta}_k, \sigma_k^2\}\right] \tag{8.9}$$

with respect to $\boldsymbol{\beta}$ and $\boldsymbol{\sigma}^2$. This approach gives rise to the *one-step backfitting estimators* of the global and local parameters (Carroll et al., 1997; Xiang & Yao, 2018). The one-step procedure is summarized in Algorithm 8.1. Notice that, in Stage 1 of Algorithm 8.1, we must define N local likelihood functions, where N is the number of local points (see (8.7)). Thereafter, the natural way to proceed is to apply the EM algorithm to each local-likelihood function in turn. That is, for each local point u_j, $j = 1, 2, \ldots, N$, the EM algorithm to implement Stage 1 of Algorithm 8.1 proceeds as follows. Define a $K-$dimensional latent variable $\mathbf{z}_i = \{z_{i1}, z_{i2}, \ldots, z_{iK}\}$ for each $i = 1, 2, \ldots, n$. The kth element of this variable is set equal to 1 to indicate that observation $(t_i, \mathbf{x}_i, y_i)$ belongs to the kth component and the rest of the elements are set equal to 0. Let $\{(t_i, \mathbf{x}_i, y_i, \mathbf{z}_i) : i = 1, 2, \ldots, n\}$

be the complete data. Then the corresponding complete-data log likelihood is given as

$$\ell^c\{\boldsymbol{\theta}(u_j)\} = \sum_{i=1}^{n}\sum_{k=1}^{K} z_{ik}\big[\log\pi_k(u_j) + \log\mathcal{N}\{y_i|\mathbf{x}_i\boldsymbol{\beta}_k(u_j), \sigma_k^2(u_j)\}\big]K_h(t_i - u_j) \tag{8.10}$$

where $\boldsymbol{\theta}(u) = (\boldsymbol{\pi}(u), \boldsymbol{\beta}(u), \boldsymbol{\sigma}^2(u))$, with $\boldsymbol{\pi}(u) = (\pi_1(u), \pi_2(u), \ldots, \pi_K(u))$, $\boldsymbol{\beta}(u) = (\boldsymbol{\beta}_1(u), \boldsymbol{\beta}_2(u), \ldots, \boldsymbol{\beta}_K(u))$ and $\boldsymbol{\sigma}^2(u) = (\sigma_1^2(u), \sigma_2^2(u), \ldots, \sigma_K^2(u))$, is a vector of the local parameters at local point u. At the $(r-1)$th iteration of the EM algorithm, we have the local parameter estimates $\boldsymbol{\theta}^{(r)}(u_j)$, for $j = 1, 2, \ldots, N$. At the E-step, we calculate the expected value of $\ell^c\{\boldsymbol{\theta}(u_j)\}$, denoted $Q\{\boldsymbol{\theta}(u_j)|\boldsymbol{\theta}^{(r-1)}(u_j)\}$, with respect to the conditional distribution of $\mathbf{z}$. This corresponds to calculating the latent variable z_{ik} for $i = 1, 2, \ldots, n$ and $k = 1, 2, \ldots, K$ using its conditional expectation $\mathbb{E}\{z_{ik}|t_i, \mathbf{x}_i, y_i, \boldsymbol{\theta}(u_j)\}$ as

$$\hat{\gamma}_{ik}^{(r)}(u_j) = \frac{\pi_k^{(r-1)}(u_j)\mathcal{N}\{y_i|\mathbf{x}_i^\top\boldsymbol{\beta}_k^{(r-1)}(u_j), \sigma_k^{2(r-1)}(u_j)\}}{\sum_{\ell=1}^{K}\pi_\ell^{(r-1)}(u_j)\mathcal{N}\{y_i|\mathbf{x}_i^\top\boldsymbol{\beta}_\ell^{(r-1)}(u_j), \sigma_\ell^{2(r-1)}(u_j)\}}, \tag{8.11}$$

for $i = 1, 2, \ldots, n$ and $k = 1, 2, \ldots, K$.

Algorithm 8.1 One-step algorithm for the SPGMRVPs model (8.4)

Stage 1: Maximize (8.7), for each $u \in \mathcal{U}$, in turn, with respect to $\boldsymbol{\pi}(u)$, $\boldsymbol{\beta}(u)$ and $\boldsymbol{\sigma}^2(u)$ and then obtain $\hat{\boldsymbol{\pi}}(t_i)$, for $i = 1, 2, \ldots, n$, by interpolation.
Stage 2: Given $\hat{\boldsymbol{\pi}}$, maximize (8.9) with respect to $\boldsymbol{\beta}$ and $\boldsymbol{\sigma}^2$ to obtain the respective estimates $\hat{\boldsymbol{\beta}}$ and $\hat{\boldsymbol{\sigma}}^2$.

$\hat{\gamma}_{ik}^{(r)}(u_j)$ is known as the responsibility that the kth component takes for explaining the ith observation (see Bishop 2006). It gives the probability that the ith observation belongs to the kth component.

Based on (8.11), $Q\{\boldsymbol{\theta}(u_j)|\boldsymbol{\theta}^{(r-1)}(u_j)\}$ is defined as follows

$$Q\{\boldsymbol{\theta}(u_j)|\boldsymbol{\theta}^{(r-1)}(u_j)\} = \sum_{i=1}^{n}\sum_{k=1}^{K} w_{ik}^{(r)}(u_j)\log\pi_k(u_j) + \sum_{i=1}^{n}\sum_{k=1}^{K} w_{ik}^{(r)}(u_j)\log\mathcal{N}\{y_i|\mathbf{x}_i^\top\boldsymbol{\beta}_k(u_j), \sigma_k^2(u_j)\}. \tag{8.12}$$

At the M-Step, we maximize $Q\{\boldsymbol{\theta}(u_j)|\boldsymbol{\theta}^{(r-1)}(u_j)\}$ to update $\boldsymbol{\theta}(u_j)$ as

$$\pi_k^{(r)}(u_j) = \frac{\sum_{i=1}^n \hat{\gamma}_{ik}^{(r)}(u_j)K_h(t_i - u_j)}{\sum_{i=1}^n K_h(t_i - u_j)}, \tag{8.13}$$

$$\boldsymbol{\beta}_k^{(r)}(u_j) = \left(\sum_{i=1}^n w_{ik}^{(r)}(u_j)\mathbf{x}_i\mathbf{x}_i^\top\right)^{-1}\left(\sum_{i=1}^n w_{ik}^{(r)}(u_j)\mathbf{x}_i y_i\right), \tag{8.14}$$

$$\sigma_k^{2(r)}(u_j) = \frac{\sum_{i=1}^n w_{ik}^{(r)}(u_j)[y_i - \mathbf{x}_i^\top \boldsymbol{\beta}_k^{(r)}]^2}{\sum_{i=1}^n w_{ik}^{(r)}(u_j)}, \tag{8.15}$$

where $w_{ik}^{(r)}(u_j) = \hat{\gamma}_{ik}^{(r)}(u_j)K_h(t_i - u_j)$.

Repeat the E-Step and M-Step until convergence.

Notice that, at the E-step, the responsibilities are calculated locally. Thus, at the M-step, the non-parametric function estimates may be non-smooth, exhibiting non-uniform behaviour, due to label-switching. This is similar to the label-switching discussed in (Stephens, 2000). By separately maximizing each local-likelihood function, the traditional local-likelihood procedure via the EM algorithm is inappropriate because it cannot guarantee that the local responsibilities will be aligned across the local points. Henceforth, we refer to Algorithm 8.1 as the *naiveEM* algorithm.

This form of label-switching problem has already been mentioned by many authors (Huang & Yao, 2012; Huang et al., 2013; Skhosana et al., 2022). Huang and Yao (2012) proposed a modified EM-type algorithm address the problem. The proposed algorithm simultaneously maximizes the complete-data local-likelihood functions (8.12) using the same (global) responsibilities $\hat{\gamma}_{ik}^{(r)} = \hat{\gamma}_{ik}^{(r)}(u)$ for $u \in \mathcal{U}$. Thus, the objective is to compute the global responsibilities. Note that, the global responsibilities $\hat{\gamma}_{ik}^{(r)}$, as calculated by Huang and Yao (2012), do not take into account the local information. As an alternative to the latter approach, Skhosana et al. (2022) proposed a novel approach in which the global responsibilities $\hat{\gamma}_{ik}^{(r)}$ are calculated using the local responsibilities $\hat{\gamma}_{ik}^{(r)}(u)$, thus taking into account the local information. In this chapter, we propose a novel approach to address the label-switching problem.

8.3 Proposed Approach

In this section, we propose an estimation strategy to address the label-switching problem by reformulating model (8.3) as a mixture of GMLRs. Estimating the mixture of GMLRs is, in effect, equivalent to simultaneously estimating all the parameters of each local GMLRs and hence the non-parametric functions.

8.3.1 The Mixture of GMLRs and Its Likelihood Estimation via the EM Algorithm

Recall that label-switching takes place when estimating the local parameters by separately maximizing each local-likelihood function (8.7). In the following, we propose to combine all the local-likelihood functions into a single likelihood function that can be maximized to obtain all the local parameter estimates simultaneously. Towards that end, we reformulate model (8.3). We assume that the parameters $\boldsymbol{\beta}_k$ and σ_k^2, for $k = 1, 2, \ldots, K$, are also non-parametric functions of the covariate t. It follows that, at each local point u_j, for $j = 1, 2, \ldots, N$, model (8.4) is a GMLRs model (8.1)

$$f_{u_j}(y|\mathbf{X} = \mathbf{x}) = \sum_{k=1}^{K} \pi_k(u_j)\mathcal{N}\{y|\mathbf{x}^\top \boldsymbol{\beta}_k(u_j), \sigma_k^2(u_j)\}. \tag{8.16}$$

One of these local GMLRs can be viewed as the conditional distribution of y given $\mathbf{X} = \mathbf{x}$. Since we do not observe the identity of this local GMLRs model, conditional on $\mathbf{X} = \mathbf{x}$, y follows a mixture of these local GMLRs models

$$\begin{aligned} f(y|\mathbf{X} = \mathbf{x}) &= \sum_{j=1}^{N} \lambda_j f_{u_j}(y|\mathbf{X} = \mathbf{x}), \\ &= \sum_{j=1}^{N} \lambda_j \left[\sum_{k=1}^{K} \pi_k(u_j)\mathcal{N}\{y|\mathbf{x}^\top \boldsymbol{\beta}_k(u_j), \sigma_k^2(u_j)\} \right], \\ &= \sum_{j=1}^{N} \sum_{k=1}^{K} \lambda_j \pi_k(u_j)\mathcal{N}\{y|\mathbf{x}^\top \boldsymbol{\beta}_k(u_j), \sigma_k^2(u_j)\}, \end{aligned} \tag{8.17}$$

where $\lambda_j > 0$ (satisfying $\sum_{j=1}^{N} \lambda_j = 1$) is the mixing proportion, probability or weight. This parameter can be interpreted in various ways. As a mixing proportion, λ_j can be viewed as the relative number of data points that were generated by the jth local GMLRs model. As a mixing probability, λ_j can be interpreted as the probability that a given data point, say $(y_i, \mathbf{x}_i)$, was generated by the jth local GMLRs model. Thus, the larger the value of λ_j, the more data will be associated with the jth local GMLRs model. Alternatively, the larger the value of λ_j, the more likely that a given data point was generated by the local model $f_{u_j}(y|\mathbf{X} = \mathbf{x})$. As a mixing weight, λ_j can be viewed as specifying the relative importance of the jth local GMLRs model. The larger the weight, the more relevant the local model is to the overall model. Estimating model (8.17) is equivalent to simultaneously maximizing the local-likelihood functions (8.7) and thus addressing label-switching.

As an alternative approach to address label-switching, we can proceed as follows. Among the resulting local models, we can choose the one with the largest weight λ_j, and use its responsibilities as the global responsibilities to simultaneously maximize the local-likelihood functions (8.12). This approach is, in principle, similar to the approach proposed by Skhosana et al. (2022).

Recall that the set of local points $\mathcal{U}$ is determined by the range $\mathcal{T}$ of the covariate t. Thus, it follows that model (8.17) represents a reformulation of model (8.4). To simplify the notation, model (8.4) can be written as

$$f(y|\mathbf{X}=\mathbf{x}, T=t) = \sum_{j=1}^{N}\sum_{k=1}^{K} \lambda_j \pi_{j,k} \mathcal{N}\Big\{y|\mathbf{x}^{\top}\boldsymbol{\beta}_{j,k}, \sigma^2_{j,k}\Big\}, \tag{8.18}$$

where $\pi_{j,k} = \pi_k(u_j)$, $\boldsymbol{\beta}_{j,k} = \boldsymbol{\beta}_k(u_j)$ and $\sigma^2_{j,k} = \sigma^2_k(u_j)$.

Note that model (8.17) is equivalent to a mixture of $N \times K$ GMLRs with weights $\lambda_j \pi_{j,k}$ and component specific parameters $\boldsymbol{\beta}_{j,k}$ and $\sigma^2_{j,k}$. This implies that it can be estimated using traditional maximum likelihood via the EM algorithm.

We now give details of the proposed estimation procedure. Consider a random sample $\{(t_i, \mathbf{x}_i, y_i) : i = 1, 2, \ldots, n\}$ from model (8.4). The corresponding likelihood with respect to model (8.17) is

$$\ell_0(\boldsymbol{\lambda}, \boldsymbol{\theta}) = \sum_{i=1}^{n} \log\Bigg[\sum_{j=1}^{N}\sum_{k=1}^{K} \lambda_j \pi_{j,k} \mathcal{N}\Big\{y|\mathbf{x}^{\top}\boldsymbol{\beta}_{j,k}, \sigma^2_{j,k}\Big\}\Bigg], \tag{8.19}$$

where $\boldsymbol{\lambda} = (\lambda_1, \lambda_2, \ldots, \lambda_N)$ and $\boldsymbol{\theta} = (\boldsymbol{\theta}(u_1), \boldsymbol{\theta}(u_2), \ldots, \boldsymbol{\theta}(u_N))$ with $\boldsymbol{\theta}(u_j) = (\boldsymbol{\pi}_{j\cdot}, \boldsymbol{\beta}_{j\cdot}, \boldsymbol{\sigma}^2_{j\cdot})$, $\boldsymbol{\pi}_{j\cdot} = (\pi_{j,1}, \pi_{j,2}, \ldots, \pi_{j,K})$, $\boldsymbol{\beta}_{j\cdot} = (\boldsymbol{\beta}_{j,1}, \boldsymbol{\beta}_{j,2}, \ldots, \boldsymbol{\beta}_{j,K})$ and $\boldsymbol{\sigma}^2_{j\cdot} = (\sigma^2_{j,1}, \sigma^2_{j,2}, \ldots, \sigma^2_{j,K})$, for $j = 1, 2, \ldots, N$.

As already mentioned, the log-likelihood function (8.19) can be maximized using the traditional EM algorithm. Note that we now have two latent variables. The first latent variable serves as an indicator variable for the identity of the local GMLRs model that generated a given data point. For each data point, we define this latent variable as $\mathbf{v}_i = (v_{i1}, v_{i2}, \ldots, v_{iN})$ where $v_{ij} = 1$ if the ith data point was generated by the jth local model and 0 otherwise. The second latent variable, $\mathbf{z}_{ij}$, serves as an indicator variable for the identity of the Gaussian component, from the jth local GMLRs model, that generated a given data point. Thus, $\mathbf{z}_{ij} = (z_{ij1}, z_{ij2}, \ldots, z_{ijK})$, where $z_{ijk} = 1$ if the ith data point was generated by the kth component from the jth local mixture model. Given the completed-data $\{(t_i, \mathbf{x}_i, y_i, \mathbf{z}_{ij}, \mathbf{v}_i) : i = 1, 2, \ldots, n; j = 1, 2, \ldots, N\}$, the corresponding (complete-data) log-likelihood is

$$\ell_0^c(\boldsymbol{\lambda}, \boldsymbol{\theta}) = \ell_0^{1c}(\boldsymbol{\lambda}) + \ell_0^{2c}(\boldsymbol{\theta}), \tag{8.20}$$

where

$$\ell_0^{1c}(\boldsymbol{\lambda}) = \sum_{i=1}^{n}\sum_{j=1}^{N} v_{ij} \log \lambda_j,$$

$$\ell_0^{2c}(\boldsymbol{\theta}) = \sum_{i=1}^{n}\sum_{j=1}^{N}\sum_{k=1}^{K} v_{ij} z_{ijk}\Bigg[\log\pi_{j,k} + \log\mathcal{N}\big\{y_i|\mathbf{x}^{\top}\boldsymbol{\beta}_{j,k}, \sigma_{j,k}^2\big\}\Bigg].$$

At the rth iteration of the E-step, we calculate the conditional expected value of $\ell_0^{1c}(\boldsymbol{\lambda})$ and $\ell_0^{2c}(\boldsymbol{\theta})$, denoted by $Q(\boldsymbol{\lambda}|\boldsymbol{\lambda}^{(r-1)})$ and $Q(\boldsymbol{\theta}|\boldsymbol{\theta}^{(r-1)})$, respectively, based on the conditional distributions of $\mathbf{v}_i$ and $\mathbf{z}_i$. This corresponds to estimating the latent variables v_{ij} and z_{ijk} by calculating $\mathbb{E}[v_{ij}|t_i, \mathbf{x}_i, y_i, \lambda_j^{(r-1)}, \boldsymbol{\theta}^{(r-1)}(u_j)]$ and $\mathbb{E}[z_{ijk}|t_i, \mathbf{x}_i, y_i, \mathbf{v}_i, \boldsymbol{\theta}^{(r-1)}(u_j)]$, respectively, as

$$\hat{v}_{ij}^{(r)} = \frac{\lambda_j^{(r-1)}\sum_{k=1}^{K}\pi_{j,k}^{(r-1)}\mathcal{N}\Big\{y_i|\mathbf{x}^{\top}\boldsymbol{\beta}_{j,k}^{(r-1)}, \sigma_{j,k}^{2(r-1)}\Big\}}{\sum_{\ell=1}^{N}\lambda_{\ell}^{(r-1)}\sum_{k=1}^{K}\pi_{j,k}^{(r-1)}\mathcal{N}\Big\{y_i|\mathbf{x}^{\top}\boldsymbol{\beta}_{j,k}^{(r-1)}, \sigma_{j,k}^{2(r-1)}\Big\}} \tag{8.21}$$

and

$$\hat{z}_{ijk}^{(r)} = \frac{\pi_{j,k}^{(r-1)}\mathcal{N}\Big\{y_i|\mathbf{x}^{\top}\boldsymbol{\beta}_{j,k}^{(r-1)}, \sigma_{j,k}^{2(r-1)}\Big\}}{\sum_{\ell=1}^{K}\pi_{j,\ell}^{(r-1)}\mathcal{N}\Big\{y_i|\mathbf{x}^{\top}\boldsymbol{\beta}_{j,\ell}^{(r-1)}, \sigma_{j,\ell}^{2(r-1)}\Big\}}. \tag{8.22}$$

Note that (8.22) is similar to (8.11), with the difference being that we now have to take into account the value of $\mathbf{v}_i$, for $i = 1, 2, \ldots, n$. Expression (8.21) $\hat{v}_{it}^{(r)}$ can be interpreted as the probability that the ith data point was generated by the jth local model. In other words, it represents the responsibility of the jth local model for the ith data point. Given that the ith data point belongs to the jth local model, $\hat{z}_{ijk}^{(r)}$ has the same interpretation as $\hat{\gamma}_{ik}(u_j)$ (8.11).

After replacing v_{ij} with $\hat{v}_{ij}$ and z_{ijk} with $\hat{z}_{ijk}$ in (8.20), we obtain $Q(\boldsymbol{\lambda}|\boldsymbol{\lambda}^{(r-1)})$ and $Q(\boldsymbol{\theta}|\boldsymbol{\theta}^{(r-1)})$.

At the rth iteration of the M-step, we update $\boldsymbol{\lambda}^{(r-1)}$, by maximizing $Q(\boldsymbol{\lambda}|\boldsymbol{\lambda}^{(r-1)})$ with respect to $\boldsymbol{\lambda}$ to obtain

$$\hat{\lambda}_j^{(r)} = \frac{\sum_{i=1}^{n}\hat{v}_{ij}^{(r)}}{n}, \qquad \text{for } j = 1, 2, \ldots, N. \tag{8.23}$$

Next, we update $\boldsymbol{\theta}^{(r-1)}$ by maximizing $Q(\boldsymbol{\theta}|\boldsymbol{\theta}^{(r-1)})$ with respect to $\boldsymbol{\theta}$. To do this, we maximize a kernel weighted version of these complete-data log-likelihood function

$$\begin{aligned} Q^{w}(\boldsymbol{\theta}|\boldsymbol{\theta}^{(r-1)}) = & \sum_{i=1}^{n}\sum_{k=1}^{K}\sum_{j=1}^{N}\hat{v}_{ij}^{(r)}\hat{z}_{ijk}^{(r)}K_h(t_i - u_j) \\ & \times\Bigg[\log\pi_{j,k} + \log\mathcal{N}\big\{y_i|\mathbf{x}^{\top}\boldsymbol{\beta}_{j,k}, \sigma_{j,k}^2\big\}\Bigg]. \end{aligned} \tag{8.24}$$

Maximizing (8.24) with respect to $\pi_{j,k}$, we get

$$\pi_{j,k}^{(r)} = \frac{\sum_{i=1}^{n} \hat{v}_{ij}^{(r)} \hat{z}_{ijk}^{(r)} K_h(x_i - u_t)}{\sum_{i=1}^{n} \hat{v}_{ij}^{(r)} K_h(x_i - u_j)}. \tag{8.25}$$

Maximizing (8.24) with respect to $\boldsymbol{\beta}_{j,k}$ and $\sigma_{j,k}^2$, we get

$$\boldsymbol{\beta}_{j,k}^{(r)} = \left(\sum_{i=1}^{n} w_{ijk}^{(r)} \mathbf{x}_i \mathbf{x}_i^{\mathsf{T}} \right)^{-1} \left(\sum_{i=1}^{n} w_{ijk}^{(r)} \mathbf{x}_i y_i \right), \tag{8.26}$$

$$\sigma_{j,k}^{2(r)} = \frac{\sum_{i=1}^{n} w_{ijk}^{(r)} \left(y_i - \mathbf{x}_i^{\mathsf{T}} \boldsymbol{\beta}_{j,k}^{(r)} \right)^2}{\sum_{i=1}^{n} w_{ijk}^{(r)}}, \tag{8.27}$$

where $w_{ijk}^{(r)} = \hat{v}_{ij}^{(r)} \hat{z}_{ijk}^{(r)} K_h(t_i - u_j)$.

We repeat the above E- and M-step until convergence.

Let $r = R$ be the iteration index at convergence. To obtain $\hat{\pi}_k(t_i)$ for $i = 1, 2, \ldots, n$ and $k = 1, 2, \ldots, K$, we linearly interpolate over $\pi_{j,k}^{(R)}$, for $k = 1, 2, \ldots, K$ and $j = 1, 2, \ldots, N$. The estimates of the other non-parametric functions can be obtained in a similar manner.

We refer to the above algorithm as the model-based EM-type algorithm. Model-based because of its mixture of mixtures structure.

Note the following properties of the model-based EM-type algorithm:

Ascent property: An important and attractive property of the classical EM algorithm is the ascent property. That is, at each iteration $\ell_0^{(r+1)}(\boldsymbol{\lambda}, \boldsymbol{\theta}) \geq \ell_0^{(r)}(\boldsymbol{\lambda}, \boldsymbol{\theta})$. Based on simulation results, we have found that the model-based EM-type algorithm possesses this property.

Convergence: The convergence of the algorithm can be evaluated in either one of the following ways: (1) Stop the algorithm when the increase in the likelihood from one iteration to the next is below some small pre-specified threshold. (2) Stop the algorithm when the change in the estimated parameters from one iteration to the next is smaller than some small value. For instance, $||\boldsymbol{\lambda}^{(r)} - \boldsymbol{\lambda}^{(r-1)}||_1 < 10^{-5}$ or $||\boldsymbol{\lambda}^{(r)} - \boldsymbol{\lambda}^{(r-1)}||_2 < 10^{-5}$, where $||\cdot||_1$ and $||\cdot||_2$ denotes the L_1 and L_2 norm, respectively, on $\mathbb{R}^N$. The latter notation, $N^{(r)}$ is used to denote the number of local grid points.

In defining model (8.17), it was assumed that the global parameters $\boldsymbol{\beta}_k$ and σ_k^2, for $k = 1, 2, \ldots, K$, were non-parametric functions. However, as mentioned, efficient estimates of the global parameters require the use of all the observed data in the estimation process.

Thus, we propose one-step backfitting estimates $(\tilde{\boldsymbol{\pi}}, \tilde{\boldsymbol{\beta}}, \tilde{\boldsymbol{\sigma}}^2)$.

Given $\hat{\boldsymbol{\pi}}$, obtained from estimating model (8.17), let $(\tilde{\boldsymbol{\beta}}, \tilde{\boldsymbol{\sigma}}^2)$ be the estimates of the global parameters obtained by maximizing the global log-likelihood function

$$\ell_1(\boldsymbol{\beta}, \boldsymbol{\sigma}^2) = \sum_{i=1}^{n} \log\left[\sum_{k=1}^{n} \hat{\pi}_k(t_i)\mathcal{N}\{y_i|\mathbf{x}_i^\top \boldsymbol{\beta}_k, \sigma_k^2\}\right]. \tag{8.28}$$

Given $(\tilde{\boldsymbol{\beta}}, \tilde{\boldsymbol{\sigma}}^2)$, let $\tilde{\boldsymbol{\pi}}$ be the estimate of the non-parametric functions $\boldsymbol{\pi}$ obtained by maximizing the following local log-likelihood function

$$\ell_2[\boldsymbol{\pi}(u)] = \sum_{i=1}^{n} \log\left[\sum_{k=1}^{K} \pi_k(u_j)\mathcal{N}\{y_i|\mathbf{x}_i^\top \tilde{\boldsymbol{\beta}}_k, \tilde{\sigma}_k^2\}\right] K_h(t_i - u). \tag{8.29}$$

for $u \in \mathcal{U}$. Note that the global parameter estimates $\tilde{\boldsymbol{\beta}}$ and $\tilde{\boldsymbol{\sigma}}^2$ are well labelled. This implies that the local log-likelihood functions (8.29) can be maximized separately without being concerned about label switching.

Let $\tilde{\boldsymbol{\theta}} = (\tilde{\boldsymbol{\pi}}, \tilde{\boldsymbol{\beta}}, \tilde{\boldsymbol{\sigma}}^2)$ be the estimate of $\boldsymbol{\theta}$. We refer to $\tilde{\boldsymbol{\theta}}$ as the one-step backfitting estimate.

In summary, the proposed approach proceeds in two stages. In the first stage, we obtain $\hat{\boldsymbol{\pi}}$. Thereafter, in the second stage, we obtain $(\tilde{\boldsymbol{\pi}}, \tilde{\boldsymbol{\beta}}, \tilde{\boldsymbol{\sigma}}^2)$. The estimator $\tilde{\boldsymbol{\pi}}$ is efficient in the sense of Huang and Yao (2012) in that the variance of $\tilde{\boldsymbol{\pi}}$ is smaller than the variance of $\hat{\boldsymbol{\pi}}$ because $\tilde{\boldsymbol{\pi}}$ is based on the global parameter estimates $\tilde{\boldsymbol{\beta}}$ and $\tilde{\boldsymbol{\sigma}}^2$.

8.3.2 *One-Step Backfitting Algorithm*

In this section, we propose an algorithm to obtain the one-step backfitting estimates $\tilde{\boldsymbol{\theta}}$. The model-based EM-type algorithm is an intermediate part of this algorithm.

8.3.2.1 Stage 0: Initializing the Algorithm

Obtain appropriate initial estimates of the global parameters and the non-parametric functions, denoted $(\boldsymbol{\beta}^{(0)}, \boldsymbol{\sigma}^{2(0)})$ and $\boldsymbol{\pi}^{(0)}$, respectively, by making use of, say a mixture of experts model.

8.3.2.2 Stage 1: Model-Based EM-Type Algorithm to Maximize (8.19)

Let $\boldsymbol{\lambda}^{(r-1)}$, $\boldsymbol{\theta}_1^{(r-1)}$ and $\boldsymbol{\theta}_2^{(r-1)}$ be the parameter estimates obtained at the $(r-1)$th iteration.

E-Step At the rth iteration, calculate $Q(\boldsymbol{\lambda}|\boldsymbol{\lambda}^{(r-1)})$ and $Q(\boldsymbol{\theta}|\boldsymbol{\theta}^{(r-1)})$, by first calculating the conditional expectations of $\mathbf{v}_i$ and $\mathbf{z}_i$, for $i = 1, 2, \ldots, n$, using (8.21) and (8.22), respectively.

M-Step Maximize $Q(\boldsymbol{\lambda}|\boldsymbol{\lambda}^{(r-1)})$, with respect to $\boldsymbol{\lambda}$, to obtain $\boldsymbol{\lambda}^{(r)}$ using (8.23). Next, maximize $Q^{w}(\boldsymbol{\theta}|\boldsymbol{\theta}^{(r-1)})$, with respect to $\boldsymbol{\theta}$ to obtain $\boldsymbol{\theta}^{(r)}$ using (8.25), (8.26) and (8.27).

Repeat the above E- and M-step until convergence.

8.3.2.3 Stage 2(a): EM Algorithm to Maximize ℓ_1

Given $\hat{\boldsymbol{\pi}}$ obtained from Stage 1, we obtain the global estimates $\tilde{\boldsymbol{\beta}}$ and $\tilde{\boldsymbol{\sigma}}^2$ of the global parameters $\boldsymbol{\beta}$ and $\boldsymbol{\sigma}^2$, respectively, by maximizing ℓ_1 in (8.28) using the usual EM algorithm.

E-Step At the rth iteration, calculate the expected value of the latent variable as

$$\gamma_{ik}^{(r)} = \frac{\hat{\pi}_k(t_i)\mathcal{N}\{y_i|\mathbf{x}^\top\boldsymbol{\beta}_k, \sigma_k^{2(r-1)}\}}{\sum_{\ell=1}^{K}\hat{\pi}_\ell(t_i)\mathcal{N}\{y_i|\mathbf{x}^\top\boldsymbol{\beta}_\ell, \sigma_\ell^{2(r-1)}\}}. \tag{8.30}$$

M-Step We obtain the global parameter estimates $\boldsymbol{\beta}^{(r)}$ and $\boldsymbol{\sigma}^{2(r)}$ using the following equations

$$\boldsymbol{\beta}_k^{(r)} = \left(\sum_{i=1}^{n}\gamma_{ik}^{(r)}\mathbf{x}_i\mathbf{x}_i^\top\right)^{-1}\left(\sum_{i=1}^{n}\gamma_{ik}^{(r)}\mathbf{x}_i y_i\right), \tag{8.31}$$

$$\sigma_k^{2(r)} = \frac{\sum_{i=1}\gamma_{ik}^{(r)}(y_i - \mathbf{x}^\top\boldsymbol{\beta}_k^{(r)})^2}{\sum_{i=1}^{K}\gamma_{ik}^{(r)}}. \tag{8.32}$$

Repeat the above E- and M-step until convergence

8.3.2.4 Stage 2(b): EM Algorithm to Maximize ℓ_2

Given $\tilde{\boldsymbol{\beta}}$ and $\tilde{\boldsymbol{\sigma}}^2$ obtained from Stage 2(a), we propose an improved estimate of the component mixing proportion functions, denoted by $\tilde{\boldsymbol{\pi}}$, obtained by maximizing each local log-likelihood function in (8.29) using the usual EM algorithm.

E-Step At the rth iteration, calculate the expected value of the latent variable as

$$\gamma_{ik}^{(r)}(u_j) = \frac{\pi_k^{(r-1)}(u_j)\mathcal{N}\{y_i|\mathbf{x}^\top\tilde{\boldsymbol{\beta}}_k, \tilde{\sigma}_k^2\}}{\sum_{\ell=1}^{K}\pi_\ell^{(r-1)}(u_j)\mathcal{N}\{y_i|\mathbf{x}^\top\tilde{\boldsymbol{\beta}}_\ell, \tilde{\sigma}_\ell^2\}}. \tag{8.33}$$

M-Step We obtain $\boldsymbol{\beta}^{(r)}$ using the following equation

$$\pi_k^{(r)}(u_j) = \frac{\sum_{i=1}^{n} \gamma_{ik}^{(r)}(u_j) K_h(t_i - u_j)}{\sum_{i=1}^{n} K_h(t_i - u_j)}. \tag{8.34}$$

Repeat the above E- and M-step until convergence.

At convergence of the EM algorithm of Stage 2(b), we obtain $\tilde{\boldsymbol{\pi}} = (\tilde{\boldsymbol{\pi}}_1, \tilde{\boldsymbol{\pi}}_2, \ldots, \tilde{\boldsymbol{\pi}}_K)$, where $\tilde{\boldsymbol{\pi}}_k = (\tilde{\pi}_k(t_1), \tilde{\pi}_k(t_2), \ldots, \tilde{\pi}_k(t_n))$, by linear interpolation over $\pi_k^{(R)}(u_j)$ for $j = 1, 2, \ldots, N$ and $k = 1, 2, \ldots, K$.

We refer to the estimates $\tilde{\boldsymbol{\pi}}$, $\tilde{\boldsymbol{\beta}}$ and $\tilde{\boldsymbol{\sigma}}^2$ as the one-step backfitting estimates. To further improve the latter estimates, we can repeat Stage 2(a) and 2(b) of the algorithm until convergence.

Remark 8.1 Note that label-switching is not a concern when obtaining the non-parametric estimates $\tilde{\boldsymbol{\pi}}$. This is because the global parameter estimates $\tilde{\boldsymbol{\beta}}$ and $\tilde{\boldsymbol{\sigma}}^2$ are the same across all the local points in Stage 2(b).

8.4 Simulations

In this section, we perform numerical experiments to demonstrate the performance of the proposed methods. The purpose of these experiments is two fold. First, we want to demonstrate the effectiveness of the proposed method towards addressing label-switching. Second, we want to demonstrate the practical utility of the proposed approach. For the rest of the chapter, we refer to the proposed one-step backfitting algorithm (see Sect. 8.3.2) as the MB-EM algorithm. The MB-EM algorithm was written using the R programming language (R Core Team, 2023). Thus, all the numerical experiments are performed on this platform. The code is available on request from the authors.

8.4.1 Choosing the Bandwidth h

In order to implement the proposed methods, we must choose a bandwidth h. In practice, a data-driven approach, such as the cross-validation (CV), is used for bandwidth selection.

Let $\mathcal{D}$ denote the observed data set. We randomly divide $\mathcal{D}$, L times, into a training data set $\mathcal{R}_l$ and test data set $\mathcal{S}_l$, for $l = 1, 2, \ldots, L$. We estimate model (8.4) using the training data set $\mathcal{R}_l$ to obtain the estimates $(\hat{\boldsymbol{\pi}}(\cdot), \hat{\boldsymbol{\beta}}, \hat{\boldsymbol{\sigma}}^2)$. We use the latter to obtain $\hat{\pi}_k(t_j)$ and

$$\gamma_{jk} = \frac{\hat{\pi}_k(t_j) \mathcal{N}\{y_j | \mathbf{x}_j^\top \hat{\boldsymbol{\beta}}_k, \hat{\sigma}_k^2\}}{\sum_{\ell=1}^{K} \hat{\pi}_\ell(t_j) \mathcal{N}\{y_j | \mathbf{x}_j^\top \hat{\boldsymbol{\beta}}_\ell, \hat{\sigma}_\ell^2\}} \tag{8.35}$$

for $(\mathbf{x}_j, t_j, y_j) \in \mathcal{S}_l$ and $k = 1, 2, \ldots, K$. Finally, we define the CV error as

$$\mathrm{CV}(h) = \sum_{l=1}^{L} \sum_{j \in \mathcal{S}_l} \sum_{k=1}^{K} \gamma_{jk}(y_j - \hat{y}_j)^2 \tag{8.36}$$

where $\hat{y}_j = \mathbf{x}_j^{\top} \hat{\boldsymbol{\beta}}_k$ is the fitted value of the test set response y_j.

We choose the bandwidth h that minimises the CV error (8.36). In our simulations we set $L = 10$.

8.4.2 Simulation Study

For each of our numerical experiments, we generate 500 data sets of sizes $n = 200, 400$ and 800. We make use of $N = 100$ local points chosen uniformly on the range of t. For illustrative purposes, in all our simulations $t = x$, where $\mathbf{x} = \begin{pmatrix} 1 & x \end{pmatrix}^{\top}$. Moreover, the covariate is generated from the uniform distribution on the interval (0, 1). We make use of the Epanechnikov kernel function. We use the proposed method (MB-EM) to obtain the model results and compare them with the results obtained using the naiveEM algorithm (Algorithm 8.1).

For initializing the algorithm, we use the mixture of experts (ME) model as implemented in the R package `mixtools` (Benaglia et al., 2009).

To evaluate the performance of the proposed method, we make use of the following measures:

Root average squared error (RASE)

$$\mathrm{RASE}^2(\pi_k) = \frac{1}{n} \sum_{i=1}^{n} \left[\tilde{\pi}_k(t_i) - \pi_k(t_i) \right]^2, \tag{8.37}$$

$$\mathrm{RASE}^2(f_{\boldsymbol{\theta}}) = \frac{1}{n} \sum_{i=1}^{n} \left[\hat{f}_{\boldsymbol{\theta}}(y_i|x_i) - f_{\boldsymbol{\theta}}(y_i|x_i) \right]^2. \tag{8.38}$$

To evaluate the goodness of the estimated component regression parameters $\tilde{\boldsymbol{\beta}}$, we make use of

$$\mathrm{ASE}(\boldsymbol{\beta}_k) = \frac{1}{D} \sum_{j=0}^{D} \left[\tilde{\beta}_{jk} - \beta_{jk} \right]^2, \tag{8.39}$$

where D is the size of the covariate vector $\mathbf{x}$ and β_{jk} is the coefficient of the covariate x_j, for $j = 0, 1, \ldots, D$ and $x_0 = 1$ for the intercept parameter.

The adjusted rand index (ARI; Hubert & Arabie 1985) is used to evaluate the clustering ability of the fitted model.

The Kolmogorov-Smirnov (KS) statistic is used to assess the goodness of the fit $F_{\boldsymbol{\theta}}$ as

$$\text{KS} = \max_i |F_{\boldsymbol{\theta}}(y_i|x_i) - \hat{F}_{\tilde{\boldsymbol{\theta}}}(y_i|x_i)|, \quad \text{for } i = 1, 2, \ldots, n. \tag{8.40}$$

where $f_{\boldsymbol{\theta}}$ and $F_{\boldsymbol{\theta}}$ are the true conditional probability distribution (8.4) and the corresponding cumulative conditional probability distribution, respectively, and $\hat{f}_{\boldsymbol{\theta}}$ and $\hat{F}_{\boldsymbol{\theta}}$ are the respective estimates.

Example 8.1 For our first example, we generate data from a $K = 2$ component SPGMRVPs model (8.4) whose parameters are given in Table 8.1. The true mixing proportion functions (left plot) and a scatter plot (right plot) of a typical sample of size 400 are shown in Fig. 8.1. It is clear from the figure that the mixing proportion functions are not constant but monotone functions of t. Figure 8.2 shows the fitted mixing proportion functions using the proposed method (left plot) and the naiveEM (right plot) for three randomly selected samples of size 200, 400 and 800, respectively. It can be seen from the figure that the fitted functions based on the naiveEM are sensitive to label-switching, characterised by wiggly and non-smooth estimated functions. Moreover, this continues to be the case as we increase the sample size. On the other hand, the estimated functions based on the proposed method exhibit stability and appear to be less sensitive to label-switching.

Table 8.1 Data generating model for Example 8.1

k	1	2
$\pi_k(t)$	$1/(1+\exp(-2t))$	$1-\pi_1(t)$
$\boldsymbol{\beta}_k$	$\begin{pmatrix}4 & -2\end{pmatrix}^{\top}$	$\begin{pmatrix}0 & 3\end{pmatrix}^{\top}$
σ_k^2	0.09	0.16

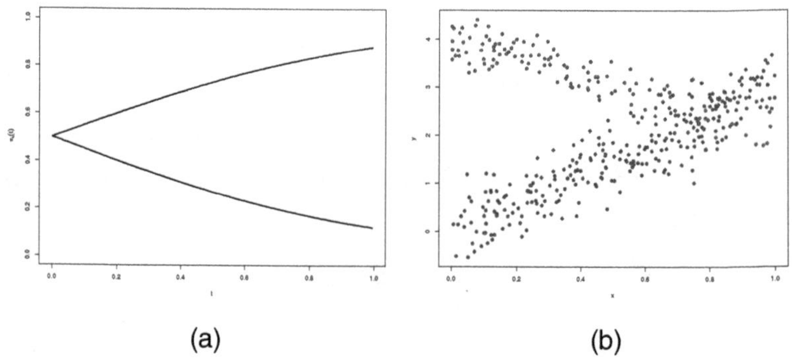

Fig. 8.1 (**a**) The $K = 2$ mixing proportion functions: a monotone decreasing function $\pi_1(t)$ and increasing function $\pi_2(t)$. (**b**) A scatter plot of a typical sample of size $n = 400$. The red data points are from component 1 and the blue data points are from component 2

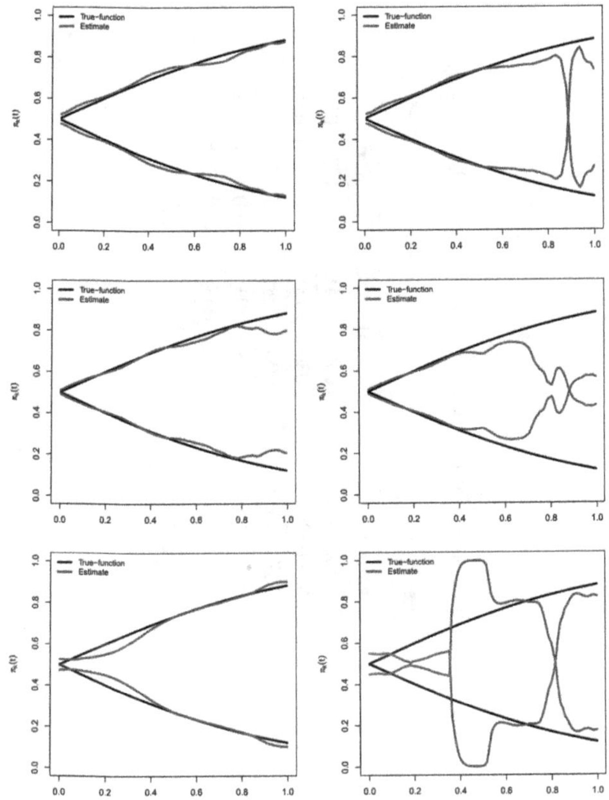

Fig. 8.2 Fitted mixing proportion functions using the proposed method (**left panel**) and the naiveEM (**right panel**) for randomly selected samples of size $n = 200$ (**first row**), 400 (**second row**) and 800 (**third row**) generated from the model in Example 1

For further evidence in support of the stability of the proposed method, Table 8.2 gives the average and standard deviation of the performance measures for the proposed method and the naiveEM calculated over the 500 simulated samples of sizes $n = 200, 400$ and 800. The tabulated results are an affirmation of the statement made based on the results shown in Fig. 8.2. Based on this results, we can say that the naiveEM is unstable, unreliable and prone to result in estimates that exhibit label-switching. On the other hand, the proposed method is stable, gives estimates that are reliable and is less sensitive to label-switching.

Example 8.2 For our next example, we generate data from the same $K = 2$ component SPGMRVPs model (8.4) as in Example 1, however, with mixing proportion functions

$$\pi_1(t) = 0.1 + 0.8\sin(\pi t) \text{ and } \pi_2(t) = 1 - \pi_1(t) \tag{8.41}$$

Table 8.2 Average (and standard deviation) of the performance measures for the proposed method and naiveEM over the 500 simulated samples of sizes $n = 200$, 400 and 800 generated from the model in Example 8.1

	$n = 200$		$n = 400$		$n = 800$	
	MB-EM	naiveEM	MB-EM	naiveEM	MB-EM	naiveEM
RASE(π_1)	0.065 (0.048)	0.138 (0.072)	0.051 (0.019)	0.145 (0.067)	0.038 (0.014)	0.143 (0.057)
RASE($f_{\tilde{\theta}}$)	0.072 (0.020)	0.099 (0.027)	0.054 (0.014)	0.090 (0.028)	0.039 (0.010)	0.082 (0.025)
ASE($\boldsymbol{\beta}_1$)	0.241 (2.105)	0.606 (3.410)	0.013 (0.019)	0.207 (1.858)	0.006 (0.008)	0.073 (0.941)
ASE($\boldsymbol{\beta}_2$)	0.217 (2.053)	0.488 (2.921)	0.006 (0.007)	0.152 (1.605)	0.003 (0.004)	0.050 (1.014)
KS	0.018 (0.009)	0.024 (0.012)	0.013 (0.007)	0.023 (0.012)	0.009 (0.005)	0.020 (0.009)
ARI	0.748 (0.066)	0.679 (0.114)	0.754 (0.043)	0.655 (0.110)	0.762 (0.033)	0.659 (0.095)

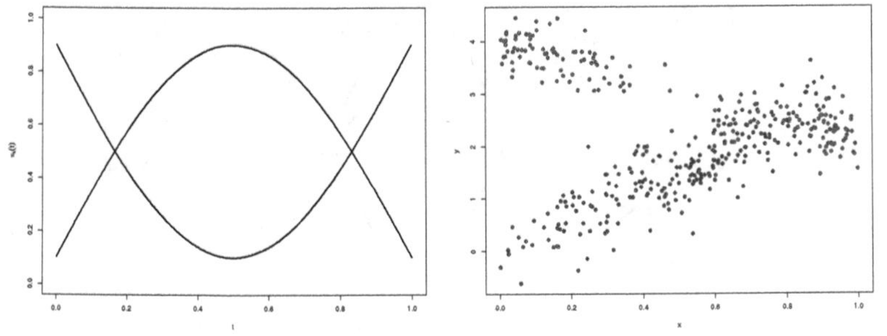

(a) (b)

Fig. 8.3 (**a**) The $K = 2$ mixing proportion functions: two parabolic functions $\pi_1(t)$ and $\pi_2(t)$. (**b**) A scatter plot of a typical sample of size $n = 400$. The red data points are from component 1 and the blue data points are from component 2

The true mixing proportion functions and a scatter plot of a typical sample of size $n = 400$ are displayed in Fig. 8.3. The mixing proportion functions are parabolic functions of t. The estimation of these functions is expected to be difficult due to the intersection between the functions.

Figure 8.4 shows the fitted mixing proportion functions using the proposed method (left plot) and the naiveEM (right plot) for three randomly selected samples of size 200, 400 and 800, respectively. As in the first example, the naiveEM is unstable and gives unreliable estimates of the mixing proportion functions. As before, this persists even if we increase the sample size. On the other hand, the proposed method exhibits stability and its estimates appear to be reliable as they are close to the true functions. To further support the latter statement, Table 8.3 gives the average and standard deviation of the performance measures for the proposed method and the naiveEM calculated over the 500 simulated samples of sizes $n = 200, 400$ and 800. Based on the tabulated results we can again say that the proposed method overcomes the challenges experienced by the naiveEM, namely, unstable and, hence unreliable, estimates and sensitivity to label-switching.

8.5 Application

In this section, we demonstrate the practical usefulness of the proposed method on real data. For real data analysis,

1. we measure the goodness-of-fit using the Bayesian information criterion (BIC)

$$\text{BIC} = -2\ell + \text{df} \times \log(n), \tag{8.42}$$

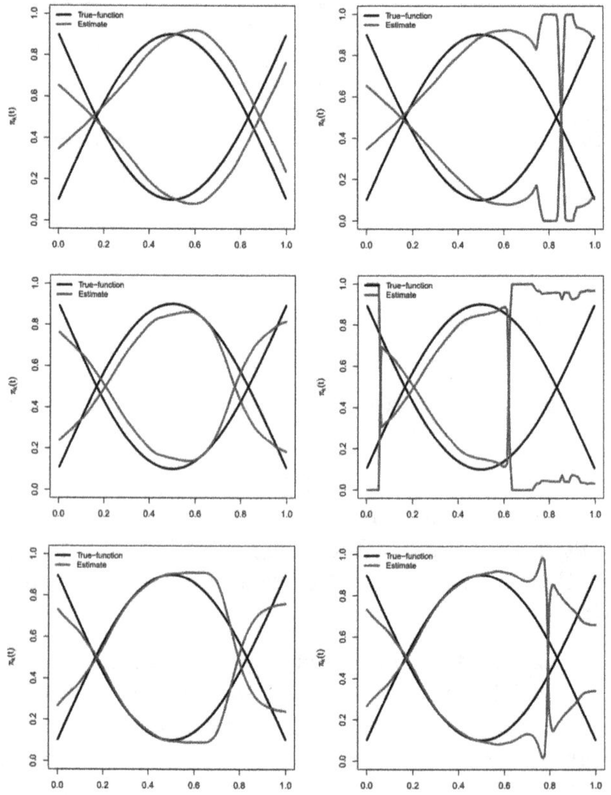

Fig. 8.4 Fitted mixing proportion functions using the proposed method (**left panel**) and using the naiveEM (**right panel**) for randomly selected samples of size $n = 200$ (**first row**), 400 (**second row**) and 800 (**third row**) generated from the model in Example 8.2

where ℓ is the maximum log-likelihood value and $\text{df} = (K-1)\text{df}_\pi + (\text{df}_{\boldsymbol{\beta}} + 1)K$ is the overall model degrees of freedom, with $\text{df}_{\boldsymbol{\beta}} = (D+1)$ and df_π is the degrees of freedom of a one-dimensional varying coefficient function (Huang et al., 2013, see page 933). Moreover, we assess the predictive ability of the fitted model using the mean squared prediction error (MPSE). Following Xiang and Yao (2018), we calculate the MSPE via a Monte Carlo cross validation (MCCV) procedure. The MCCV procedure randomly partitions the data into a training set with size $n(1-r)$ and a test set with size nr, where r is the proportion of data in the test set. The model is estimated using the data in the training set and then validated using data in the test set. The procedure is repeated T times and we take the average of the MSPEs. We take $r = 0.1, 0.2$ and 0.3. Moreover, we set $T = 200$; and

2. lastly, we use a conditional bootstrap approach to calculate the pointwise 95% confidence intervals of the fitted mixing proportion functions and the 95%

Table 8.3 Average (and standard deviation) of the performance measures for the proposed method and naiveEM over the 500 simulated samples of sizes $n = 200, 400$ and 800 generated from the model in Example 8.2

	$n = 200$		$n = 400$		$n = 800$	
	MB-EM	naiveEM	MB-EM	naiveEM	MB-EM	naiveEM
RASE(π_1)	0.126 (0.027)	0.161 (0.058)	0.082 (0.017)	0.132 (0.046)	0.071 (0.011)	0.117 (0.035)
RASE($f_{\bar{\theta}}$)	0.117 (0.020)	0.126 (0.023)	0.081 (0.014)	0.091 (0.020)	0.064 (0.010)	0.074 (0.012)
ASE($\boldsymbol{\beta}_1$)	0.008 (0.011)	0.251 (2.199)	0.004 (0.005)	0.045 (0.900)	0.002 (0.002)	0.002 (0.003)
ASE($\boldsymbol{\beta}_2$)	0.024 (0.032)	0.264 (2.149)	0.012 (0.017)	0.076 (1.017)	0.005 (0.007)	0.007 (0.010)
KS	0.026 (0.009)	0.031 (0.012)	0.018 (0.007)	0.026 (0.013)	0.014 (0.005)	0.021 (0.010)
ARI	0.637 (0.071)	0.619 (0.079)	0.654 (0.048)	0.629 (0.057)	0.656 (0.034)	0.634 (0.042)

confidence intervals of the component regression coefficients and variances. That is, for a given value of x, we sample the corresponding value of the response, denoted by y^*, from the fitted SPGMRVPs model $\sum_{k=1}^{K} \hat{\pi}_k(t)\mathcal{N}\{y|\mathbf{x}^\mathsf{T}\hat{\boldsymbol{\beta}}_k, \hat{\sigma}_k^2\}$. We repeat this sampling process n times to get a bootstrap sample $\mathcal{S} = \{(x_i, y_i^*) : i = 1, 2, \ldots, n\}$. We generate B bootstrap samples $\mathcal{S}^{(1)}, \mathcal{S}^{(2)}, \ldots, \mathcal{S}^{(B)}$ in the above manner. We fit the SPGMRVPs model (8.4) on each of these bootstrap samples, thus generating a sampling distribution of $\hat{\pi}_k(t)$, $\hat{\boldsymbol{\beta}}_k$ and $\hat{\sigma}_k^2$. To compute the 95% confidence intervals, we take the 2.5th and 97.5th percentiles of the sampling distributions as the lower and upper limits, respectively, of the interval. We set $B = 200$.

8.5.1 CO_2 data

In this application, we consider data on per capita gross national product (GNP) and carbon dioxide (CO_2) emissions in 1996 for a group of 28 OECD countries. The data was obtained from the R package `mixtools` (Benaglia et al., 2009). A scatter plot of the data is given in Fig. 8.5. Each data point on the figure is accompanied by the corresponding country's code. For instance, NOR is for Norway and MEX is for Mexico.

Previous studies of this data have found two linear regression components corresponding to two developmental paths (Hurn et al., 2003; Young & Hunter, 2010). The first linear regression function is steep and has a positive slope (higher GNP, higher CO2 emissions) whereas the second function is almost flat with a negative slope (higher GNP, lower CO2 emissions). According to Hurn et al. (2003), analysis of this data could assist in identifying the development paths of lower GNP countries in order to ensure that they do not pursue economic growth (high GNP) at an expense to the environment (high CO_2). This is an important step towards

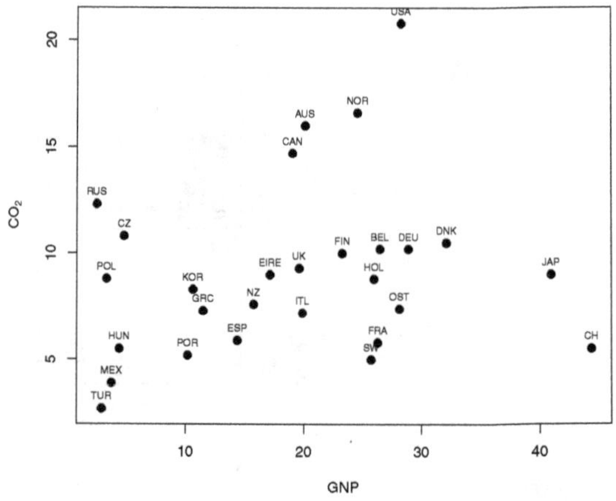

Fig. 8.5 Scatter plot of the CO_2 data. Each data point is accompanied by the corresponding country's code. For instance, *NOR* Norway and *MEX* Mexico

achieving the 13th Sustainable Development Goal (SDG) of taking urgent action against climate change and its impacts.

From Fig. 8.5, it does seem like most high GNP countries have low per-capita CO_2 emissions. This implies that, as per capita GNP increases so does the probability of a country belonging to the higher GNP, lower CO2 component. In other words, the mixing proportion is not constant. It is instead a function of per capita GNP. Thus, we fit the $K = 2$ component SPGMRVPs

$$f(y|X = x, T = t) = \pi(t)\mathcal{N}\{y|\mathbf{x}^{\mathsf{T}}\boldsymbol{\beta}_1, \sigma_1^2\} + (1 - \pi(t))\mathcal{N}\{y|\mathbf{x}^{\mathsf{T}}\boldsymbol{\beta}_2, \sigma_2^2\} \quad (8.43)$$

where $y = CO_2$, $\mathbf{x} = \begin{pmatrix}1 & x\end{pmatrix}^{\mathsf{T}}$ and $x = t = \text{GNP}$.

We fit model (8.43) using the proposed method. The optimal bandwidth was chosen to be $h = 22.7$. We compare the fitted model with the $K = 2$ component GMLRs (8.1) model which assumes that the mixing proportion function is constant. Table 8.4 gives the estimated parameters along with the boostrap standard errors. The estimated parameters based on the two models are similar. To evaluate the goodness of fit and the prediction performance of the fitted models, Tables 8.4 and 8.5 gives the BIC and MSPE, respectively. As can be seen from the table, the fitted model based on the proposed method gives the best fit to the data.

The first and second column of Fig. 8.6 shows the fitted component regression lines (first row) and the fitted mixing proportion functions for the first (second row) and second (third row) regression components obtained from the fitted model (8.43) and the fitted GMLRs model. For the fitted component regression plots, we also include the hard clustered data points based on the responsibilities $\hat{\gamma}_{ik}$. An interpretation of the results on the figure suggests that, in 1996, low GNP countries such as Turkey (TUR) and Mexico (MEX) were on a higher GNP and higher CO_2 emissions developmental path. On the other hand, low GNP countries

Table 8.4 Estimated parameters (and the bootstrap standard errors) for the fitted models. The BIC is also provided to assess the goodness-of-fit of each model

	$\boldsymbol{\beta}_1$		$\boldsymbol{\beta}_2$		σ_1^2	σ_2^2	BIC
SPGMRVPs	1.457 (1.586)	0.675 (0.092)	8.778 (1.408)	−0.027 (0.089)	0.664 (0.367)	4.137 (1.848)	**153.418**
GMLRs	1.415 (1.449)	0.677 (0.123)	8.679 (0.940)	−0.023 (0.042)	0.655 (0.832)	4.200 (1.868)	157.205

The bold values are used to show which method between (SPGMRVPs and GMLRs) performs the best.

Table 8.5 The average (and standard deviation) of the mean square prediction error (MSPE) of the fitted models for values of $r = 0.1, 0.2$ and 0.3, where r is as defined in the text

$r = 0.1$		$r = 0.2$		$r = 0.3$	
SPGMRVPs	GMLRs	SPGMRVPs	GMLRs	SPGMRVPs	GMLRs
13.420 (7.847)	15.951 (10.580)	14.926 (8.536)	15.494 (9.367)	15.801 (7.063)	16.474 (7.972)

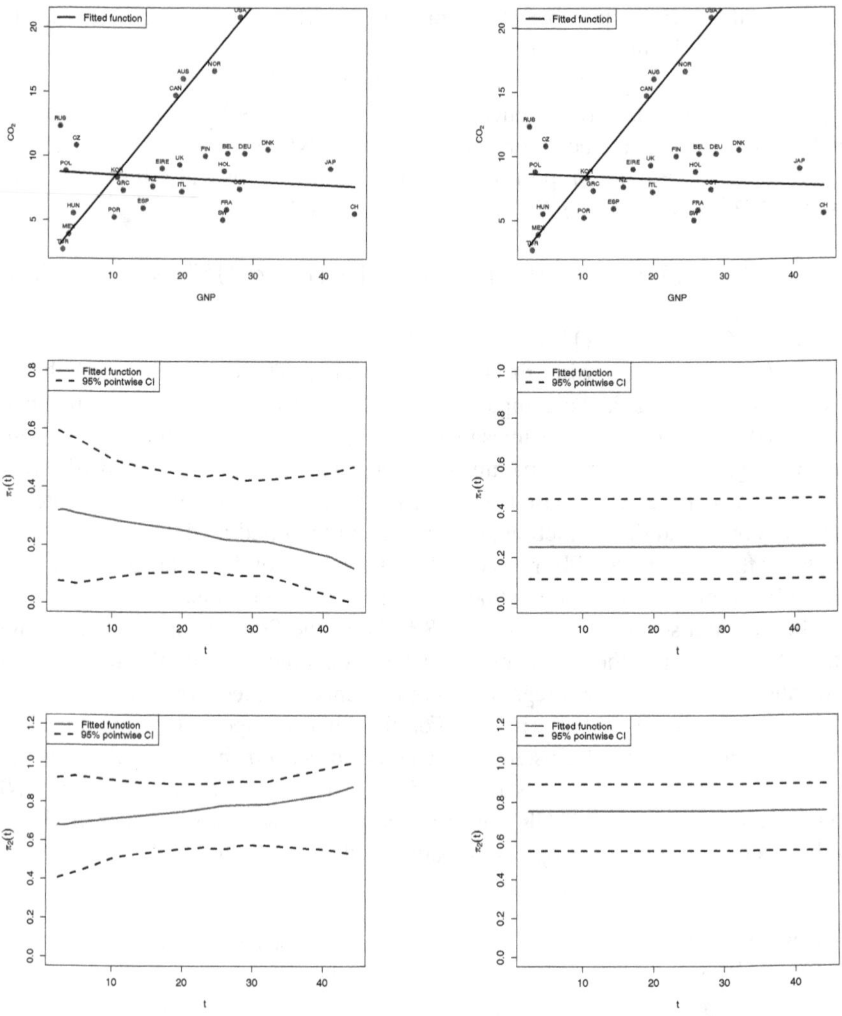

Fig. 8.6 Fitted component linear regression functions (first row) with hard clustered data points, fitted component mixing proportion functions for the first component (second row) and the second component (third row) based on model (8.43) (first column) and the GMLRs model (second column). Also included are the 95% bootstrap pointwise confidence intervals

like Czech-Republic (CZ) and Poland (POL) were on a higher GNP and lower CO_2 developmental path. According to the Our World in Data (Accessed on 12 December 2023), per capita CO_2 emissions in Turkey and Mexico have increased over the period 1996–2022, whereas per capita CO2 emissions in Czech-Republic and Poland have decreased over the same period. This is an indication that, indeed, these countries have been on their respective development paths, as determined by the model.

For the fitted mixing proportion functions, we also include the 95% bootstrap pointwise confidence intervals. The fitted mixing proportion functions shows a gradual decrease in the probability that a country belongs to the high GNP and high CO2 emissions development path as per capita GNP increases. Consequently, the probability of belonging to the high GNP and low CO2 development path is increasing as per capita GNP increases.

8.6 Conclusion

This chapter was concerned with estimating semi-parametric Gaussian mixtures of regressions with varying mixing proportions (SPGMRVPs), where the mixing proportions are non-parametric functions of the covariate (s) and the component variances and regression coefficients are parametric. To estimate the latter functions, we must define a set of local-likelihood functions. Applying the EM algorithm to maximize each local-likelihood function separately does not guarantee that the component labels on the local parameter estimates will be aligned. In the event of a misalignment, the result is label-switching. We proposed a novel two-stage estimation procedure to: (1) address label-switching and (2) obtain good estimates of the parametric and non-parametric terms of the model. In the first-stage, we reformulate the SPGMRVPs as a mixture of mixtures of linear regression models. Estimating the latter model is equivalent to simultaneously maximizing the local-likelihood functions. To improve the first-stage estimates, in the second-stage, we propose one-step backfitting estimates of the parametric and non-parametric terms. We demonstrated the effectiveness and practical usefulness of the proposed method on simulated datasets and a real dataset, respectively.

The proposed method can be easily extended to other semi- and non-parametric Gaussian mixtures of regressions. For instance, it could be of practical value for future studies to investigate the effectiveness of the proposed approach in estimating the non-parametric Gaussian mixtures of regressions (Huang et al., 2013) and the Gaussian mixtures of single index models (Xiang & Yao, 2020). Unlike the former, the latter can take multivariate covariates without suffering from the curse-of-dimensionality. Other interesting avenues for future research include an extension of the proposed approach to robust mixtures of regressions, where the component error distributions are skewed or heavy-tailed (Kanfer & Millard, 2020; Naderi et al., 2023) or even non-parametric (Ma et al., 2021; Millard & Kanfer, 2022) and mixtures of mixed-effects models for longitudinal data analysis (Bai et al., 2016) and censored data analysis (Lachos et al., 2017). Finally, considering robust longitudinal data analysis, the present work can be extended to estimate a mixture of mixed-effects model that combines variable selection and robust estimation as in Taavoni and Arashi (2022).

Acknowledgments The authors would like to thank Dr. Jannie Pretorius from the Centre for the advancement of Scholarship at the University of Pretoria, South Africa for assistance with the high-performance computer used to generate the empirical results reported in this chapter.

References

Bai, X., Chen, K., & Yao, W. (2016) Mixture of linear mixed models using multivariate t distribution. *Journal of Statistical Computation and Simulation, 86*(4), 771–787.

Benaglia, T., Chauveau, D., Hunter, D. R., & Young, D. (2009). mixtools: An R package for analyzing finite mixture models. *Journal of Statistical Software, 32*(6), 1–29. URL https://www.jstatsoft.org/v32/i06/.

Bishop, C. M. (2006). *Pattern recognition and machine learning*. Springer.

Carroll, R. J., Fan, J., Gijbels, I., & Wand, M. P. (1997). Generalized partially linear single-index models. *Journal of the American Statistical Association, 92*(438), 477–489.

Dempster, A. P., Laird, N. M., & Rubin, D. B. (1977). Maximum likelihood from incomplete data via the EM algorithm. *Journal of the Royal Statistical Society: Series B (Methodological), 39*(1), 1–22.

DeSarbo, W. S., & Cron, W. L. (1988). A maximum likelihood methodology for clusterwise linear regression. *Journal of Classification, 5*, 249–282.

Fan, J., & Gijbels, I. (1996). *Local polynomial modelling and its applications: Monographs on statistics and applied probability* (Vol. 66). CRC Press.

Frühwirth-Schnatter, S. (2006). *Finite mixture and Markov switching models*. Springer Series in Statistics. Springer.

Fruhwirth-Schnatter, S., Celeux, G., & Robert, C. P. (2019). *Handbook of mixture analysis*. CRC Press.

Huang, M., & Yao, W. (2012). Mixture of regression models with varying mixing proportions: a semiparametric approach. *Journal of the American Statistical Association, 107*(498), 711–724.

Huang, M., Li, R., & Wang, S. (2013). Nonparametric mixture of regression models. *Journal of the American Statistical Association, 108*(503), 929–941.

Hubert, L., & Arabie, P. (1985). Comparing partitions. *Journal of Classification, 2*, 193–218.

Hurn, M., Justel, A., & Robert, C. P. (2003). Estimating mixtures of regressions. *Journal of Computational and Graphical Statistics, 12*(1), 55–79.

Jacobs, R. A., Jordan, M. I., Nowlan, S. J., & Hinton, G. E. (1991) Adaptive mixtures of local experts. *Neural Computation, 3*(1), 79–87.

Kanfer, F., & Millard, S. (2020). A scale mixture approach to t-distributed mixture regression. In *Computational and methodological statistics and biostatistics: Contemporary essays in advancement* (pp. 329–347). Springer.

Lachos, V. H., Moreno, E. J. L., Chen, K., & Cabral, C. R. B. (2017). Finite mixture modeling of censored data using the multivariate student-t distribution. *Journal of Multivariate Analysis, 159*, 151–167.

Lindley, D. V. (2000). The philosophy of statistics. *Journal of the Royal Statistical Society Series D: The Statistician, 49*(3), 293–337.

Ma, Y., Wang, S., Xu, L., & Yao, W. (2021). Semiparametric mixture regression with unspecified error distributions. *Test, 30*, 429–444.

McLachlan, G. J., & Peel, D. (2000). *Finite mixture models*. Wiley Series in Probability and Statistics. Wiley.

Millard, S. M., & Kanfer, F. H. J. (2022). Mixtures of semi-parametric generalised linear models. *Symmetry, 14*(2), 409.

Naderi, M., Mirfarah, E., Wang, W., & Lin, T. (2023). Robust mixture regression modeling based on the normal mean-variance mixture distributions. *Computational Statistics & Data Analysis, 180*, 107661.

Quandt, R. E. (1972). A new approach to estimating switching regressions. *Journal of the American Statistical Association*, *67*(338), 306–310.
Quandt, R. E., & Ramsey, J. B. (1978). Estimating mixtures of normal distributions and switching regressions. *Journal of the American Statistical Association*, *73*(364), 730–738.
R Core Team. (2023). *R: A language and environment for statistical computing*. R Foundation for Statistical Computing. https://www.R-project.org/.
Schlattmann, P. (2009). *Medical applications of finite mixture models*. Springer.
Skhosana, S. B., Kanfer, F. H. J., & Millard, S. M. (2022). Fitting non-parametric mixture of regressions: Introducing an EM-type algorithm to address the label-switching problem. *Symmetry*, *14*(5), 1058.
Stephens, M. (2000). Dealing with label switching in mixture models. *Journal of the Royal Statistical Society: Series B (Statistical Methodology)*, *62*(4), 795–809.
Taavoni, M., & Arashi, M. (2022). Estimation in multivariate linear mixed models for longitudinal data with multiple outputs: Application to pbcseq data analysis. *Biometrical Journal*, *64*(3), 539–556.
Tibshirani, R., & Hastie, T. (1987). Local likelihood estimation. *Journal of the American Statistical Association*, *82*(398), 559–567.
Titterington, D. M., Smith, A. F. M., & Makov, U. E. (1985). *Statistical analysis of finite mixture distributions*. Wiley.
Xiang, S., & Yao, W. (2018). Semiparametric mixtures of nonparametric regressions. *Annals of the Institute of Statistical Mathematics*, *70*, 131–154.
Xiang, S., & Yao, W. (2020). Semiparametric mixtures of regressions with single-index for model based clustering. *Advances in Data Analysis and Classification*, *14*(2), 261–292.
Xue, J., & Yao, W. (2022). Machine learning embedded semiparametric mixtures of regressions with covariate-varying mixing proportions. *Econometrics and Statistics*, *22*, 159–171.
Young, D. S., & Hunter, D. R. (2010). Mixtures of regressions with predictor-dependent mixing proportions. *Computational Statistics & Data Analysis*, *54*(10), 2253–2266.

Chapter 9
Computational Comparisons of Two-Component Mixtures Using Lindley-Type Models

O. van Heerden, S. Makgai, A. Bekker, and J. T. Ferreira

9.1 Introduction

The Lindley distribution, as firstly introduced by Lindley (1958), has made a resurgence since the seminal paper by Ghitany et al. (2008), which illustrated its capabilities as an alternative model to the exponential distribution. The Lindley is a mixture of a gamma distribution with a shape parameter equal to 1 (i.e. an exponential distribution) and a gamma distribution with a shape parameter equal to 2. This simplicity in the construction of the Lindley has inspired many variations that have been proposed over the years.

One of these variations was developed by Shanker and Mishra (2013), who proposed a two-parameter Lindley distribution with an additional parameter α that added more flexibility to the value of the mixing parameter—and will here be referred to as the Shanker distribution. The same motivation is true for the model proposed by Shanker et al. (2013), which also added an additional parameter to the well-known Lindley model, allowing for more flexibility. This distribution will be referred to as the second Shanker (SeS) distribution. The three parameter

O. van Heerden · S. Makgai (✉)
Department of Statistics, University of Pretoria, Pretoria, South Africa
e-mail: u20439530@tuks.co.za; seite.makgai@up.ac.za

A. Bekker
Department of Statistics, University of Pretoria, Pretoria, South Africa

Centre for Environmental Studies, Department of Geography, Geoinformatics and Meteorology, University of Pretoria, Pretoria, South Africa
e-mail: andriette.bekker@up.ac.za

J. T. Ferreira
Department of Statistics, University of Pretoria, Pretoria, South Africa

National Institute for Theoretical and Computational Sciences (NITheCS), South Africa

C. A. Coelho, D.-G. Chen (eds.), *Statistical Modeling and Applications*, Emerging Topics in Statistics and Biostatistics, https://doi.org/10.1007/978-3-031-69622-0_9

Lindley (TLP) distribution was proposed by Shanker et al. (2017), which also allows more flexibility to the mixing parameter. The XLindley model, which has a mixing parameter of $\frac{1}{2}$, was proposed by Nawel et al. (2023), and demonstrated that this distribution adequately represented survival time data. Zakerzadeh and Dolati (2009) proposed a generalised Lindley (GL) distribution, which is a mixture between a gamma distribution with scale θ and shape α and a gamma distribution with scale θ and shape $\alpha + 1$. Ekhosuehi et al. (2020) proposed a mixture of gamma and exponential distributions similar to the Lindley distribution, but with the shape parameter of the gamma distribution not fixed and the mixing parameter being dependent on the value of the shape parameter. Some authors have considered Lindley-type models with mixtures between an exponential and a gamma distribution with different shape parameters (Shanker et al., 2017). For other literature which includes interesting developments of other Lindley-type models, see Ferreira and van der Merwe (2022), van der Merwe and Ferreira (2022), Alrasheedi et al. (2022), Iqbal and Iqbal (2020), Maya and Irshad (2017), Nadarajah et al. (2011), and Tomy (2018). The relevance for the investigation of finite mixtures of the Lindley distribution is emphasized by the recent research done by Abushal et al. (2023) and Al-Moisheer et al. (2021). It is for the first time, to the authors' knowledge, that an ensemble of Lindley-type models are considered as components for finite mixture modeling.

According to Punzo et al. (2018), finite mixture models aid in modeling multi-modal or multi-modal-appearing data. A finite mixture model is a convex combination of probability distributions that provides a versatile method for analyzing heterogeneous subpopulations. Mixtures have a crucial role in not only facilitating the identification or modeling of distinct clusters, but they can also serve as a catalyst for the development of novel and adaptable models that were not previously considered. Assuming that the component unimodal probability density functions (PDFs) should all belong to the same parametric family, a general finite mixture $p(x; \cdot)$ with k components assumes the following form:

$$p(x; \boldsymbol{\pi}, \boldsymbol{\xi}) = \sum_{i=1}^{k} \pi_i f(x; \xi_i), \qquad x > 0. \tag{9.1}$$

In (9.1):

- $f(x; \xi_i)$ is the PDF of the ith component with parameters ξ_i;
- $\boldsymbol{\pi} = (\pi_1, \ldots, \pi_k)$ is the vector of mixture weights, with $\pi_i \in (0, 1]$ and $\sum_{i=1}^{k} \pi_i = 1$;
- $\boldsymbol{\xi} = (\xi_1, \ldots, \xi_k)$ is the vector of component parameters ξ_i;

Consider the one-parameter Lindley distribution, which as mentioned above, has been proven to be a good model of lifetime data; this distribution can be a considered candidate in finite mixtures for modeling lifetime data that is either multi-modal or may have the appearance of multi-modality. A few mixture models of the Lindley and its variations have been proposed in literature: Al-Moisheer et al. (2021) proposed a mixture of the Lindley distribution; Abushal et al. (2023) proposed a mixture based on a special case of Shanker distribution in which $\alpha = \theta$, whereas Khan and Jan (2015) suggested a mixture based on the SeS distribution.

Note the difference between the mixture that creates the Lindley distribution and the finite mixture defined in (9.1). In the mixture construction that creates the Lindley distribution, the two different models only have one common parameter value, i.e. the same value of θ, whereas this is not the case for the finite mixture models. In the finite mixture models, the parameter values of the models being mixed are not the same. In the finite mixture model defined in (9.1), it follows that the value of π is independent of ξ, whereas in the mixtures that create the Lindley and some of its derivatives, the value of the mixing probability is tied to the value of the parameters of the models being mixed. Consider the Lindley distribution, where the mixing parameter is defined as $\frac{\theta}{\theta+1}$ and is clearly dependent on the value of θ. Also, when considering the XLindley distribution, the mixing parameter is fixed. There are some exceptions where the values of the mixing parameter that creates some of the Lindley derivatives are not necessarily dependent on the value of θ. This is seen when considering the Shanker distribution with mixing parameter $\frac{\alpha\theta}{\alpha\theta+1}$; even though the mixing parameter is a function of θ, α can be chosen such to negate θ.

In this chapter, we will focus on the two component mixture ($k = 2$ in Eq. (9.1)). In Sect. 9.2 a review will be given of the models that will be considered for finite mixture modeling. We focus on parsimonious models to avoid overfitting, and we will refer to these two-component mixtures as of the Lindley-type. The latter is to indicate the resemblance with the Lindley model, where some new candidates are also added. In Sect. 9.3, estimation inference will receive attention, followed by simulation studies in Sect. 9.4. Real data is analysed in Sect. 9.5, and final thoughts conclude the chapter in Sect. 9.6.

9.2 Mixtures of Lindley-Type Distributions

In this section, an overview of considered Lindley-type candidates in this chapter is presented and briefly discussed. In addition to the probability density function (PDF) of each considered model, each corresponding hazard function will also be considered. Leemis (2003) p, i.e., p. 214–219 defines and explains the hazard function, in particular also mentioning possible reasons for an increasing-, decreasing-, and with particular focus to the models investigated in this chapter, a bathtub shaped hazard model. The bathtub shaped hazard model being of interest considering many of the data being modelled in the application section of the chapter will include bimodal and contaminated data and this may result in a bathtub shaped hazard function. See for example Reyes et al. (2021) who introduced a bimodal exponential distribution for which the hazard function is bathtub shaped. Leemis (2003) p. 218 describes the reason for the bathtub shaped hazard function through the example of a mortality rate for different ages, where the infant mortality rate is higher than the mortality rate for teenagers, but the mortality rate increases from adulthood onwards.

9.2.1 *Mixture of Lindley Distributions*

The probability density function (PDF) of the Lindley distribution is given by Ghitany et al. (2008):

$$f_L(x;\theta) = \frac{\theta}{\theta+1}(1+x)e^{-\theta x}\ , x > 0, \theta > 0. \tag{9.2}$$

Subsequently, the PDF for the two component mixture of Lindley distributions is described by Al-Moisheer et al. (2021):

$$f_{mix}(x;\pi,\theta_1,\theta_2) = \pi f_L(x,\theta_1) + (1-\pi) f_L(x,\theta_2), x > 0, \theta_1, \theta_2 > 0. \tag{9.3}$$

Figure 9.1a and b display the PDF of the two component mixture of Lindley distributions (Eq. (9.3)). The parameter vector is defined as $\underline{\Theta} = (\pi, \theta_1, \theta_2)$. The second component in the mixture model changes the decay of the tail. The greater the parameters θ_1 and θ_2, the closer the mode will be to zero. Note that if π is fairly large and θ_1 has a larger value that brings the mode closer to zero and θ_2 has a smaller value then the tail will be heavier as can be seen in the functions highlighted in yellow in Figs. 9.1a and 9.15b. The inverse is also true, when looking at the functions highlighted in blue in Figs. 9.1a and 9.15b, when θ_1 and θ_2 are the

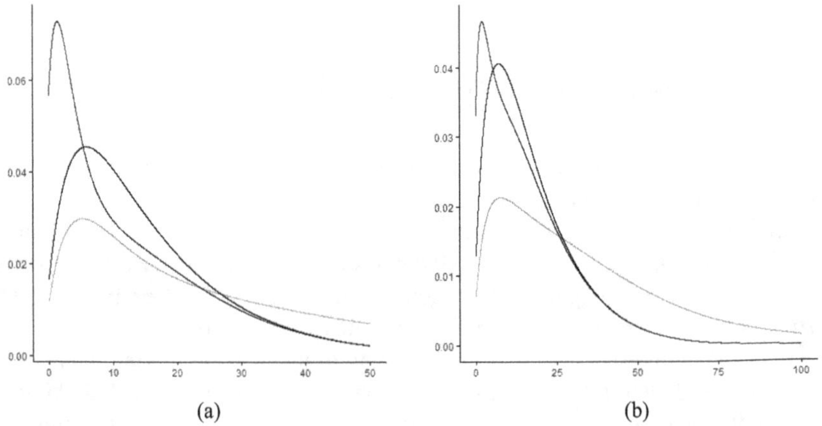

Fig. 9.1 PDF of Lindley mixture model for different parameter values. (**a**) yellow: (0.7, 0.05, 0.2) red: (0.7, 0.1, 0.2) blue: (0.7, 0.1, 0.5). (**b**) yellow: (0.7, 0.25, 0.1) red: (0.7, 0.5, 0.1) blue: (0.7, 0.5, 0.2)

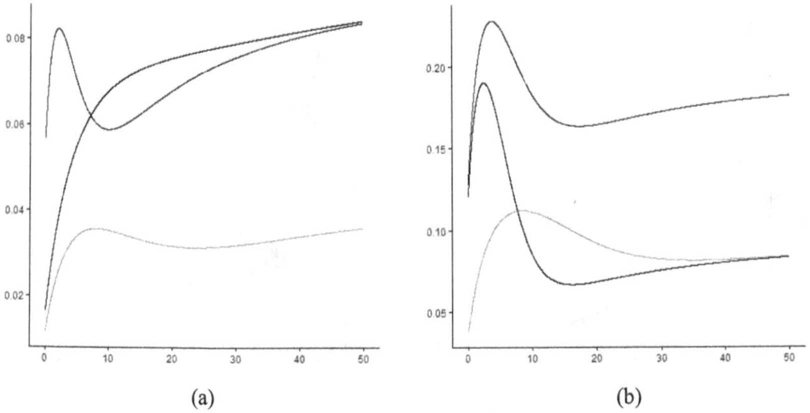

Fig. 9.2 Lindley mixture hazard function for different parameter values. (**a**) yellow: (0.7, 0.05, 0.2) red: (0.7, 0.1, 0.2) blue: (0.7, 0.1, 0.5). (**b**) yellow: (0.7, 0.25, 0.1) red: (0.7, 0.5, 0.1) blue: (0.7, 0.5, 0.2)

same as described in the previous case, but the value of π is small then the peak of the model will be stronger with a light tail.

Figure 9.2a and b display the hazard functions of the two component mixture of Lindley distributions (Eq. (9.3)) for different parameter values. For certain parameter values, the hazard function is bathtub shaped, implying that for survival data there is an initial period where failure is likely to occur, however if this stage is survived, the object (typical examples of survival data would involve components of machines or lifetime data of individuals with disease) is expected to last much longer before the object decays, considering the hazard increases as time continues.

9.2.2 *Mixture of XLindley Distributions*

The PDF of the XLindley as described by Nawel et al. (2023) is given by:

$$f_{XL}(x;\theta) = \frac{\theta}{2}(1+\theta x)e^{-\theta x}, \; x > 0, \theta > 0. \tag{9.4}$$

It follows that the PDF of the two component mixture of the XLindley distribution is given by:

$$f_{mix}(x;\pi,\theta_1,\theta_2) = \pi f_{XL}(x,\theta_1)+(1-\pi)f_{XL}(x,\theta_2), x > 0, \theta_1, \theta_2 > 0. \tag{9.5}$$

Figure 9.3a and b display the PDF of the two component mixture of XLindley distributions (Eq. (9.5)). The parameter vector is defined as $\underline{\Theta} = (\pi, \theta_1, \theta_2)$. The mode for all the parameter sets plotted are all zero or very close to zero. Similar

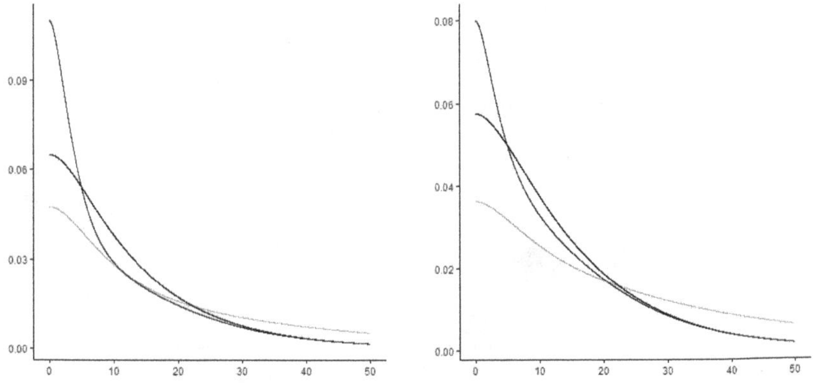

(a) (b)

Fig. 9.3 PDF of XLindley mixture model for different parameter values. (**a**) yellow: (0.7, 0.05, 0.2) red: (0.7, 0.1, 0.2) blue: (0.7, 0.1, 0.5). (**b**) yellow: (0.7, 0.25, 0.1) red: (0.7, 0.5, 0.1) blue: (0.7, 0.5, 0.2)

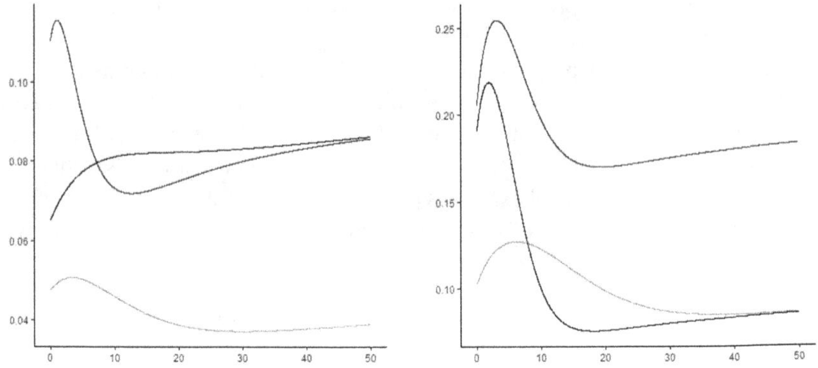

(a) (b)

Fig. 9.4 XLindley mixture hazard function for different parameter values. (**a**) yellow: (0.7, 0.05, 0.2) red: (0.7, 0.1, 0.2) blue: (0.7, 0.1, 0.5). (**b**) yellow: (0.7, 0.25, 0.1) red: (0.7, 0.5, 0.1) blue: (0.7, 0.5, 0.2)

to the Lindley mixture model, a large value of π and a larger value of θ_1 with a smaller value of θ_2 produces a model with a greater mode that decays quickly into a lighter tail—see the blue functions in Fig. 9.3a and b. Compared to the Lindley, the inverse is not as noticeably true, i.e. for a large π with small θ_1 and large θ_2, the tail is indeed heavy, but not as heavy compared to the case mentioned in the previous section on the Lindley mixture model—see the yellow functions in Fig. 9.3a and b.

Figure 9.4a and b display the hazard functions of the two component mixture of XLindley distributions (Eq. (9.5)) for different parameter values. The parameter vector is defined as $\underline{\Theta} = (\pi, \theta_1, \theta_2)$. Again there is a bathtub shape for some parameter values, implying there are many failures early on. If an object does indeed survive this bump, it may well survive much longer until it does eventually fail. This eventual failure being highlighted by the increasing hazard function as time goes on, after the initial bump. The bathtub shape for the set of parameter values $\underline{\Theta} = (0.7, 0.5, 0.1)$ is severe, as is shown in Fig. 9.4b. There is also a hazard function that is ever increasing, implying that there is no initial bump that an object can survive.

9.2.3 *Mixture of Shanker Distributions*

The Shanker distribution was proposed by Shanker and Mishra (2013) with PDF given by:

$$fs(x; \theta, \alpha) = \frac{\theta^2}{\alpha\theta + 1}(\alpha + x)e^{-\theta x}, \ x > 0, \theta > 0, \alpha\theta > -1. \tag{9.6}$$

Subsequently, the PDF of the mixture of Shanker models is given by,

$$\begin{aligned} &f_{mix}(x; \pi, \theta_1, \theta_2, \alpha_1, \alpha_2) \\ &\quad = \pi fs(x, \theta_1, \alpha_1) + (1 - \pi) fs(x, \theta_2, \alpha_2), x > 0, \theta_1, \theta_2 > 0, \alpha_1, \alpha_2 \geq 0. \end{aligned} \tag{9.7}$$

The value for α in the mixture was restricted to $\alpha \geq 0$, since the EM algorithm tended to a value of α such that $\alpha\theta$ was very close to -1 while fitting certain datasets. With $\alpha \geq 0$, the value of $\pi = \frac{\alpha\theta}{\alpha\theta+1} \in [0, 1)$. This mixture is also similar to the mixture of Second Shanker models proposed by Khan and Jan (2015), since the Shanker can be seen as a re-parameterisation of the Second Shanker. The difference between the mixture proposed by Khan and Jan (2015) and the mixture defined above (Eq. (9.7)), is that in the mixture above the values of α_1 and α_2 are bounded to be greater than 0, but this is not the case of the mixture proposed by Khan and Jan (2015).

Figure 9.5a and b display the PDF of the two component mixture of Shanker distributions (Eq. (9.7)). The parameter vector is defined as $\underline{\Theta} = (\pi, \theta_1, \theta_2, \alpha_1, \alpha_2)$. For certain parameter values the mode of the model is zero and for some it is greater than 0. It should be noted that the mode of the Shanker distribution only exists for $|\alpha\theta| < 1$ (Shanker & Mishra, 2013). It is also clear that for some combination of parameter values, the model decays quickly into a heavy tail—see the yellow function in Fig. 9.6b. This is likely due to one component having a mode of zero, while the other has a mode greater than zero. The tail of the Shanker mixture model can be altered significantly by manipulating the parameters.

Figure 9.6a and b display the hazard functions of the two component mixture of Shanker distributions (Eq. (9.7)) for different parameter values. The hazard function

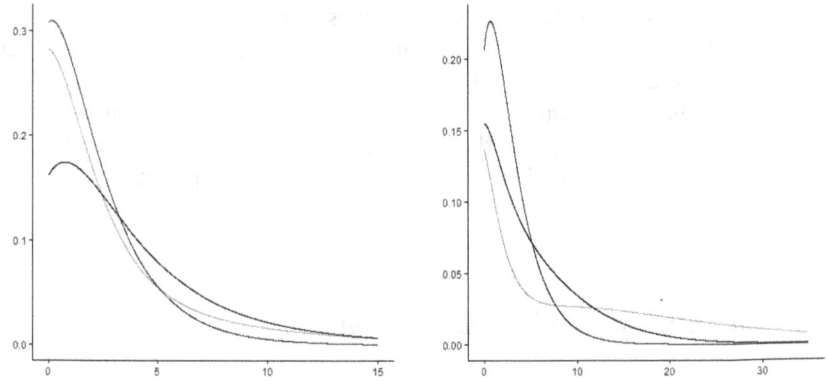

(a) (b)

Fig. 9.5 PDF of Shanker mixture model for different parameter values. (**a**) yellow: (0.3, 0.25, 0.75, 0.25, 1.5) red: (0.3, 0.25, 0.5, 0.75, 1.5) blue: (0.3, 0.5, 0.75, 0.5, 1.5). (**b**) yellow: (0.7, 0.1, 0.75, 0.25, 2) red: (0.7, 0.25, 0.75, 0.5, 2) blue: (0.7, 0.5, 0.75, 0.5, 2)

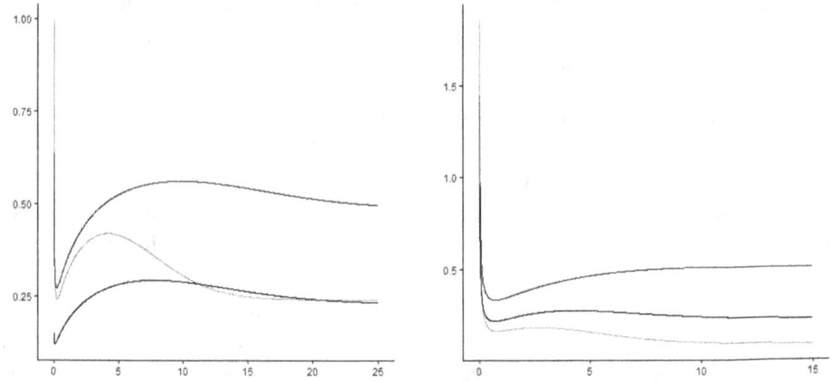

(a) (b)

Fig. 9.6 Shanker mixture hazard function for different parameter values. (**a**) yellow: (0.3, 0.25, 0.75, 0.25, 1.5) red: (0.3, 0.25, 0.5, 0.75, 1.5) blue: (0.3, 0.5, 0.75, 0.5, 1.5). (**b**) yellow: (0.7, 0.1, 0.75, 0.25, 2) red: (0.7, 0.25, 0.75, 0.5, 2) blue: (0.7, 0.5, 0.75, 0.5, 2)

in Fig. 9.6a has a bathtub shape, but rather than increasing continually after the initial dip, the hazard function appears to eventually level out, almost as though it reduces to an exponential in terms of hazard rate. The same appears to be true for the hazard functions in Fig. 9.6b which appear to also have a bathtub shape that levels out.

The mixture of the Second Shanker (SeS) and the three parameter Lindley (TLP) distributions, proposed by Shanker et al. (2013) and Shanker et al. (2017)

respectively, will not be investigated in this chapter. However, their construction is relevant when considering other models of interest later in this chapter. The construction and the PDF of the SeS and TLP distributions given respectively as follows:

$$f(x,\theta,\alpha)=\frac{\theta^2}{\theta+\alpha}(1+\alpha x)e^{-\theta x}, x>0, \theta>0, \alpha>-\theta \text{ (Shanker et al., 2013).} \tag{9.8}$$

and

$$f(x,\theta,\alpha,\beta)=\frac{\theta^2}{\alpha\theta+\beta}(\alpha+\beta x)e^{-\theta x}, x>0, \theta>0, \beta>0, \alpha\theta+\beta>0 \text{ (Shanker et al., 2017).} \tag{9.9}$$

9.2.4 *Mixture of Special Shanker Distributions*

The special Shanker (SpecShank) distribution is defined as a Shanker distribution with $\alpha=\theta$. The PDF of the special Shanker distribution is given by:

$$f_{SpecS}(x;\theta)=\frac{\theta^2}{\theta^2+1}(\theta+x)e^{-\theta x}. \tag{9.10}$$

Subsequently, the PDF of the two component mixture of the special Shanker distribution proposed by Abushal et al. (2023) is given by:

$$f_{mix}(x;\pi,\theta_1,\theta_2)=\pi f_{SpecS}(x,\theta_1)+(1-\pi)f_{SpecS}(x,\theta_2), x>0, \theta_1,\theta_2>0. \tag{9.11}$$

Figure 9.7a and b display the PDF of the two component mixture of special Shanker distributions (Eq. (9.11)). The parameter vector is defined as $\underline{\Theta}=(\pi,\theta_1,\theta_2)$. The same goes for the Special Shanker mixture model as does for the XLindley and Lindley mixture models. The PDF for the Special Shanker does, for the same set of parameters, look the same as the PDF for the Lindley mixture model, however the difference is in the modes of the Special Shanker mixture model, which has a greater peak than that of the Lindley mixture model.

Figure 9.8a and b display the hazard functions of the two component mixture of Special Shanker distributions (Eq. (9.11)) for different parameter values. Bathtub shapes and increasing hazard functions are again observed. Some of these hazard functions have a strong bathtub shape to them, however none are as strongly defined as for a certain set of parameter values of the XLindley mixture model as can be observed in Fig. 9.4b.

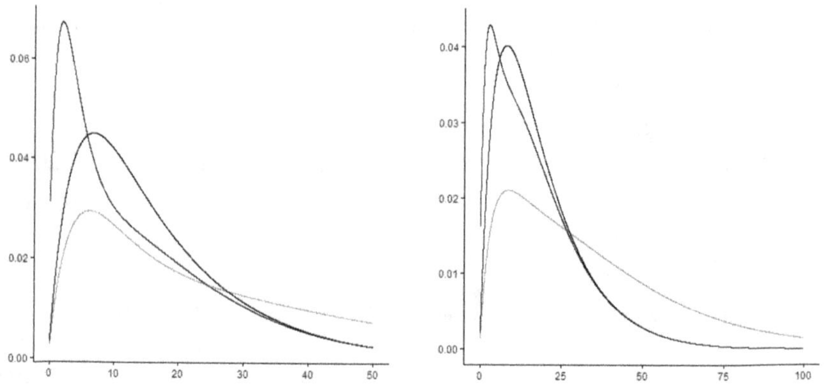

(a) (b)

Fig. 9.7 PDF of the Special Shanker mixture model for different parameter values. (**a**) yellow: (0.7, 0.05, 0.2) red: (0.7, 0.1, 0.2) blue: (0.7, 0.1, 0.5). (**b**) yellow: (0.7, 0.25, 0.1) red: (0.7, 0.5, 0.1) blue: (0.7, 0.5, 0.2)

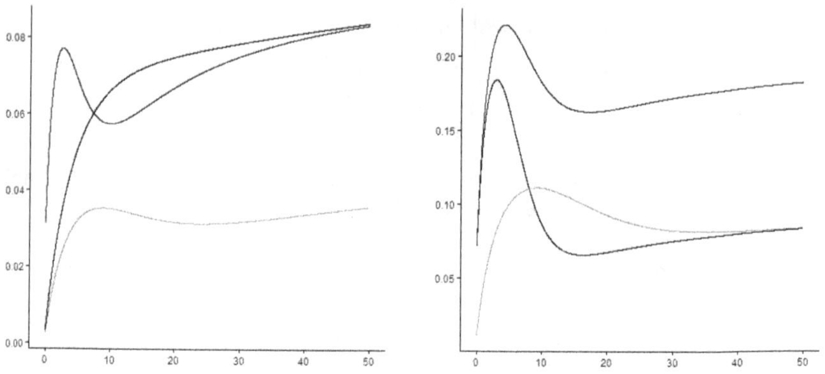

(a) (b)

Fig. 9.8 Special Shanker mixture hazard function for different parameter values. (**a**) yellow: (0.7, 0.05, 0.2) red: (0.7, 0.1, 0.2) blue: (0.7, 0.1, 0.5). (**b**) yellow: (0.7, 0.25, 0.1) red: (0.7, 0.5, 0.1) blue: (0.7, 0.5, 0.2)

9.2.5 Mixture of Gamma Distributions

The (usual) gamma distribution is given by Bain and Engelhardt (1992) as having two parameters; a shape parameter α and scale parameter θ, with PDF given by:

$$f_G(x;\alpha,\theta) = \frac{\theta^\alpha}{\Gamma(\alpha)} x^{\alpha-1} e^{-\theta x}, \quad x > 0, \alpha > 0, \theta > 0. \tag{9.12}$$

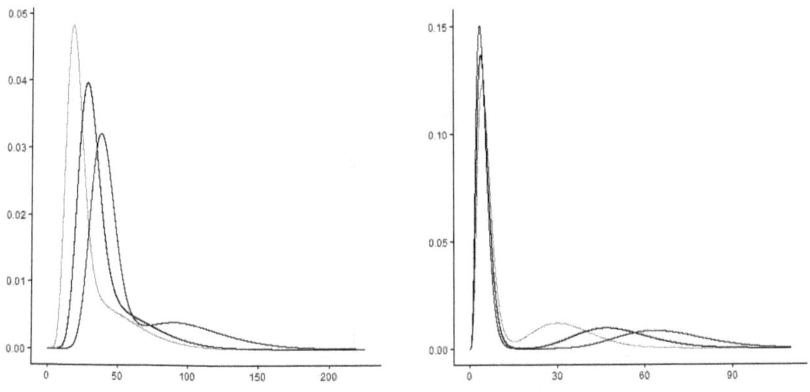

(a) (b)

Fig. 9.9 PDF of Gamma mixture model for different parameter values. (**a**) yellow: (0.7, 0.5, 0.1, 10, 5) red: (0.7, 0.5, 0.1, 15, 6) blue: (0.7, 0.5, 0.1, 20, 10). (**b**) yellow: (0.7, 0.9, 0.3, 5, 10) red: (0.7, 1, 0.3, 5, 15) blue: (0.7, 1.1, 0.3, 5, 20)

The PDF of a two component mixture of gamma distributions is given by Young et al. (2019):

$$f_{mix}(x; \underline{\Theta}) = \pi f_G(x, \alpha_1, \theta_1) + (1 - \pi) f_G(x, \alpha_2, \theta_2), x > 0, \theta_1, \theta_2, \alpha_1, \alpha_2 > 0 \tag{9.13}$$

where $\underline{\Theta} = (\pi, \theta_1, \theta_2, \alpha_1, \alpha_2)$.

Figure 9.9a and b display the PDF of the two component mixture of gamma distributions (Eq. (9.13)). The five parameters of the two component gamma mixture model may make it more cumbersome, however it does allow for the modelling of bimodal data. The mode of the gamma model can very easily be altered by manipulating the parameters to get the desired result and one can even re-parameterise the gamma in order to make the mode one of the parameters—see Tomarchio and Punzo (2020) for example who re-parameterised the gamma in terms of the mode.

Figure 9.10a and b display the hazard functions of the two component mixture of Gamma distributions (Eq. (9.13)) for different parameter values. Some very interesting hazard functions are observed. In Fig. 9.10a, there are some bathtub shapes with a strong increase after the initial dip. Much stronger than was observed for the more parsimonious models, such as the Lindley, XLindley and Special Shanker models. The gamma mixture model may be more cumbersome since it has five parameters, however this added flexibility does allow for more interesting shapes of the hazard function. Figure 9.10b displays a bump in some of the functions, similar as is seen in Fig. 9.8b for example, however in the case of the gamma mixture, the function does not increase after the bump, but rather flatlines, almost as if it reduces to an exponential, considering the exponential distribution has a constant hazard rate (Leemis, 2003, p. 220).

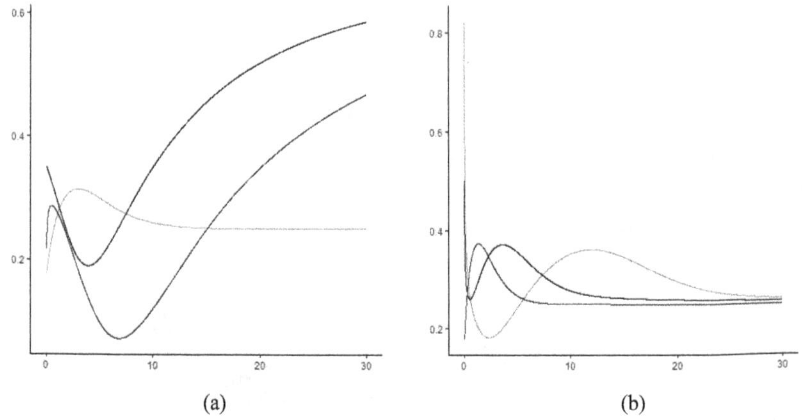

Fig. 9.10 Gamma mixture hazard function for different parameter values. (**a**) yellow: (0.7, 0.25, 0.75, 1, 2) red: (0.7, 0.5, 0.75, 1, 6) blue: (0.7, 0.5, 0.75, 1.1, 10). (**b**) yellow: (0.7, 0.25, 0.75, 0.7, 5) red: (0.7, 0.25, 1, 0.8, 3) blue: (0.7, 0.25, 1.25, 1, 2)

9.2.6 Mixture of Exponential and Gamma Distributions

The Lindley distribution is a mixture of a gamma distribution with shape parameter 1 and scale parameter θ and a gamma distribution with shape parameter 2 and scale parameter θ (Alrasheedi et al., 2022), with mixing parameter $\pi = \frac{\theta}{\theta+1}$. It may be of value to consider a general mixture between a gamma distribution with shape parameter equal to 2 and an exponential distribution (i.e., $\pi \in [0, 1]$) when also considering a mixture between Lindley distributions. This mixture is a special case of the gamma mixture. This mixture will be referred to as EG2 and has PDF given by:

$$f_{EG2}(x; \underline{\Theta}) = \pi f_G(x, 1, \theta_1) + (1 - \pi) f_G(x, 2, \theta_2), \ \ x > 0, \theta_1, \theta_2 > 0, \tag{9.14}$$

where $\underline{\Theta} = (\pi, \theta_1, \theta_2, \alpha_2)$. Figure 9.11a and b display the PDF of the EG2 model (Eq. (9.14)).

Another special case of the mixture of the gamma distributions is proposed in this section. This special case is to be constructed as a mixture of an exponential distribution and a gamma distribution, however this time the value of the shape parameter will not be fixed as it is in the model above (Eq. (9.14)). This mixture is similar to the distribution proposed by Ekhosuehi et al. (2020), which consists of an exponential distribution and a gamma distribution where the shape parameter is not fixed as is the case with the Lindley distribution. The difference between the model proposed by Ekhosuehi et al. (2020) and the model given in Eq. (9.15) is that the mixing parameter of the model in Eq. (9.15) is independent of any other parameters and the parameter values of the exponential and gamma also differ. This model will

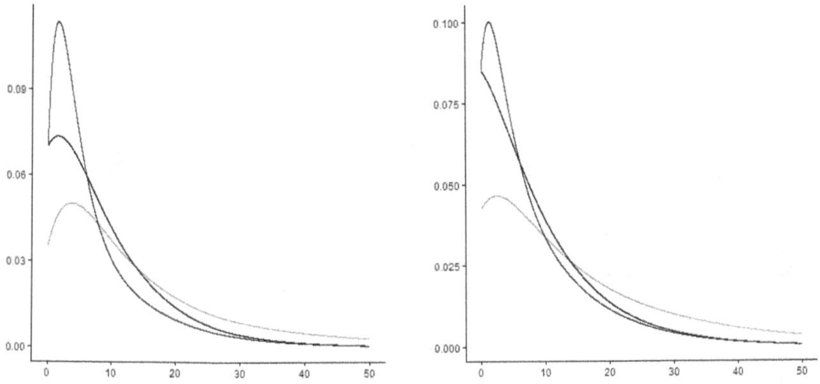

(a) (b)

Fig. 9.11 PDF of EG2 model for different parameter values. (**a**) yellow: (0.7, 0.05, 0.2) red: (0.7, 0.1, 0.2) blue: (0.7, 0.1, 0.5). (**b**) yellow: (0.7, 0.25, 0.1) red: (0.7, 0.5, 0.1) blue: (0.7, 0.5, 0.2)

be referred to as EG. The PDF of this model is given as,

$$f_{EG}(x; \underline{\Theta}) = \pi f_G(x, 1, \theta_1) + (1 - \pi) f_G(x, \alpha_2, \theta_2), \; x > 0, \theta_1, \theta_2, \alpha_2 > 0, \tag{9.15}$$

where $\underline{\Theta} = (\pi, \theta_1, \theta_2)$.

Figure 9.11a and b display the PDFs for different sets of parameter values of the EG2 model. The same is true for the EG2 as is the case for the Lindely and Special Shanker mixture models—large π-value, large θ_1-value, and small θ_2-value results in a high peak, which quickly decays into a light tail. Also, large π-value, small θ_1-value, and large θ_2-value results in a lower peak, which has a heavier tail.

Figure 9.12a and b display the hazard functions of the EG2 model (Eq. (9.14)) for different parameter values. Figure 9.12a shows a bump followed by a flatline, similar to the gamma mixture model in Fig. 9.10b. Figure 9.12b shows a bathtub shape, with one set of parameter values producing an extreme drop in the bathtub shape.

Figure 9.13a and b display the PDFs for different sets of parameter values of the EG model. The PDFs are oddly shaped, which is a result of the shape parameter being bendable and not fixed in the gamma component as is the case with the EG2 model. This added flexibility could help account for contamination (contamination as defined by Punzo (2019) being "outliers" or "bad" points), since the shape parameter in the gamma component can be used to account for x-values where more observations may have been sampled than expected. Figure 9.13a shows that for a fixed π-value with θ_2 greater than θ_1, increasing the shape parameter will move the notch in the exponential further along the x-axis, accounting for contaminated values. Figure 9.13b shows that for a fixed π-value and a fixed θ_1-value, where θ_2 is greater than θ_1, when θ_2 and α_2 increase, the notch will move down the x-axis (closer to zero). In short, the added shape parameter, α_2, in the model may make it

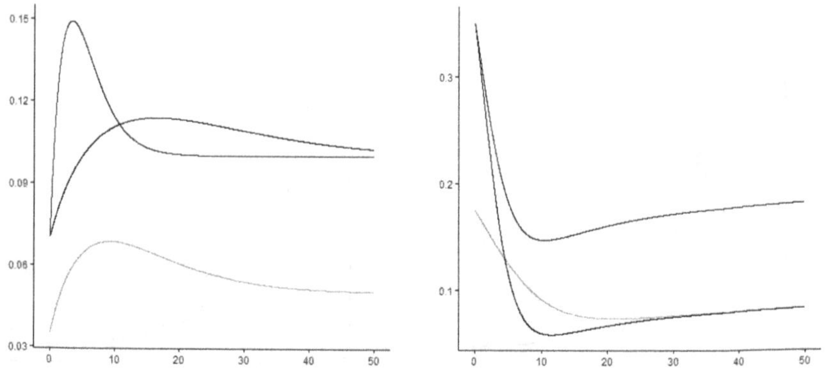

(a) (b)

Fig. 9.12 EG2 hazard function for different parameter values. (**a**) yellow: (0.7, 0.05, 0.2) red: (0.7, 0.1, 0.2) blue: (0.7, 0.1, 0.5). (**b**) yellow: (0.7, 0.25, 0.1) red: (0.7, 0.5, 0.1) blue: (0.7, 0.5, 0.2)

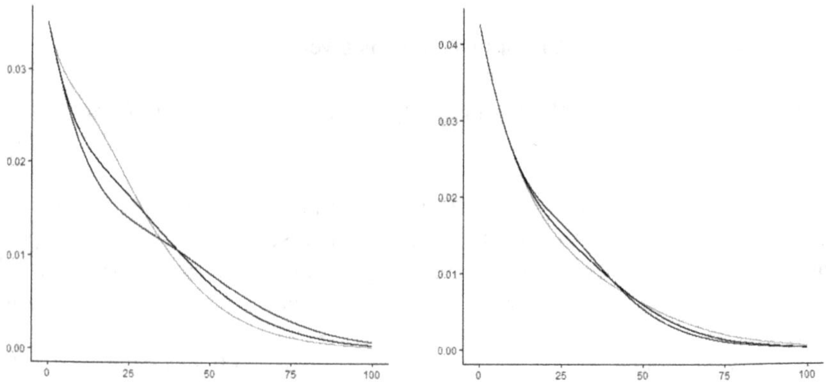

(a) (b)

Fig. 9.13 PDF of EG model for different parameter values. (**a**) yellow: (0.7, 0.05, 0.1, 3) red: (0.7, 0.05, 0.1, 4) blue: (0.7, 0.05, 0.1, 5). (**b**) yellow: (0.85, 0.05, 0.1, 5) red: (0.85, 0.05, 0.15, 6) blue: (0.85, 0.05, 0.2, 7)

useful to address any contamination or possible "bad data points" as referred to by Punzo (2019).

Figure 9.14a and b display the hazard functions of the EG model (Eq. (9.15)) for different parameter values. Figure 9.14b clearly displays a bathtub shaped hazard function, whereas Fig. 9.14a displays a function with a distinct bump (one has a bathtub shape before the bump), which then flatlines almost as if the model decays into an exponential.

The two models introduced above are both special cases of the gamma mixture defined in Sect. 9.2.5. The inspiration for these two models come from the Lindley

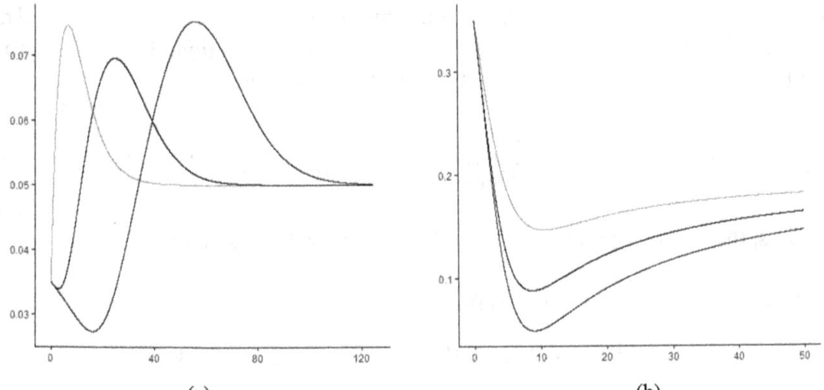

Fig. 9.14 EG hazard function for different parameter values. (**a**) yellow: (0.7, 0.05, 0.25, 2) red: (0.7, 0.05, 0.25, 5) blue: (0.7, 0.05, 0.25, 10). (**b**) yellow: (0.7, 0.5, 0.2, 2) red: (0.7, 0.5, 0.2, 3) blue: (0.7, 0.5, 0.2, 4)

and Lindley-type models mentioned in the introduction (for example see Shanker and Mishra 2013, Ekhosuehi et al. 2020, Iqbal and Iqbal 2020, Maya and Irshad 2017, Nadarajah et al. 2011, and Bakoban 2010 among others), the idea being that these models are all a mixture of gamma distributions, however for many of them, the mixing proportion is dependent on one or more of the parameter values. Bakoban (2010) proposed a very similar mixture to the ones described in this section (Eqs. (9.14) and (9.15)), which is a mixture between an exponential and an exponentiated gamma distribution where the mixing proportion is independent of any of the parameters. The two distributions defined in this section also bear a resemblance to the model proposed by Ekhosuehi et al. (2020), which is a mixture of an exponential distribution and a gamma distribution where the mixing proportion is dependent on the value of the shape parameter in the gamma component.

9.2.7 Mixture of Generalised Lindley Distributions

The generalised Lindley (GL) distribution was proposed by Zakerzadeh and Dolati (2009). The generalised Lindley is a mixture of a gamma distribution with shape α and scale θ and a gamma distribution with shape $\alpha+1$ and scale θ where the mixing parameter is $\pi = \frac{\theta}{\theta+\gamma}$. This also happens to be the mixing parameter of the Second Shanker distribution (Eq. (9.8)). The PDF is given by:

$$f(x, \pi, \theta, \alpha, \gamma) = \frac{\theta^2}{(\theta+\gamma)\Gamma(\alpha+1)}(\theta x)^{\alpha-1}(\alpha+\gamma x)e^{-\theta x}, \; x > 0, \theta, \alpha, \gamma > 0. \tag{9.16}$$

Two special cases of GL will be introduced in this section. The first special case of GL is where $\gamma = 1$, which leads the mixing parameter to become $\pi = \frac{\theta}{\theta+1}$. This is the mixing parameter which creates the Lindley distribution. This special case will be referred to as GLLind or GLL, with PDF given by:

$$f_{GLL}(x, \pi, \theta, \alpha) = \frac{\theta^2}{(\theta+1)\Gamma(\alpha+1)}(\theta x)^{\alpha-1}(\alpha+x)e^{-\theta x}, \ x > 0, \theta, \alpha > 0. \tag{9.17}$$

Thus, the two component mixture of GLLind has PDF given by:

$$f_{mix}(x, \underline{\Theta}) = \pi f_{GLL}(x, \theta_1, \alpha_1) + (1-\pi) f_{GLL}(x, \theta_2, \alpha_2), \ x > 0, \theta_1, \theta_2, \alpha_1, \alpha_2 > 0, \tag{9.18}$$

where $\underline{\Theta} = (\pi, \theta_1, \theta_2, \alpha_1, \alpha_2)$.

Figure 9.15a and b displays the PDF of the two component mixture of GLLind (Eq. (9.18)). The PDF of the GLLind mixture model looks the same as for the gamma mixture model, however this does not mean that they are the same model, as can be seen in the differences in their fit to data in the application section—see Table 9.7 for example.

Figure 9.16a and b display the hazard functions of the two component mixture of GLLind distributions (Eq. (9.18)) for different parameter values. The differences between the GLLind and the gamma are clear in the hazard function, with the hazard functions in Fig. 9.16b having a bathtub shape that rises compared to those that flatline in Fig. 9.10b. Also, the bathtub shapes in Fig. 9.10a, which displays the hazard functions for the gamma mixture are clearly more pronounced than the bathtub shapes in Fig. 9.16a, which displays the hazard function for the GLLind mixture.

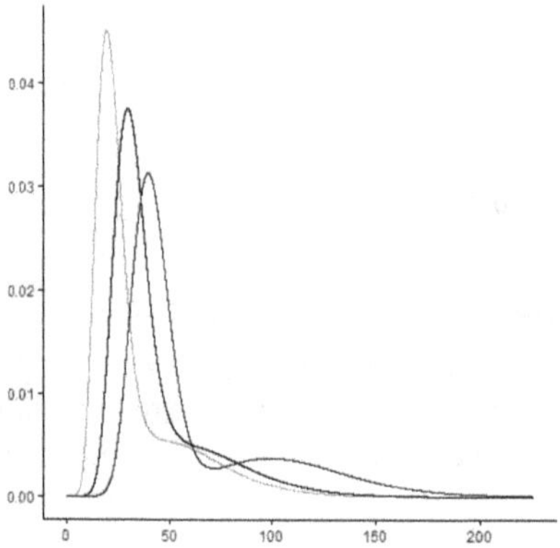

(a)

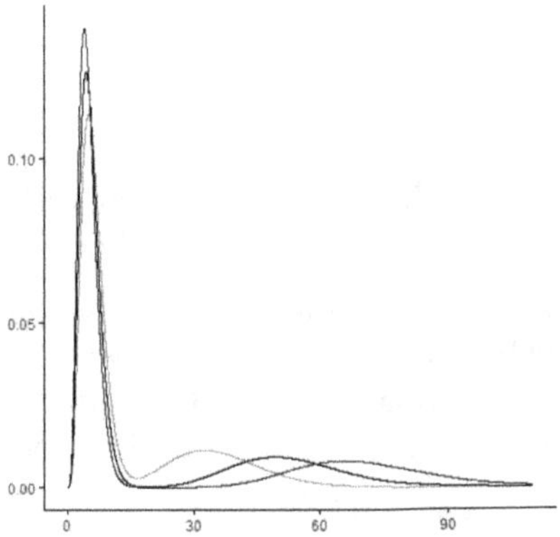

(b)

Fig. 9.15 PDF of GLLind mixture model for different parameter values. (**a**) yellow: (0.7, 0.5, 0.1, 10, 5) red: (0.7, 0.5, 0.1, 15, 6) blue: (0.7, 0.5, 0.1, 20, 10). (**b**) yellow: (0.7, 0.9, 0.3, 5, 10) red: (0.7, 1, 0.3, 5, 15) blue: (0.7, 1.1, 0.3, 5, 20)

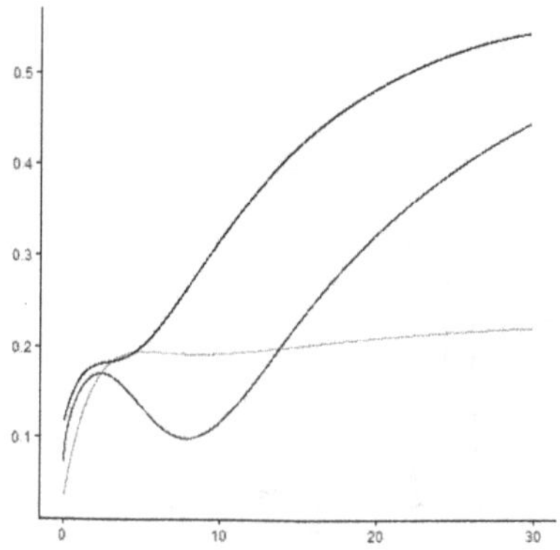

(a)

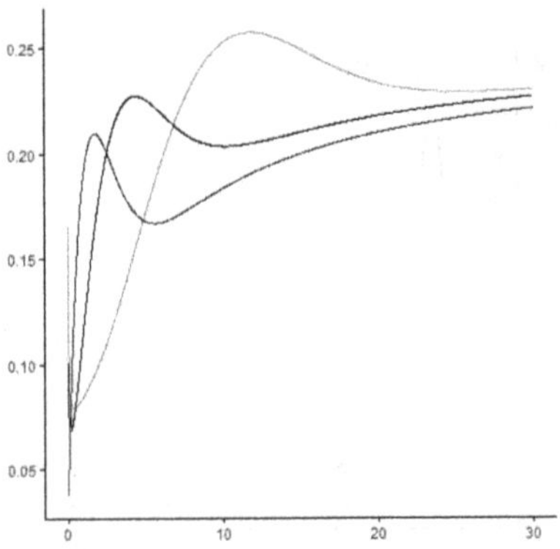

(b)

Fig. 9.16 GLLind mixture hazard function for different parameter values. (**a**) yellow: (0.7, 0.25, 0.75, 1, 2) red: (0.7, 0.5, 0.75, 1, 6) blue: (0.7, 0.5, 0.75, 1.1, 10). (**b**) yellow: (0.7, 0.25, 0.75, 0.7, 5) red: (0.7, 0.25, 1, 0.8, 3) blue: (0.7, 0.25, 1.25, 1, 2)

The second special case of GL is where $\gamma = \theta$. This distribution will be referred to as generalised XLindley (GXL) distribution. It follows that the mixing probability becomes $\pi = \frac{\theta}{2\theta} = \frac{1}{2}$. This is also the mixing probability of the XLindley distribution. The PDF of GXL is then given by:

$$f_{GXL}(x, \pi, \theta, \alpha) = \frac{\theta}{2\Gamma(\alpha+1)}(\theta x)^{\alpha-1}(\alpha+\theta x)e^{-\theta x}, \ x > 0, \theta, \alpha > 0. \quad (9.19)$$

The two component mixture of GXL has PDF given by:

$$\begin{aligned} f_{mix}(x, \underline{\Theta}) &= \pi f_{GXL}(x, \theta_1, \alpha_1) \\ &\quad + (1-\pi) f_{GXL}(x, \theta_2, \alpha_2), \ x > 0, \theta_1, \theta_2, \alpha_1, \alpha_2 > 0, \end{aligned} \quad (9.20)$$

where $\underline{\Theta} = (\pi, \theta_1, \theta_2, \alpha_1, \alpha_2)$.

Figure 9.17a and b display the PDF of the two component mixture of GXL (Eq. (9.20)). Same as for the GLLind mixture model, the PDFs of the GXL mixture model looks the same as the PDF of the gamma mixture model for the same parameter values, however this does not mean that they are in fact the same model, as there are differences in the fit of the GXL, GLLind and gamma distributions—see Table 9.7 for example.

Figure 9.18a and b display the hazard functions of the two component mixture of GXL distributions (Eq. (9.20)) for different parameter values. The differences between the GXL and the gamma are clear in the hazard function, with the hazard functions in Fig. 9.18b having a bathtub shape that rises compared to those that flatline in Fig. 9.10b. Also, the bathtub shapes in Fig. 9.10a, which displays the hazard functions for the gamma mixture are clearly more pronounced than the

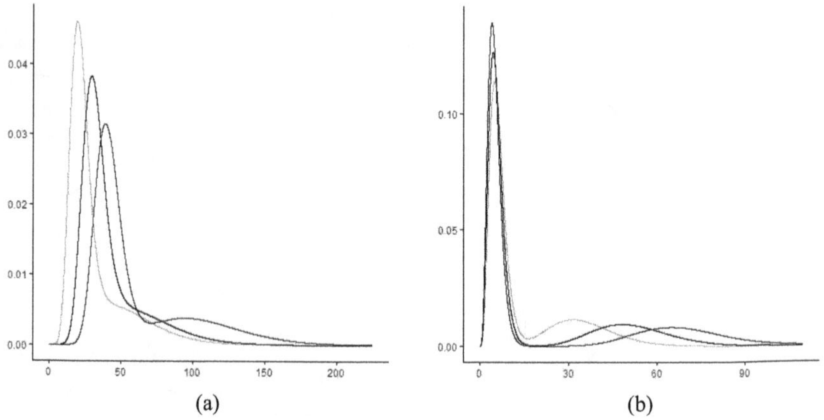

Fig. 9.17 PDF of GXL mixture model for different parameter values. (**a**) yellow: (0.7, 0.5, 0.1, 10, 5) red: (0.7, 0.5, 0.1, 15, 6) blue: (0.7, 0.5, 0.1, 20, 10). (**b**) yellow: (0.7, 0.9, 0.3, 5, 10) red: (0.7, 1, 0.3, 5, 15) blue: (0.7, 1.1, 0.3, 5, 20)

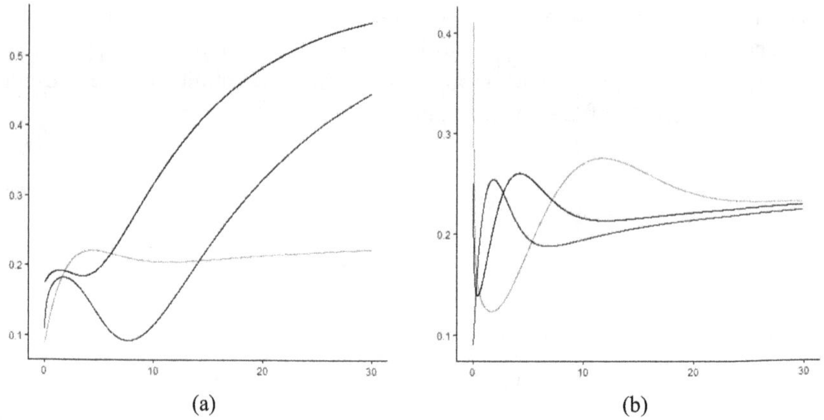

Fig. 9.18 GXL mixture hazard function for different parameter values. (**a**) yellow: (0.7, 0.25, 0.75, 1, 2) red: (0.7, 0.5, 0.75, 1, 6) blue: (0.7, 0.5, 0.75, 1.1, 10). (**b**) yellow: (0.7, 0.25, 0.75, 0.7, 5) red: (0.7, 0.25, 1, 0.8, 3) blue: (0.7, 0.25, 1.25, 1, 2)

bathtub shapes in Fig. 9.18a, which displays the hazard function for the GXL mixture.

9.2.8 Summary of Considered Distributions

A summary of the considered models in this section, is given in Tables 9.1 and 9.2. The last two rows of Table 9.1 proposes a mixture model of an exponential and a

Table 9.1 Summary of the different models

Models	Component 1	Component 2	π	Citation
Lindley (9.2)	$Exp(\theta)$	$Gamma(2, \theta)$	$\frac{\theta}{\theta+1}$	Lindley (1958) and Ghitany et al. (2008)
XLindley (9.4)	$Exp(\theta)$	$Gamma(2, \theta)$	$\frac{1}{2}$	Nawel et al. (2023)
Shanker (9.6)	$Exp(\theta)$	$Gamma(2, \theta)$	$\frac{\alpha\theta}{\alpha\theta+1}$	Shanker and Mishra (2013)
Second Shanker (9.8)	$Exp(\theta)$	$Gamma(2, \theta)$	$\frac{\theta}{\theta+\alpha}$	Shanker et al. (2013)
Three Parameter Lindley (9.9)	$Exp(\theta)$	$Gamma(2, \theta)$	$\frac{\alpha\theta}{\alpha\theta+\beta}$	Shanker et al. (2017)
Generalised Lindley (9.16)	$Gamma(\alpha, \theta)$	$Gamma(\alpha + 1, \theta)$	$\frac{\theta}{\theta+\gamma}$	Zakerzadeh and Dolati (2009)
EG2 (9.14)	$Exp(\theta_1)$	$Gamma(2, \theta_2)$	$\pi \in [0, 1]$	Proposed
EG (9.15)	$Exp(\theta_1)$	$Gamma(\alpha_2, \theta_2)$	$\pi \in [0, 1]$	Proposed

Table 9.2 Special cases of the generalised Lindley models

Models	Component 1	Component 2	γ value	π
GLLind (9.17)	$Gamma(\alpha, \theta)$	$Gamma(\alpha + 1, \theta)$	1	$\frac{\theta}{\theta+1}$
GXL (9.19)	$Gamma(\alpha, \theta)$	$Gamma(\alpha + 1, \theta)$	θ	$\frac{1}{2}$

gamma with shape 2 (Eq. (9.14)). In this case, the value of the mixing parameter is independent of the value of the other parameters. Considering the GL model, two special cases were given attention in this section and are summarized in Table 9.2.

9.3 Estimation and Inference

9.3.1 *Overview of the EM Algorithm*

The parameter estimation in this chapter is carried out via the Expectation Maximisation (EM) algorithm (Algorithm 9.1), which is a widely used approach for maximum likelihood (ML) estimation in mixture modelling and when data is incomplete. Within the formulation of the mixture models, data is viewed as incomplete, since for each observation one would not know its component membership. The EM algorithm for fitting a mixture of distributions to data is described by Punzo (2019). It is briefly summarised in Algorithm 9.1 for a two component mixture model.

9.3.2 *Computational Aspects of the EM Algorithm*

In the EM algorithm, the quantity L_{EM} needs to be maximised and for the majority of the models defined in Sect. 9.2, this is done using optimisation functions in the

Algorithm 9.1 The Expectation Maximisation Algorithm

1. Choose appropriate initial values for parameter estimates.
2. Calculate for each data point x_j:

$$v_{ij} = \pi_i \frac{f(x_j; \underline{\Theta}_i)}{\sum_{i=1}^{2} \pi_i f(x_j; \underline{\Theta}_i)} \tag{9.21}$$

3. Calculate π_i using:

$$\pi_i = \frac{\sum_{j=1}^{n} v_{ij}}{\sum_{i=1}^{2} \sum_{j=1}^{n} v_{ij}}. \tag{9.22}$$

4. Maximize:

$$L_{EM} = \sum_{j=1}^{n} [v_{1j} lnf(x_j; \underline{\Theta}_1) + v_{2j} lnf(x_j; \underline{\Theta}_2)] \tag{9.23}$$

5. Repeat steps 2–4 until the parameter values that maximise L_{EM} converge.

---

coding language R. The PDF of each model is coded as a function in R, thereafter the negative log-likelihood function for a model with respect to a given dataset will be coded as a function as well. The base R `nlm` function will then be used to minimise this function to provide the maximum likelihood estimates. This will be a subroutine within each iteration of the EM algorithm for each component of the mixture model. The EM algorithm was coded in R from first principles.

These optimisation functions require an initial estimate for the function to be used. The initial estimates in each iteration of the EM algorithm will be the previous solution of the EM algorithm. However, for the two component gamma mixture, the initial estimates will be obtained via the method of moments as defined by Young et al. (2019):

$$\alpha_j^{(0)} = \frac{\tilde{x}_j^2}{\tilde{s}_j^2}, \quad \beta_j^{(0)} = \frac{\tilde{x}_j}{\tilde{s}_j^2}, \tag{9.24}$$

where the values $\tilde{x}_j$ and $\tilde{s}_j^2$ are calculated as follows,

$$\tilde{x}_j = \frac{\sum_{i=1}^{n} x_i v_{ij}}{\sum_{i=1}^{n} v_{ij}}, \quad \tilde{s}_j^2 = \frac{n \sum_{i=1}^{n} v_{ij}(x_i - \tilde{x}_j)^2}{(n-1) \sum_{i=1}^{n} v_{ij}}. \tag{9.25}$$

For the special case of the gamma mixture where $\alpha_1 = 1$ and $\alpha_2 = 2$ (EG2 model, Eq. (9.14)), the maximum likelihood estimates in the EM algorithm exist in closed form. The estimate for θ_1 in the EM algorithm is,

$$\theta_1 = \frac{\sum_{i=1}^{n} v_i}{\sum_{i=1}^{n} v_i x_i}. \tag{9.26}$$

The estimate for θ_2 in the EM algorithm is,

$$\theta_2 = \frac{2\sum_{i=1}^{n} \nu_i}{\sum_{i=1}^{n} \nu_i x_i}. \tag{9.27}$$

9.3.3 *Performance Measures*

The performance measures will be used to evaluate the fitted models are the Akaike Information Criterion (AIC) (Akaike, 1998) and the Bayesian Information Criterion (BIC) (Schwarz, 1978). These will be determined for each of the finite mixture models, where the model with the lowest AIC and BIC will be preferred. The following formulae are used to calculate the AIC and BIC measures (Burnham & Anderson, 2004):

$$AIC = -2lnL(\underline{\Theta}) + 2K, \tag{9.28}$$

and

$$BIC = -2lnL(\underline{\Theta}) + Kln(n), \tag{9.29}$$

with the number of different parameters K, n the size of the sample and $lnL(\underline{\Theta})$ the log-likelihood value with respect to the estimated parameters $\underline{\Theta}$.

Another performance tool that will be considered is the Q-Q plot. The Q-Q plot as described by Wicklin (2011) is a useful tool that visually illustrates the effectiveness of each model.

9.4 Simulated Data Analysis

9.4.1 *Simulating from the Lindley Distribution*

The Lindley distribution (Eq. (9.2)) is described earlier in the document as being a mixture between an exponential with parameter θ (with mixture weight $\pi = \frac{\theta}{\theta+1}$) and a gamma with shape 2 and scale θ (with mixture weight $1 - \pi$) (Alrasheedi et al., 2022; Shanker & Mishra, 2013). This construction is important to take note of when trying to simulate values from the Lindley distribution, as there is no built-in function in R that can simulate directly from the Lindley distribution.

Algorithm 9.2 describes as shown in Ghitany et al. (2008) an algorithm from which to simulate Lindley random variates. The same concept as outlined by Algorithm 9.2 was adapted to create the algorithm for simulating EG2 random variates (Eq. (9.14)). The steps of this simulation is given in Algorithm 9.3.

Algorithm 9.2 Simulating from a Lindley distribution (Eq. (9.3))

1. Generate a vector U with n values where $U_i \sim Unif(0, 1)$.
2. Generate a vector E with n values where $E_i \sim Exp(\theta)$.
3. Generate a vector G with n values where $G_i \sim gamma(2, \theta)$.
4. Calculate the mixing probability, $\pi = \frac{\theta}{\theta+1}$.
5. Suppose the vector containing the simulated Lindley values is the vector L. If $U_i \leq \pi$, then $L_i = E_i$. Otherwise, $L_i = G_i$.

Algorithm 9.3 Simulating from the EG2 model (Eq. (9.14))

1. Generate a vector U with n values where $U_i \sim Unif(0, 1)$.
2. Generate a vector E with n values where $E_i \sim Exp(\theta)$.
3. Generate a vector G with n values where $G_i \sim gamma(2, \theta)$.
4. Suppose the vector containing the simulated EG2 values is the vector $EG2$. If $U_i \leq \pi$, then $EG2_i = E_i$. Otherwise, $EG2_i = G_i$.

9.4.2 *Simulated Data Analysis: EG2*

Algorithm 9.3 is used to simulate EG2 random variates (Eq. (9.14)) across a range of different parameter values and sample sizes, where the EG2 model is fitted to the simulated data and the bias and mean squared error (MSE) are calculated. This procedure is repeated 1000 times for each set of sample sizes and parameter values. The bias and MSE values are calculated using the formulae given by Abushal et al. (2023);

$$\begin{aligned} Bias &= \frac{1}{1000}\sum_{i=1}^{1000}(\underline{\Theta} - \hat{\underline{\Theta}}) \\ MSE &= \frac{1}{1000}\sum_{i=1}^{1000}(\underline{\Theta} - \hat{\underline{\Theta}})^2. \end{aligned} \tag{9.30}$$

Tables 9.3 and 9.4 contain the results of the simulation study for the EG2 model for various parameter values. The EG2 model will be shown to be a useful model in the data application section, Sect. 9.5. For illustrative purposes, Fig. 9.19a and b display the box-plots of the bias and the MSE of selected parameter values of the EG2 model. The bias values appear to be close to zero and the MSE values are also fairly small for larger sample sizes.

Figure 9.19a displays a box-plot of the simulated bias of the parameter values for a given set of these parameter values of the EG2 model, and Fig. 9.19b displays a box-plot of the simulated MSE of the parameter values for a given set of these parameter values of the EG2 model. This was done to give an indication of the spread of the individual bias and MSE values that were generated.

Table 9.3 Simulated bias and MSE values for the EG2 model for $\theta_1 = 0.1, \theta_2 = 0.25$ and various values of π

		Bias				MSE			
		n=30	n=75	n=100	n=200	n=30	n=75	n=100	n=200
$\pi = 0.1$	π	−0.01037	−0.00527	−0.00605	−0.00301	0.00365	0.00124	0.00115	0.00034
	θ_1	−0.04177	−0.02257	−0.01811	−0.01086	0.00544	0.00229	0.00168	0.00084
	θ_2	−0.00141	0.00018	0.00071	0.00026	0.00132	0.00052	0.00038	0.00019
$\pi = 0.25$	π	0.01359	0.00762	0.00347	0.00230	0.01437	0.00564	0.00390	0.00116
	θ_1	−0.02129	−0.00983	−0.00835	−0.00319	0.00219	0.00093	0.00073	0.00028
	θ_2	−0.00266	0.00027	−0.00090	0.00055	0.00225	0.00086	0.00061	0.00028
$\pi = 0.5$	π	0.06717	0.02229	0.01919	0.00601	0.03679	0.01338	0.00873	0.00286
	θ_1	−0.00972	−0.00582	−0.00486	−0.00169	0.00127	0.00048	0.00033	0.00013
	θ_2	−0.01423	−0.00217	−0.00105	0.00148	0.00658	0.00217	0.00144	0.00055
$\pi = 0.75$	π	0.13302	0.04665	0.03413	0.01033	0.05575	0.01383	0.00837	0.00161
	θ_1	−0.00487	−0.00088	−0.00105	−0.00082	0.00077	0.00023	0.00017	0.00007
	θ_2	−0.03927	−0.02764	−0.01127	−0.00465	0.02212	0.01014	0.00464	0.00126
$\pi = 0.9$	π	0.14835	0.04815	0.03230	0.00545	0.06076	0.01126	0.00637	0.00026
	θ_1	−0.00105	−0.00007	−0.00026	−0.00032	0.00055	0.00016	0.00012	0.00005
	θ_2	−0.08532	−0.04734	−0.04437	−0.02105	0.06022	0.02125	0.02010	0.00604

Table 9.4 Simulated bias and MSE values for the EG2 model for $\theta_1 = 0.1, \theta_2 = 0.75$ and various values of π

		Bias				MSE			
		n=30	n=75	n=100	n=200	n=30	n=75	n=100	n=200
$\pi = 0.1$	π	−0.03768	−0.01742	−0.01055	−0.00369	0.01122	0.00355	0.00234	0.00078
	θ_1	−0.11754	−0.03864	−0.02198	−0.00859	0.03908	0.00715	0.00313	0.00082
	θ_2	−0.01295	−0.00781	−0.00234	−0.00347	0.01645	0.00606	0.00385	0.00188
$\pi = 0.25$	π	−0.01765	−0.00925	−0.00740	−0.00421	0.01858	0.00681	0.00514	0.00183
	θ_1	−0.03232	−0.00673	−0.00668	−0.00202	0.00665	0.00094	0.00079	0.00026
	θ_2	−0.02247	−0.00775	−0.00627	−0.00321	0.02636	0.00869	0.00608	0.00267
$\pi = 0.5$	π	0.00672	−0.00359	−0.00204	−0.00135	0.02683	0.01012	0.00701	0.00328
	θ_1	−0.00670	−0.00308	−0.00309	−0.00082	0.00129	0.00039	0.00030	0.00012
	θ_2	−0.07557	−0.02103	−0.02356	−0.00590	0.08971	0.02084	0.01481	0.00589
$\pi = 0.75$	π	0.06247	0.02341	0.00746	0.00570	0.02858	0.00996	0.00747	0.00286
	θ_1	−0.00145	0.00055	−0.00059	−0.00063	0.00069	0.00022	0.00016	0.00008
	θ_2	−0.19778	−0.06456	−0.06302	−0.02321	0.50475	0.09347	0.09229	0.02313
$\pi = 0.9$	π	0.11036	0.03856	0.03116	0.00733	0.03842	0.00805	0.00538	0.00146
	θ_1	0.00209	0.00099	0.00076	−0.00019	0.00050	0.00017	0.00012	0.00006
	θ_2	−0.27251	−0.18880	−0.16564	−0.13417	0.99529	0.48401	0.36297	0.19117

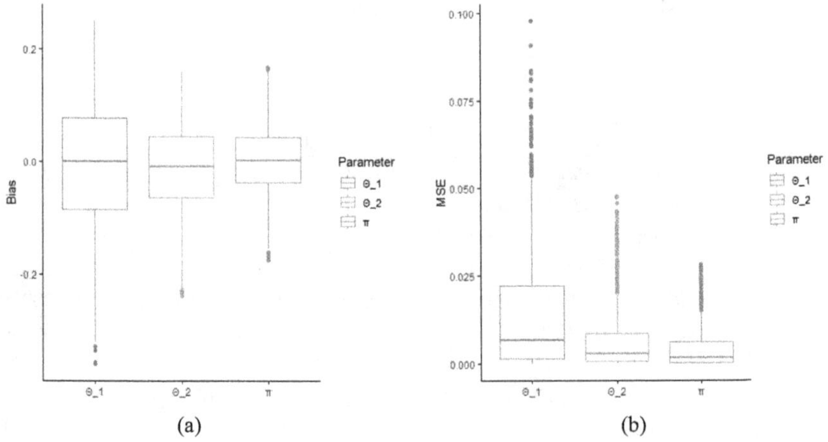

Fig. 9.19 Simulated from EG2 for $\pi = 0.25$, $\theta_1 = 0.5$, $\theta_2 = 0.75$ and $n = 100$. (**a**) Box-Plots for the Bias of the EG2 model. (**b**) Box-Plots for MSE of the EG2 model

9.5 Real Data Analysis

In this section, the performances of the seven mixture models given in Sect. 9.2 are illustrated using Q-Q plots and calculated AIC and BIC measures for three real life data sets (Boeing, Monoclonal Gamopathy, and Kidney respectively). Values in bold indicate the better performing model based on AIC and BIC values.

9.5.1 Boeing Dataset

In this dataset the number of successive failures of the air-conditioning system on a Boeing 720 aircraft is modelled. This dataset is used by and can be obtained in the paper of Al-Moisheer et al. (2021) who proposed a mixture of Lindley models (Eq. (9.3)). Figure 9.20a and b displays the histogram, as well as the possible modes of the data, respectively. The `locmodes` function in the `multimode` package in R was applied for Fig. 9.20b. The data appear bimodal when observing Fig. 9.20a and b and thus it follows that a finite mixture model(s) may be fitted to this dataset. The observation $x = 603$, considered as a possible outlier, was removed from the dataset before the data fitting.

Tables 9.5 and 9.6 gives the estimated parameter values for the considered models fitted to the Boeing dataset. Table 9.6 gives specifically the estimated parameter values for the mixture models fitted to the dataset. From the results, the EG2 model (Eq. (9.14)) with the lowest AIC and BIC values, indicates to be a suitable contender for this data set. Figures 9.21a and b and 9.22a and b display the Q-Q plots for selected models fitted to the Boeing dataset. From these figures, it is illustrated that the EG2 model is a suitable model for this dataset.

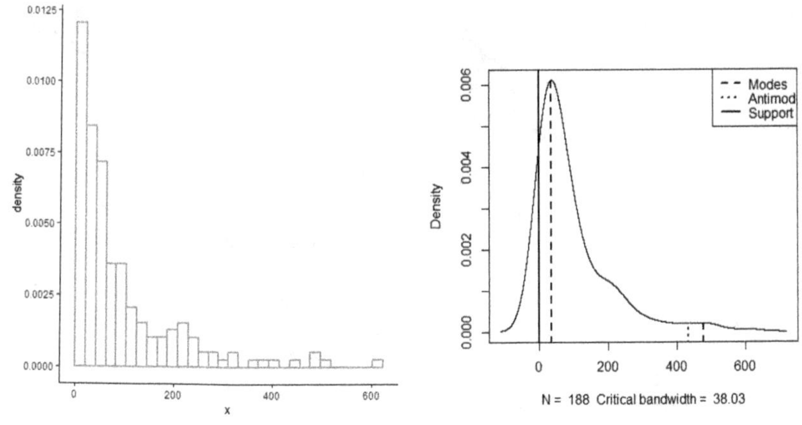

(a) (b)

Fig. 9.20 Plots displaying the characteristics of the Boeing dataset: Histogram and a density plot displaying the modes of the data. (**a**) Histogram of the Boeing dataset. (**b**) Graph showing modes using `locmodes` function

Table 9.5 Models fitted to the Boeing dataset

	Exponential	Lindley	XLindley	SpecShank	Shanker	Gamma	GLLind	GXL
θ	0.01119	0.02214	0.01631	0.00755	0.01128	0.01036	0.01143	0.01407
α	–	–	–	–	10000	0.92600	0.00001	0.78015
AIC	**2056.20**	2138.50	2070.70	2464.40	2058.20	2057.40	2062.50	2066.80
BIC	**2059.40**	2141.70	2073.90	2467.60	2064.70	2063.90	2069.00	2073.30
$ln(L)$	1027.10	1068.20	1034.30	1231.20	1027.10	1026.70	1029.30	1031.40

Table 9.6 Two-component mixture models fitted to the Boeing dataset

	Lindley	XLindley	SpecShank	EG2	EG	Shanker	Gamma	GLLind	GXL
π	0.52054	0.48216	0.50544	0.71344	0.68499	0.31741	0.36518	0.40371	0.48051
θ_1	0.06509	0.04097	0.07011	0.00879	0.00859	0.06880	0.01280	0.01255	0.01236
θ_2	0.01290	0.01086	0.01320	0.06972	0.05792	0.01066	0.03743	0.04470	0.05421
α_1	–	–	–	–	–	0.00002	2.29336	1.13820	1.35353
α_2	–	–	–	–	1.77122	279.790	1.41171	0.61282	1.30752
AIC	**2051.60**	2052.00	2053.20	**2050.00**	2051.90	2053.90	2052.80	2053.90	2053.30
BIC	**2061.30**	2061.70	2062.90	**2059.70**	2064.80	2070.00	2069.00	2070.00	2069.50
$ln(L)$	1022.80	1023.00	1023.60	1022.00	1021.90	1021.90	1021.40	1021.90	1021.70

9.5.2 MGUS2 Dataset

This dataset can be obtained from the `survival` package in R. Contained in the dataset is the time in months of patients with monoclonal gammopathy until death or last contact. For data analysis purposes, this data was scaled by a factor of 10^{-1}.

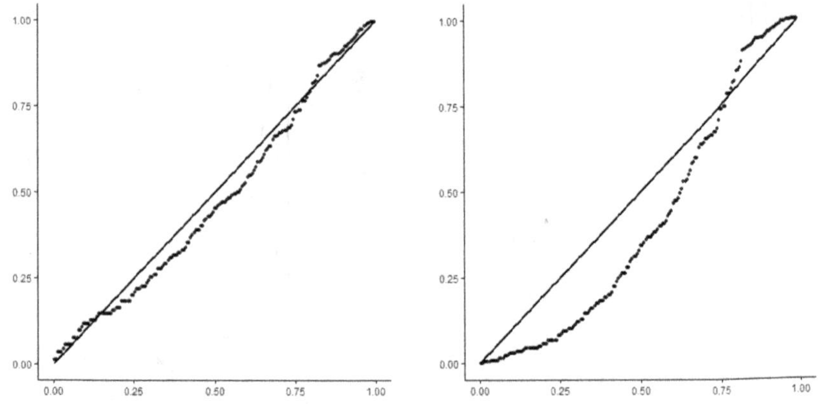

(a) (b)

Fig. 9.21 Q-Q plots for the exponential and Lindley distributions fitted to the Boeing dataset. (**a**) Exponential Q-Q plot. (**b**) Lindley Q-Q plot

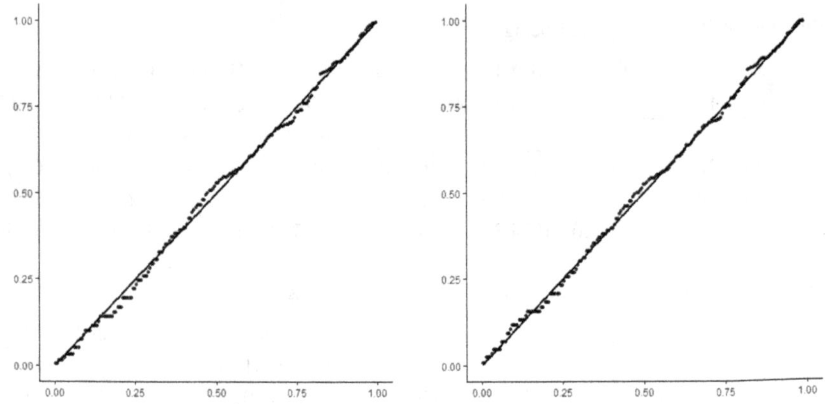

(a) (b)

Fig. 9.22 Q-Q plots for the Lindley mixture model and the EG2 model fitted to the Boeing dataset. (**a**) Lindley Mixture Q-Q plot. (**b**) EG2 Q-Q plot

Figure 9.23a and b displays the histogram, as well as the possible modes of the data, respectively. The `locmodes` function in the `multimode` package in R was used here for Fig. 9.23b. The data appear bimodal when observing Fig. 9.23b, and thus it follows that a finite mixture model(s) can be fitted to this dataset.

Tables 9.7 and 9.8 gives the estimated parameter values for the considered models fitted to the MGUS2 dataset. Table 9.8 gives specifically the estimated parameter values for the mixture models fitted to the dataset.

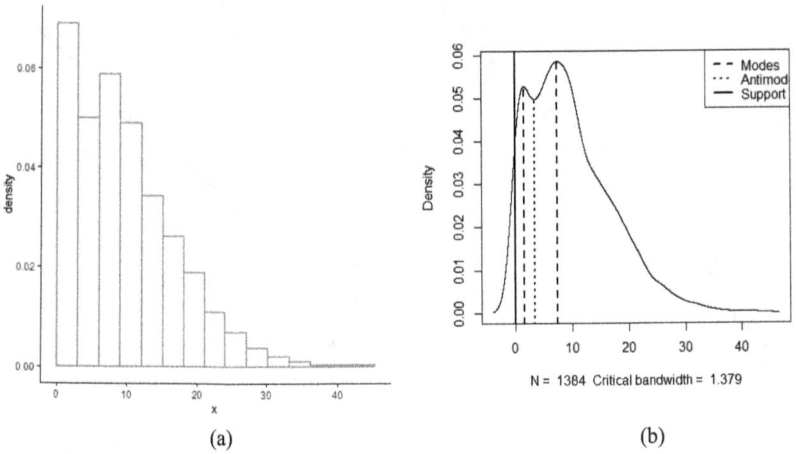

Fig. 9.23 Plots displaying the characteristics of the MGUS2 dataset: Histogram and a density plot displaying the modes of the data. (**a**) Histogram of the MGUS2 dataset. (**b**) Graph showing modes using `locmodes` function

Table 9.7 Models fitted to the MGUS2 dataset

	Exponential	Lindley	XLindley	SpecShank	Shanker	Gamma	GLLind	GXL
θ	0.10439	0.19196	0.15776	0.20808	0.16728	0.11892	0.15367	0.15354
α	–	–	–	–	3.94509	1.13918	0.60529	0.95605
AIC	9024.70	9011.40	**8948.20**	9181.70	**8943.20**	9012.40	**8943.60**	8918.90
BIC	9029.90	9016.60	**8953.40**	9187.00	**8953.70**	9022.80	**8954.00**	8959.40
$ln(L)$	4511.30	4504.70	4473.10	4589.90	4469.60	4504.20	4469.80	4472.50

Table 9.8 Mixture models fitted to the MGUS2 dataset

	Lindley	XLindley	SpecShank	EG2	EG	Shanker	Gamma	GLLind	GXL
π	0.08187	0.10161	0.10361	0.11060	0.13994	0.88939	0.12021	0.10628	0.10778
θ_1	3.38425	0.15776	2.55194	2.12034	1.45576	0.18668	2.81419	4.10775	4.52100
θ_2	0.17777	0.15776	0.18485	0.18670	0.22927	2.12000	0.21546	0.22311	0.23149
α_1	–	–	–	–	–	0.00000	1.39049	1.51730	1.38136
α_2	–	–	–	–	2.52812	10000	2.33153	1.56280	1.97303
AIC	8836.40	8952.20	8825.40	8823.30	8812.00	8827.30	**8809.10**	**8806.20**	**8809.50**
BIC	8852.10	8967.90	8841.10	8839.00	8832.90	8853.40	**8835.30**	**8832.40**	**8835.70**
$ln(L)$	4415.20	4473.10	4409.70	4408.60	4402.00	4408.60	4399.60	4398.10	4399.80

Figures 9.24a and b, 9.25a and b, and 9.26a and b display the Q-Q plots for selected models fitted to the MGUS2 dataset. Figure 9.27a and b displays the histogram of the MGUS2 dataset overlaid with two selected fitted mixture models, illustrating that the fitted models account for the bimodality in the data.

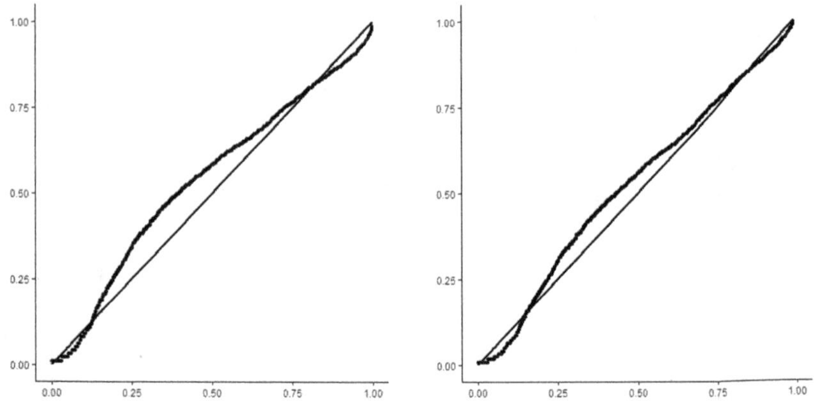

(a) (b)

Fig. 9.24 Q-Q plots for the exponential distribution and the gamma mixture model fitted to the MGUS2 dataset. (**a**) Exponential Q-Q plot. (**b**) XLindley Q-Q plot

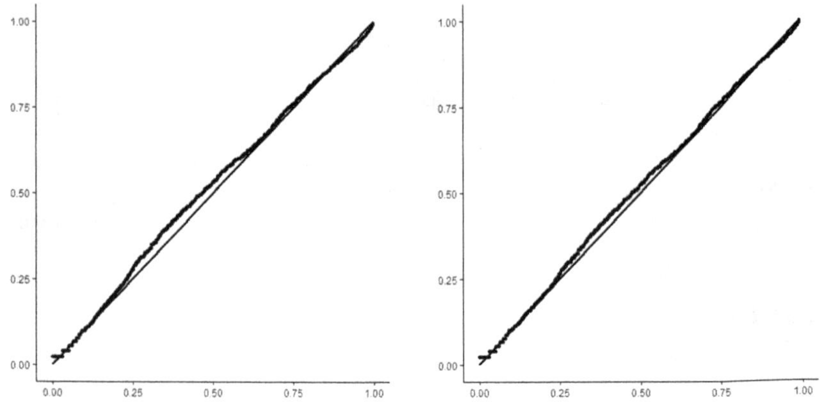

(a) (b)

Fig. 9.25 Q-Q plots for the Lindley mixture model and the EG2 model fitted to the MGUS2 dataset. (**a**) Lindley Mixture Q-Q plot. (**b**) EG2 Q-Q plot

The Q-Q plots illustrate that the EG, EG2 and the Gamma Mixture models are suitable contenders for this dataset. Results from Tables 9.7 and 9.8 indicate that the Shanker (Eq. (9.6)), XLindley (Eq. (9.4)) and the GLLind mixture distributions (Eq. (9.18)) can also be considered as contenders, when analysing the AIC and BIC values. The added flexibility of the EG and EG2 mixture distributions proves to be advantageous, as these models outperformed the parsimonious Lindley distribution (Eq. (9.2)).

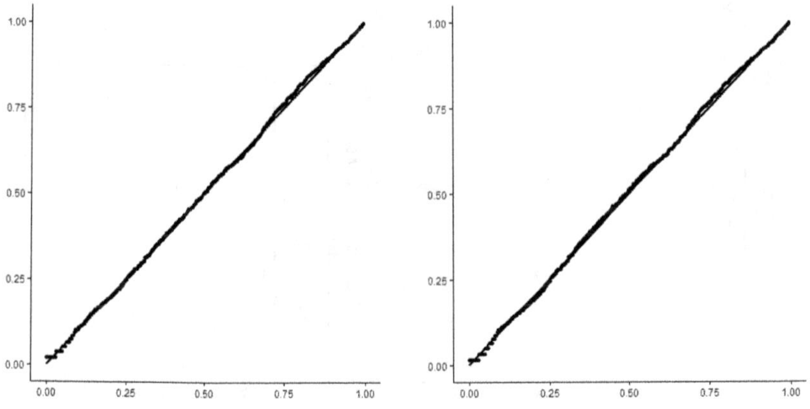

(a) (b)

Fig. 9.26 Q-Q plots for the EG model and the gamma mixture model fitted to the MGUS2 dataset. (**a**) EG Q-Q plot. (**b**) Gamma Mixture Q-Q plot

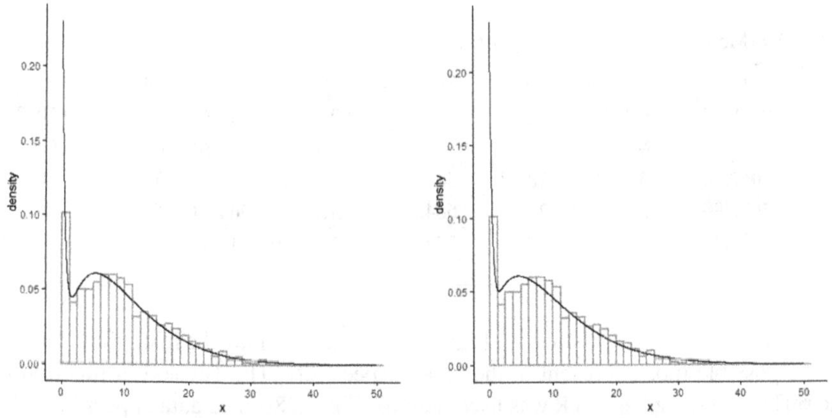

(a) (b)

Fig. 9.27 Histograms of the MGUS2 dataset overlaid with the EG model and the gamma mixture model. (**a**) EG model. (**b**) Gamma mixture model

9.5.3 Kidney Dataset

This dataset can be obtained from the `survival` package in R. Contained in the dataset is the time until the recurrence of infection for kidney patients after the insertion of a catheter. For data analysis purposes, this data was scaled by

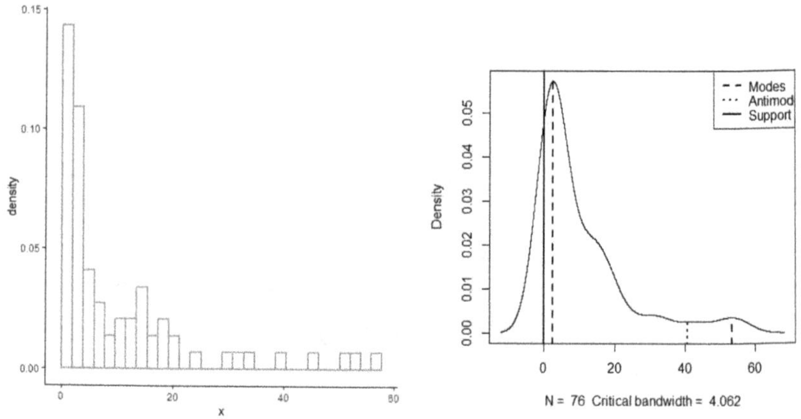

(a) (b)

Fig. 9.28 Plots displaying the characteristics of the Kidney dataset: Histogram and a density plot displaying the modes of the data. (**a**) Histogram of the Kidney dataset. (**b**) Graph showing modes using `locmodes` function

Table 9.9 Models fitted to the Kidney dataset

	Exponential	Lindley	XLindley	SpecShank	Shanker	Gamma	GLLind	GXL
θ	0.09839	0.18166	0.1397	0.19713	0.09849	0.0736	0.10640	0.10631
α	–	–	–	–	10000	0.7480	0.17757	0.62619
AIC	**506.450**	545.160	518.17	561.950	508.450	**503.72**	519.780	510.650
BIC	**508.780**	547.490	520.50	564.280	513.110	**508.39**	524.440	515.310
$ln(L)$	252.230	271.580	258.09	279.980	252.230	249.86	257.890	253.320

multiplying it with the constant $\frac{1}{10}$. Figure 9.28a and b display the histogram, as well as the possible modes present in the data, respectively. The `locmodes` function in the `multimode` package in R was used here for Fig. 9.28b. The data appear bimodal when observing Fig. 9.28a and b and thus it follows that a finite mixture model(s) may be fitted to this dataset.

Table 9.9 gives the estimated parameter values of the considered models fitted to the Kidney dataset. Table 9.10 specifically gives the estimated parameter values of the mixture models fitted to the Kidney dataset.

Figures 9.29a and b and 9.30a and b display the Q-Q plots for selected models fitted to the Kidney dataset. Figure 9.30a illustrates that the EG2 (Eq. (9.14)) and the gamma mixture models can be considered as contenders for this dataset. When considering both the AIC and BIC values, the proposed EG2 model proves to be the most suitable mixture model for the dataset.

Table 9.10 Mixture models fitted to the Kidney dataset

	Lindley	XLindley	SpecShank	EG2	EG	Shanker	Gamma	GLLind	GXL
π	0.53416	0.45372	0.54844	0.59706	0.60974	0.39718	0.51459	0.49827	0.47777
θ_1	0.68613	0.69237	0.70728	0.06381	0.06485	1.01063	0.08669	1.14097	1.17190
θ_2	0.09968	0.08956	0.10111	0.99924	1.17153	0.06454	1.01221	0.09734	0.10078
α_1	–	–	–	–	–	0.00000	1.54512	1.90842	1.89869
α_2	–	–	–	–	2.28561	10000	2.06637	0.85913	1.27438
AIC	494.160	495.140	493.440	**491.140**	492.960	495.140	494.070	494.390	494.340
BIC	501.150	502.130	500.430	**498.140**	502.280	506.800	505.730	506.040	505.990
$ln(L)$	244.080	244.570	243.720	242.570	242.480	242.570	242.040	242.190	242.170

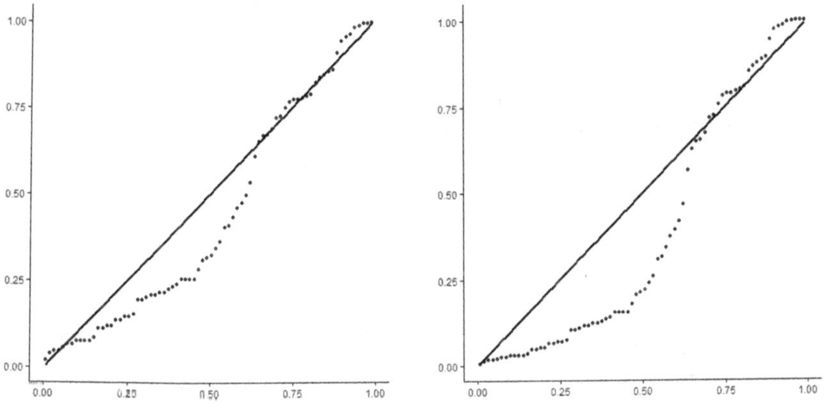

(a) (b)

Fig. 9.29 Q-Q plots for the exponential and Lindley distributions fitted to the Kidney dataset. (**a**) Exponential Q-Q plot. (**b**) Lindley Q-Q plot

9.5.4 *Fitting Models to Bootstraps of the Data*

Here we consider a resampling (with replacement) of the Boeing dataset, which describes the number of successive failures of the air-conditioning system on a Boeing 720 passenger jet. The idea of resampling or bootstrapping is summarised by Wehrens et al. (2000). The EG model (Eq. (9.15)), EG2 model (Eq. (9.14)) and the Lindley mixture model (Eq. (9.3)) are fitted to the 1000 (70%) resampled Boeing dataset. Figure 9.31a and b displays the boxplots for the AIC and BIC values for each of the three models fitted to the 1000 datasets resampled from the Boeing dataset. The boxplots illustrate that the EG2 model offers more flexibility as compared to the other models.

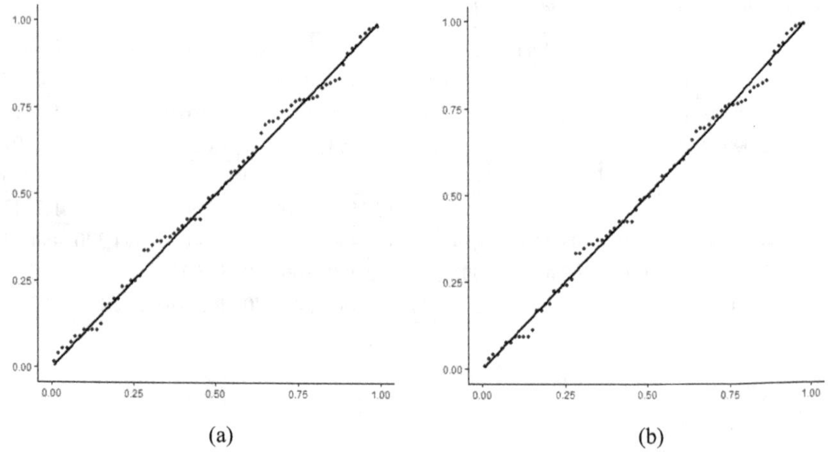

Fig. 9.30 Q-Q plots for the EG2 model and gamma mixture model fitted to the Kidney dataset. (**a**) EG2 Q-Q plot. (**b**) Gamma mixture Q-Q plot

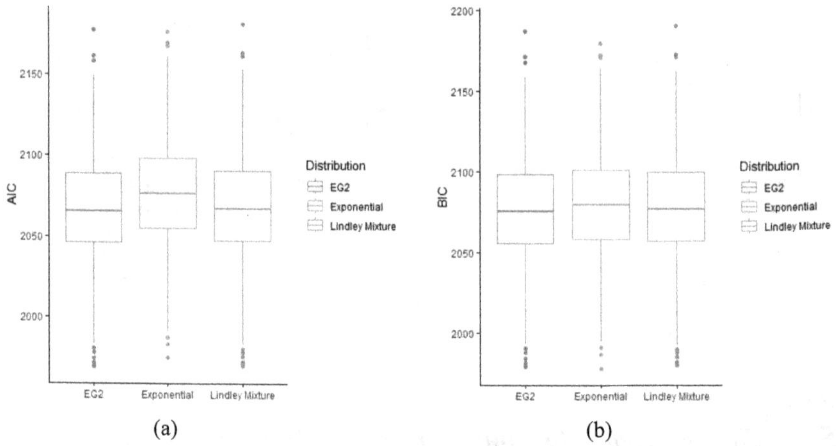

Fig. 9.31 Boxplots of bootstrapped AIC and BIC values for selected models fitted to the Boeing dataset. (**a**) Boxplot of AIC values. (**b**) Boxplot of BIC values

In addition to this, a resampling (with replacement) of the MGUS2 dataset is considered. The MGUS2 dataset in this section is resampled 250 times and the AIC and BIC values are calculated for three models. Figure 9.32a and b displays the boxplots of the AIC and BIC values, respectively, of the three mixture models. The boxplots illustrate that the proposed EG2 model offers again more flexibility as compared to the other models.

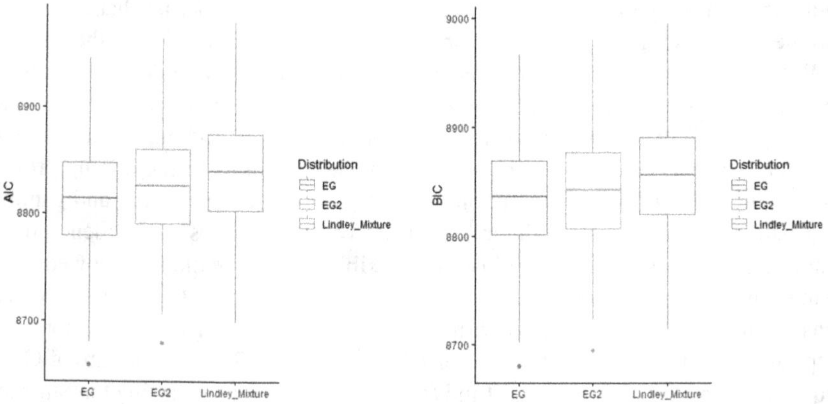

(a) (b)

Fig. 9.32 Boxplots of bootstrapped AIC and BIC values for selected models fitted to the MGUS2 dataset. (**a**) Boxplot of AIC values. (**b**) Boxplot of BIC values

9.6 Conclusion

In this chapter, seven parsimonious models, including two component ($k = 2$ in Eq. (9.1)) mixture models are presented, where the Lindley-type models are the focus of the study. The relevance for this investigation and look into finite mixtures of the Lindley distribution is emphasized by resent research, where it is for the first time that an ensemble of Lindley-type models are considered as components for finite mixture modeling.

A review, together with some properties of the considered models are presented in Sect. 9.2. We investigate the effect of the parameters on the PDFs of the two component mixtures of Lindley, XLindley, special Shanker, gamma, Generalised Lindley, as well as a mixture of exponential and gamma distributions. Two special cases of the Generalised Lindley distributions are given attention and two mixtures of exponential and gamma distributions are proposed (i.e. the EG and EG2 models). A general mixture between a gamma distribution with shape parameter equal to 2 and an exponential distribution (i.e., $\pi \in [0, 1]$) is the basis of the construction of the EG and EG2 models.

A simulation study of the EG2 model is also presented, where the bias and MSE values decreased for increasing sample sizes. The flexibility of the EG and EG2 models are proven and illustrated via three real life data applications. This flexibility and feasibility of the models are due to the fact that the mixing proportion is independent from the parameter values. In general, Lindley-type mixture models appear to be good contenders for lifetime data. In the data application of the MGUS2 (monoclonal gammopathy) dataset, it is observed that the Q-Q plots of the more complex mixture models did fit the data best, while the simpler models, such as the EG and EG2 showed a good fit as well. When looking at the Q-Q plots, one could ask if more parameters in a model are necessary, considering that the added parameters (for some of the models) only give a slightly improved fit to the data.

With the exception of the Shanker mixture model and EG model, the Lindley-type mixture models were all parsimonious with only three parameters, which is the mixing probability and the scale parameters for each of the two components. For the Shanker distribution, the added parameter α added flexibility in the mixing probability value, allowing for more option in the degree of the exponential and gamma distribution contained within the model. The two special cases of the Generalised Lindley, GLLind and GXL also illustrated flexibility, as these models allowed for a change in the shape parameter of the mixture of the gamma distributions. When considering the more cumbersome models, the mixture of gamma distributions outperformed the mixture of Shanker distributions (Eq. (9.7)). The mixture of GXL (Eq. (9.20)) and the mixture of GLLind (Eq. (9.18)) also outperforming the Shanker mixture, while being competitive with the Gamma mixture distribution.

This perspective and comparisons of finite mixtures of Lindley-type models can be expanded upon for other datasets for an extensive overview of the power of parsimonious Lindley models. It may be worthwhile to consider mixture models of the variations of the Lindley or Lindley-type distributions in various other survival analysis data sets.

Acknowledgments We thank the editors as well as two anonymous reviewers for their constructive feedback and allowing us to present an improved manuscript. This work was based upon research supported in part by the National Research Foundation (NRF) of South Africa (SA), grant RA201125576565 & RA211204653274, nr 145681 & 151035; NRF ref. SRUG2204203865, and the Centre of Excellence in Mathematical and Statistical Sciences, based at the University of the Witwatersrand (SA). The opinions expressed and conclusions arrived at are those of the authors and are not necessarily to be attributed to the NRF.

References

Abushal, T. A., Sindhu, T. N., Lone, S. A., Hassan, M. K., & Shafiq, A. (2023). Mixture of Shanker distributions: Estimation, simulation and application. *Axioms, 12*(3), 231.

Akaike, H. (1998). Information theory and an extension of the maximum likelihood principle. In *Selected papers of hirotugu akaike* (pp. 199–213) Springer.

Al-Moisheer, A., Daghestani, A., & Sultan, K. (2021). Mixture of two one-parameter Lindley distributions: Properties and estimation. *Journal of Statistical Theory and Practice*, *15*, 1–21.

Alrasheedi, A., Abouammoh, A., & Kayid, M. (2022). A new flexible extension of the Lindley distribution with applications. *Journal of King Saud University-Science*, *34*(1), 101714.

Bain, L. J., & Engelhardt, M. (1992). *Introduction to probability and mathematical statistics* (Vol. 4). Duxbury Press.

Bakoban, R. (2010). A study on mixture of exponential and exponentiated gamma distributions. *Advances and Applications in Statistical Sciences*, *2*(1), 101–127.

Burnham, K. P., & Anderson, D. R. (2004). Multimodel inference: Understanding AIC and BIC in model selection. *Sociological Methods & Research*, *33*(2), 261–304.

Ekhosuehi, N., Lawrence, N., & Opone, F. (2020). A new mixture of exponential-gamma distribution. *Gazi University Journal of Science*, *33*(2), 548–564.

Ferreira, J., & van der Merwe, A. (2022). A noncentral Lindley construction illustrated in an INAR (1) environment. *Stats, 5*(1), 70–88.

Ghitany, M. E., Atieh, B., & Nadarajah, S. (2008). Lindley distribution and its application. *Mathematics and Computers in Simulation, 78*(4), 493–506.

Iqbal, T., & Iqbal, M. Z. (2020). On the mixture of weighted exponential and weighted gamma distribution. *International Journal of Analysis and Applications*, *18*(3), 396–408.

Khan, A. H., & Jan, T. (2015). Estimation of stress-strength reliability model using finite mixture of two parameter Lindley distributions. *Journal of Statistics Applications and Probability, 4*(1), 147–159.

Leemis, L. (2003). Statistical modeling: Regression, survival analysis, and time series analysis. *Open Education Resources, 3*. https://scholarworks.wm.edu/cgi/viewcontent.cgi?article=1003&context=oer

Lindley, D. V. (1958). Fiducial distributions and Bayes' theorem. *Journal of the Royal Statistical Society Series B (Methodological)*, 102–107.

Maya, R., & Irshad, M. (2017). Extended version of generalised Lindley distribution. *South African Statistical Journal, 51*(1), 19–44.

van der Merwe, A., & Ferreira, J. T. (2022). An adapted discrete Lindley model emanating from negative binomial mixtures for autoregressive counts. *Mathematics, 10*(21), 4141.

Nadarajah, S., Bakouch, H. S., & Tahmasbi, R. (2011). A generalized Lindley distribution. *Sankhya B, 73*, 331–359.

Nawel, K., Gemeay, A. M., Zeghdoudi, H., Karakaya, K., Alsahangiti, A. M., Bakr, M. E., Balogun, O. S., Muse, A. H., & Hussam, E. (2023). Modeling voltage real data set by a new version of Lindley distribution. *IEEE Access, 11*, 67220–67229.

Punzo, A. (2019). A new look at the inverse Gaussian distribution with applications to insurance and economic data. *Journal of Applied Statistics, 46*(7), 1260–1287.

Punzo, A., Mazza, A., & Maruotti, A. (2018). Fitting insurance and economic data with outliers: a flexible approach based on finite mixtures of contaminated gamma distributions. *Journal of Applied Statistics, 45*(14), 2563–2584.

Reyes, J., Gómez-Déniz, E., Gómez, H. W., & Calderín-Ojeda, E. (2021). A bimodal extension of the exponential distribution with applications in risk theory. *Symmetry, 13*(4), 679.

Schwarz, G. (1978). Estimating the dimension of a model. *The Annals of Statistics*, 461–464.

Shanker, R., & Mishra, A. (2013). A two-parameter Lindley distribution. *Statistics in Transition New Series, 14*(1), 45–56.

Shanker, R., Sharma, S., & Shanker, R. (2013). A two-parameter Lindley distribution for modeling waiting and survival times data. *Applied Mathematics, 4*(2), 28215. https://doi.org/10.4236/am.2013.42056

Shanker, R., Shukla, K. K., Shanker, R., & Tekie, A. (2017). A three-parameter Lindley distribution. *American Journal of Mathematics and Statistics, 7*(1), 15–26.

Tomarchio, S. D., & Punzo, A. (2020). Dichotomous unimodal compound models: Application to the distribution of insurance losses. *Journal of Applied Statistics, 47*(13–15), 2328–2353.

Tomy, L. (2018). A retrospective study on Lindley distribution. *Biometrics and Biostatistics International Journal, 7*(3), 163–169.

Wehrens, R., Putter, H., & Buydens, L. M. (2000). The bootstrap: a tutorial. *Chemometrics and Intelligent Laboratory Systems, 54*(1), 35–52.

Wicklin, R. (2011). Modeling the distribution of data? Create a Q-Q plot. Online, Accessed Oct 21, 2023, Available at https://blogs.sas.com/content/iml/2011/10/28/modeling-the-distribution-of-data-create-a-qq-plot.html.

Young, D. S., Chen, X., Hewage, D. C., & Nilo-Poyanco, R. (2019). Finite mixture-of-gamma distributions: Estimation, inference, and model-based clustering. *Advances in Data Analysis and Classification, 13*, 1053–1082.

Zakerzadeh, H., & Dolati, A. (2009). Generalized Lindley distribution. *Journal of Mathematical Extension, 3*(2), 13–25.

Chapter 10
Baranchik-Type Estimators Under Modified Balanced Loss Functions

Lahoucine Hobbad, Idir Ouassou, and Mohamed Alahiane

10.1 Introduction

Baranchik (1970) introduced a class of estimators to estimate the mean of a multivariate normal distribution. These estimators can be represented as follows $\delta_{a,b,S}(X) = \left(1 - aS(\|X\|^2)/(b + \|X\|^2)\right) X$. This class of estimators encompasses various options, including the straightforward choice of $S(t) = t/(t+b)$ with $b \geq 0$. Notably, when $b = 0$, these estimators correspond to the James-Stein estimators, (see Stein 1956 and James and Stein 1961).

The Balanced Loss Functions (BLFs), proposed by Zellner (1994), aim to meet two key criteria in the estimation: goodness of fit and accuracy. Traditional goodness of fit criteria, such as the sum of squared residuals in regression problems, tend to produce estimates that provide a good fit and are unbiased. However, these estimates may not be as accurate as the biased estimators. So, we need a framework that balances these two considerations. BLFs offer an attractive solution by incorporating the proximity of an estimator denoted as δ, to both a target estimator δ_0 and the unknown parameter θ to be estimated. They can be formulated as $L_{\omega,\delta_0}(\delta, \theta) = \omega\|\delta - \delta_0\|^2 + (1-\omega)\|\delta - \theta\|^2$ where, $\theta \in \mathbb{R}^d$ denotes the unknown parameter vector, while $\omega \in [0, 1)$ denotes the weight assigned to the proximity of δ relatively to δ_0, where δ represents a given estimator of θ.

Marchand and Strawderman (2020), have extended the concept of balanced loss functions by introducing MBLFs in both cases **(i)** and **(ii)**. These MBLFs integrate the functions $\rho \geq 0$ and $\ell \geq 0$, which are increasing, concave functions and satisfy a completely monotone property. They showed that Baranchik-type estimators

L. Hobbad · I. Ouassou (✉) · M. Alahiane
Cadi Ayyad University, National School of Applied Sciences, Marrakesh, Morocco
e-mail: i.ouassou@uca.ac.ma

C. A. Coelho, D.-G. Chen (eds.), *Statistical Modeling and Applications*, Emerging Topics in Statistics and Biostatistics, https://doi.org/10.1007/978-3-031-69622-0_10

$\delta_a(X) = \left(1 - aS(\|X\|^2)/\|X\|^2\right) X$ dominate the usual estimator $\delta_0(X) = X$, when X has a scale mixture of normal distributions.

Hobbad and Ouassou (2021), generalized this result by showing that the Baranchik-type estimator $\delta_{a,S}(X) = \left(1 - a(1-\omega)S(\|X\|^2)/\|X\|^2\right) X$ dominates the estimator $\delta_0(X) = X$ for spherically symmetric distributions with a density. This result holds for MBLFs of types **(i)** and **(ii)**, where ρ and ℓ need only to be increasing and concave functions.

In this chapter, we will study the problem of estimating a d-dimensional parameter $\theta = (\theta_1, \ldots, \theta_d)$ when the observation is the vector $X = (X_1, X_2, \ldots, X_d)$ of a general spherically symmetric distribution around $\theta = (\theta_1, \ldots, \theta_d)$ without necessarily needing a probabilitydensity under MBLFs **(i)** and **(ii)**. Our objective is to find a sufficient condition on the constant a for which the Baranchik-type estimator:

$$\delta_{a,S}(X) = \left(1 - a(1-\omega)S(\|X\|^2)/\|X\|^2\right) X \tag{10.1}$$

dominates the estimator δ_0, where the function S is a twice differentiable almost everywhere function, and satisfying:

$$0 \leq S(.) \leq 1, \;\; S(.) \neq 0, \;\; S'(.) \geq 0 \text{ and } S''(.) \leq 0. \tag{10.2}$$

This chapter is organized as follows. In Sect. 10.2, we give some conditions for the Baranchik-type estimator (10.1) to dominate δ_0 under MBLFs **(i)**, when the random vector X has a spherically symmetric distribution without necessarily needing a probability density. In Sect. 10.3, under the same model, we provide a condition for which the Baranchik-type estimator (10.1) dominates $\delta_0(X) = X$ under MBLFs **(ii)**. A simulation study is performed in Sect. 10.4.

10.2 Minimax Estimators of the Location Parameters with Respect to the Loss $\omega\rho(\|\delta - \delta_0\|^2) + (1-\omega)\rho(\|\delta - \theta\|^2)$

In this section, we will study the minimax Baranchik-type estimators of a location vector θ when X has a spherically symmetric distribution without necessarily needing a density under the MBLFs **(i)**. We assume that the function ρ satisfies:

$$\textbf{H1}: \rho(0) = 0, 0 < \rho'(0) < +\infty, \text{ and } \rho \text{ is concave.}$$

Examples of ρ functions satisfying hypothesis **H1**: $(i)\ \rho(t) = t$, $(ii)\ \rho(t) = 1 - \exp(-t/\alpha)$ with $\alpha > 0$, $(iii)\ \rho(t) = \log(1+t)$, $(iv)\ \rho(t) = (1 + t/\gamma)^\beta$ with $\gamma > 0$, $\beta \in (0, 1)$, $(v)\ \rho(t) = r^2 t/(rt+1)$ with $r > 0$, $(vi)\ \rho(t) = \arctan(t)$ and $(vii)\ \rho(t) = \tanh(t)$. With the exception of (vi) and (vii), these last two examples are not completely monotones i.e., $(-1)^k \rho^k(t) \geq 0$ for $t > 0$ and $k \in \mathbb{N}$.

We propose the following lemma to prove the domination and the minimaxity results.

Lemma 10.1 *Let X be a random vector with a spherically symmetric distribution around θ, and let $g(.)$ be a weakly differentiable function such that $E_\theta\left[|(X-\theta)^\top g(X)|\right] < \infty$. Then*

$$E_\theta\left[\rho'(\|X-\theta\|^2)(X-\theta)^\top g(X)\right] = \frac{1}{d}E\left[r^2\rho'(r^2)\int_{B_{r,\theta}} \nabla g(X)dV_{r,\theta}(X)\right],$$

where E_θ denotes the expectation with respect to the radius distribution and where $V_{r,\theta}(.)$ is the uniform distribution on $B_{r,\theta}$ (the ball of radius r and centered at θ).

We introduce also the following lemma to prove the Theorem 10.1. The importance of this lemma in the present study stems from the fact that if $X \sim S.S.(\theta)$, then $X|R \sim U(\|X-\theta\|^2 = R^2)$, where $S.S(\theta)$ denote the Spherically Symmetric distribution around θ.

Lemma 10.2 *Let $X \sim S.S(\theta)$. For $d \geq 4$ and $g(x) = -a\left(S(\|x\|^2)/\|x\|^2\right)x$, we have*

$$E_\theta\left[(1-\omega)\rho'(\|X-\theta\|^2)[2(1-\omega)g(X).(X-\theta)+(1-\omega)^2\|g(X)\|^2]\right]$$
$$\leq E\left[(1-\omega)^2 a\rho'(R^2)\left(\frac{a(1-\omega)}{R^2}-\frac{2(d-2)}{d}\right)\right.$$
$$\left.\times E_{R,\theta}[R^2\frac{S(\|X\|^2)}{\|X\|^2}/\|X-\theta\| = R]\right]$$

Then, we give the following main dominance theorem.

Theorem 10.1 *Assume that $d \geq 4$, X has a spherically symmetric distribution around θ ($X \sim SS_d(\theta)$) and that the function ρ satisfies **H1**. For estimating θ under the loss **(i)** with $\delta_0(X) = X$, the estimator $\delta_{a,S}(X)$ in (10.1) where the function S satisfies (10.2), dominates $\delta_0(X)$ provided:*

$$0 < a < \frac{d-2}{d} \times \frac{E\left[\rho'(R^2)R^2\right]}{\omega\rho'(0)E\left[R^{-2}\right]+(1-\omega)E\left[\rho'(R^2)R^{-2}\right]}, \tag{10.3}$$

where E denotes the expectation with respect to the distribution of the radius R.

Remark 10.1 Suppose that X is spherically symmetric distributed about θ with a probability density $f(\|x-\theta\|^2)$. The domination condition (see Hobbad and Ouassou 2021) is:

$$0 < a < \frac{2\,K^2\,(d-2)/d}{\{\omega\rho'(0)+K(1-\omega)\}\;E\left(\rho'(R^2)R^{-2}\right)}, \tag{10.4}$$

where $K = E(\rho'(R^2))$. In the case where $\rho(t) = t$, the domination condition in Theorem 10.1 becomes $0 < a < (d-2)E\left[R^2\right]/dE\left[R^{-2}\right]$ and the condition in Hobbad and Ouassou (2021) is $0 < a < 2(d-2)/dE\left[R^{-2}\right]$. We deduce that if $E\left[R^2\right] < 2$ then the bound on the risk without a probability density is smaller than that with a probability density, otherwise we obtain the opposite result.

Example 10.1 (The Student Distribution) Suppose that $X \in \mathbb{R}^d$ is distributed according to a Multivariate Student distribution with an unknown location parameter θ, a scale I_d and $\nu > 0$ degrees of freedom: $X \hookrightarrow t_d(\theta, I_d, \nu)$, with the probability density function $f(x) = \dfrac{\Gamma(\frac{d+\nu}{2})}{(\pi\nu)^{\frac{d}{2}}\Gamma(\frac{\nu}{2})}\left(1 + \|x\|^2/\nu\right)^{-\frac{d+\nu}{2}}$. The distribution is the mixture of multivariate normal distributions with the inverse gamma distribution as the weighting function $G(t) = \dfrac{\nu^{\frac{\nu}{2}} \times t^{\frac{\nu}{2}-1} \times exp(-\frac{\nu t}{2})}{2^{\frac{\nu}{2}}\Gamma(\frac{\nu}{2})}$.

For $\rho(t) = 1 - \exp(-t/\alpha)$, we have $\rho'(t) = \alpha^{-1}\exp(-t/\alpha)$. Then from Theorem 10.1, we obtain

$$a_0 = \frac{d-2}{d}\,\frac{\nu(\nu-1)}{r^2\left[\omega\,(1/\alpha + 1/r)^{\nu+1} + (1-\omega)\,(1/\alpha + 1/r)^2\right]}.$$

For the distribution with a probability density (see Hobbad and Ouassou 2021), we have :

$$a_0 = \frac{2(d-2)}{d}\,\frac{\alpha(\nu-1)/(r+\alpha)}{\omega\left(\frac{1}{\alpha} + \frac{1}{r}\right)^{\nu} + (1-\omega)}.$$

Example 10.2 (The Symmetric Kotz Type Distribution) Let X be the random vector having a symmetric Kotz type distribution with unknown location parameter θ, scale I_d, with the probability density function:

$$f(x) = C_d(\|x\|^2)^{N-1}\exp(-r(\|x\|^2)^s),\ \text{where } r, s > 0 \text{and} 2N + d > 2, \tag{10.5}$$

and C_d is a normalization constant. For $s = 1$ and $\nu = N + d/2 - 1 > 0$, the radial density of $\|X - \theta\|^2 = R^2$ can be expressed as $h(u) = \dfrac{r^\nu}{\Gamma(\nu)}u^{\nu-1}e^{-ru}$, $u > 0$ which is a gamma distribution $\Gamma(\nu, r)$. From (10.1)

(continued)

Example 10.2 (continued)
we obtain

$$a_0 = \frac{d-2}{d}\frac{\nu(\nu-1)}{r^2\left[\omega\left(1/\alpha+1/r\right)^{\nu+1}+(1-\omega)\left(1/\alpha+1/r\right)^2\right]}.$$

For the distribution with a probability density (see Hobbad and Ouassou 2021), we have

$$a_0 = \frac{2(d-2)}{d}\frac{\alpha(\nu-1)/(r+\alpha)}{\omega\left(\frac{1}{\alpha}+\frac{1}{r}\right)^{\nu}+(1-\omega)}.$$

Example 10.3 (The Uniform Distribution on a Sphere) For the uniform distribution on the sphere $S_{m,\theta}$ ($X \sim U(S_{m,\theta})$) and $\rho(t) = \log(1+t)$, we have $\rho'(t) = \dfrac{1}{1+t}$ and $R^2 = \|X-\theta\|^2 = m^2$. Hence, the cut-off point a_0 is

$$a_0(m) = \frac{d-2}{d}\frac{m^4}{\omega(1+m^2)+(1-\omega)}.$$

Example 10.4 (The Uniform Distribution on a Ball) Let X be the random vector having the uniform distribution on the ball $B_{m,\theta}$ i.c, $X \sim U(B_{m,\theta})$, with $d \geq 4$ and $R^2 = \|X-\theta\|^2$ having a probability density $h(t) = \dfrac{d}{2m^d}t^{\frac{d}{2}-1}\mathrm{i}_{(0,m^2)}(t)$. According to the choice $\rho(t) = \log(1+t)$, $d = 4$ and from (10.1), we obtain:

$$a_0(m) = \frac{1}{2}\frac{m^4/2 - m^2 + \log(1+m^2)}{m^2\omega + (1-\omega)\log(1+m^2)}.$$

For a distribution with a probability density (see Hobbad and Ouassou 2021), we have

$$a_0(m) = \frac{\left\{m^2-\log(1+m^2)\right\}^2}{\left(m^4\omega/2+(1-\omega)\left\{m^2-\log(1+m^2)\right\}\right)\log(1+m^2)}$$

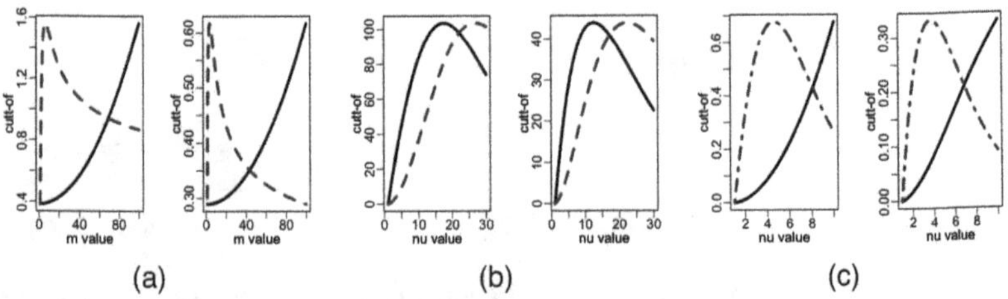

Fig. 10.1 (**a**) Uniform in the ball cases (**b**) Student distribution cases (**c**) Kotz distribution cases: without density (–), with density (- -) for $\omega = 1/4$ and $\omega = 3/4$

In the Fig. 10.1, we compare the limit found in this paper from (10.1) for spherically symmetric variables (without a probability density) with the limit found in Hobbad and Ouassou (2021)(with a probability density) under the MBLFs (**i**).

10.3 Minimax Estimators of Location Paramters with Respect to Loss $\ell(\omega\|\delta - \delta_0\|^2 + (1-\omega)\|\delta - \theta\|^2)$

In this section, we consider a random vector X with a spherically symmetric distribution that does not necessarily have a probability density. We study the minimax Baranchik-type estimators of the location vector θ of X under the MBLFs (**ii**) framework. We assume that the function ℓ satisfies **H2**: $\ell(\cdot) \geq 0,\ \ \ell'(\cdot) > 0$ and ℓ is a concave function. As with the choice of ρ, in the previous section, the condition of complete monotonicity over ℓ' in Marchand and Strawderman (2020) is relaxed here. The examples of ℓ satisfying the **H2** condition are the examples of functions ρ satisfying the **H1** condition in the previous section. But, we do not require finiteness of $\ell'(0)$, so many other losses, such as $L^{\frac{q}{2}}$ losses with $\ell(t) = t^q; 0 < q < 1$, satisfy **H2**. Another interesting choice is (see Kubokawa and Strawderman 2017) in the following form: $\ell(t) = 2Q(\sqrt{t}) - 1$ with Q is a cumulative distribution function on $\mathbb{R}$ of even and unimodal density Q'.

The following lemma establishes a relation between the loss differences by exploiting the concavity of ℓ.

Lemma 10.3 *Suppose that X has a spherically symmetric distribution. For the problem of estimating θ under the MBLFs* (**ii**)*, we have*

$$\Delta L_{\omega,\delta_0,\ell}(\theta,\delta) \leq (1-\omega)^2 \ell'((1-\omega)\|\delta_0 - \theta\|)\Delta L_{0,\delta_0,\ell}(\theta,\delta).$$

The following theorem gives the minimaxity conditions for the estimators $\delta_{a,S}(X)$ given by (10.1) for general spherically symmetric distributions.

Theorem 10.2 *Suppose that X has a spherically symmetric distribution ($X \sim S.S(\theta)$) and the function ℓ satisfies* **H2**. *For $d \geq 4$, the estimator $\delta_{a,S}(X)$ given* *by (10.1) dominates $\delta_0(X) = X$ under the MBLFs* **(ii)** *for all θ with a strict inequality for some θ provided*

$$0 < a \le 2\frac{d-2}{d} \cdot \frac{E\left[R^2\ell'((1-\omega)R^2)\right]}{(1-\omega)E\left[\ell'((1-\omega)R^2)R^{-2}\right] \times E\left[R^2\right]}.$$

Remark 10.2 Suppose that X is spherically symmetric distribution around θ with the density $f(\|x-\theta\|^2)$. The condition of domination found by Hobbad and Ouassou (2021) is:

$$0 < a < 2\frac{d-2}{d} \cdot \frac{E\left[\ell'((1-\omega)R^2)\right]}{(1-\omega)E\left[\ell'((1-\omega)R^2)R^{-2}\right]}. \tag{10.6}$$

In the case when $\ell(t) = t$, the domination condition in Theorem 10.2 and in Hobbad and Ouassou (2021) becomes $0 < a < (d-2)/d \times 2/((1-\omega)E\left[R^{-2}\right])$ and for $\omega = 0$, it becomes $0 < a < (d-2)/d/E\left[R^{-2}\right]$.

Since $\ell'((1-\omega)R^2)$ is decreasing in R, according to **H2** and using the covariance inequality, we obtain $E\left[R^2\ell'((1-\omega)R^2)\right] \le E\left[R^2\right] \times E\left[\ell'((1-\omega)R^2)\right]$. So,

$$\begin{aligned} &2\frac{d-2}{d} \cdot \frac{E\left[R^2\ell'((1-\omega)R^2)\right]}{(1-\omega)E\left[\ell'((1-\omega)R^2)R^{-2}\right] \times E\left[R^2\right]} \\ &\le 2\frac{d-2}{d} \cdot \frac{E\left[\ell'((1-\omega)R^2)\right]}{(1-\omega)E\left[\ell'((1-\omega)R^2)R^{-2}\right]} \end{aligned}$$

Example 10.1 (The Uniform Distribution on a Sphere) For the uniform distribution on the sphere $S_{m,\theta}$ i.e, $X \sim U(S_{m,\theta})$) and $\ell(t) = t^q$ with $0 < q < 1$, we have $\ell'(t) = qt^{q-1}$ and $R^2 = \|X-\theta\|^2 = m^2$. Hence, the cut-off point from (10.2) is given by: $a_0(m) = \frac{d-2}{d}\frac{2m^2}{1-\omega}$

Example 10.2 (The Uniform Distribution on a Ball) Let X be the random vector having the uniform distribution on the ball $B_{m,\theta}$ i,e, $X \sim U(B_{m,\theta})$, with $d \ge 4$ and $R^2 = \|X-\theta\|^2$ having the probability density $h(t) = \frac{d}{2m^d}t^{\frac{d}{2}-1}\mathrm{i}_{(0,m^2)}(t)$. From the choice $\ell(t) = t^q$ with $0 < q < 1$ and from (10.2), we obtain the cut-off point a_0 is

$$a_0(m) = \frac{2(d-2)}{d}\frac{d(2q+d-4)m^2}{(1-\omega)(d+2)(2q+d)}.$$

(continued)

Example 10.2 (continued)
For the distribution with a probability density (see paper Hobbad and Ouassou (2021)), we obtain:

$$a_0(m) = \frac{2(d-2)}{d} \frac{(2q+d-4)m^2}{(1-\omega)(2q+d-2)}.$$

Example 10.3 (The Symmetric Kotz Type Distribution) Suppose that X possesses a symmetric Kotz type distribution with unknown location parameter θ, scale I_d. For $s = 1$, the probability density of the radius R^2 is $h(u) = \frac{r^\nu}{\Gamma(\nu)} u^{\nu-1} e^{-ru}$ for$u > 0$. When $\ell(t) = t^q$ with $0 < q < 1$ and from Theorem 10.2, we obtain the cut-off point a_0 as follows

$$a_0 = \frac{2(d-2)}{d} \frac{(\nu+q-1)(\nu+q-2)}{(1-\omega)(\nu+1)}.$$

For the distribution with a probability density (see Hobbad and Ouassou 2021), we obtain: $a_0 = \frac{2(d-2)}{d} \frac{\nu+q-2}{r(1-\omega)}$.

Example 10.4 (The Student Distribution) Suppose that $X \in \mathbb{R}^d$ is distributed according to a multivariate student distribution with unknown location parameter θ, scale I_d and $\nu > 0$ degrees of freedom. From $\ell(t) = t^q$ with $0 < q < 1$ and (10.2), we get the cut-off point a_0 as follows

$$a_0 = \frac{2(d-2)}{d} \frac{4(\nu/2+q-1)(\nu/2+q-2)}{\nu^2(1-\omega)}.$$

For the distribution with a probability density (see Hobbad and Ouassou 2021), we obtain:

$$a_0 = \frac{2(d-2)}{d} \frac{2(\nu/2+q-2)}{\nu(1-\omega)}.$$

The Fig. 10.2 illustrates a comparison between the bound found in this chapter of (10.2) for spherically symmetric random vectors (without a probability density) and the bound found in Hobbad and Ouassou (2021) (with a probability density) under the MBLFs **(ii)** .

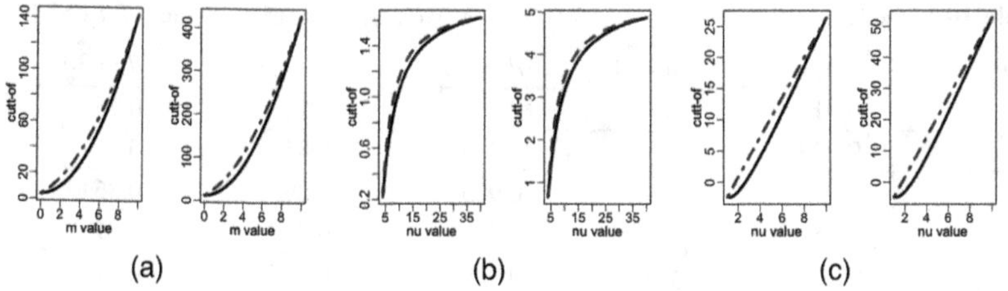

Fig. 10.2 (**a**) Uniform on the ball cases, (**b**) Student distribution cases, (**c**) Kotz distribution cases, for $\omega = 1/4$ and $\omega = 3/4$. (–) without a probability density, (- -) with a density

10.4 Simulation

We consider a random vector X having a spherically symmetric distribution without necessarily needing a probability density. We know that we can write $X - \theta = RU$ with R is a positive real random variable (the radius distribution), U has a uniform distribution on the unit sphere and R and U are independent. In order to simulate X, we assume that R is a discrete variable with a Poisson distribution, so it's not continuous with respect to the Lebesgue measure which indicates that X has no probability density. The following lemma allows us to simulate U. Finally, we obtain a simulation of $X = \theta + RU$.

Lemma 10.4 *Let $N_1, \ldots, N_d$ be independent and identifically distributed random variables as $N(0, 1)$. we consider :*

$$T = \left(N_1^2 + \cdots + N_d^2\right)^{1/2}.$$

Then the random vector $(N_1/T, \cdots, N_d/T)$ has the uniform distribution on a the sphere $S(0, 1)$.

Table 10.1 gives the normalized risk of $\delta_{a,S}(X)$ ie, $\mathcal{R}(\theta, \ell, \delta_{a,S}(X))/\mathcal{R}(\theta, \ell, \delta_0(X)$ for various values of the dimension d and parameter θ using the MBLFs (ii). We obtain these results by considering (i) $\omega = 1/2$, (ii) $\ell(t) = t^q, q = 1/2$, (iii) R has a Poisson distribution of parameter $\lambda = 8$, (iv) $S(t) = 1, a = 1$ and (v) $\theta = \|\theta\|(1, 0, \cdots, 0)$. Table 10.2 gives the values of the normalized risk of $\delta_{a,S}(X)$ i.e., $\mathcal{R}(\theta, \rho, \delta_{a,S}(X))/\mathcal{R}(\theta, \rho, \delta_0(X))$ for various values of the dimension d and of the parameter θ. The experimental values found under the following assumptions (i) $\theta = \|\theta\|(1, 0, \cdots, 0)$, (ii) $\rho(t) = \ln(1 + t)$, (iii) $\omega = 1/3$, (iv) $s(t) = t/(1 + t)$, (v) $a = 5$ and (vi), R has the Poisson distribution of parameter $\lambda = 8$. Using the normalized risk $(\mathcal{R}(\theta, \rho, \delta_{a,S}(X))/\mathcal{R}(\theta, \rho, \delta_0(X)))$ and $(\mathcal{R}(\theta, \ell, \delta_{a,S}(X))/\mathcal{R}(\theta, \ell, \delta_0(X)))$, the improvement of $\delta_{a,S}$ over δ_0 is the more significant (respectively non-significant) the closer this quantity is to 0 (respectively almost 1).

Table 10.1 the Risk ratios $(\mathcal{R}(\theta, \ell, \delta_{a,S}(X))/\mathcal{R}(\theta, \ell, \delta_0(X)))$, for various values of d and $\|\theta\|$

$\|\theta\|\backslash d$	$d = 6$	$d = 10$	$d = 15$	$d = 20$	$d = 30$
0	0.91	0.89	0.861	0.81	0.76
2	0.923	0.9	0.883	0.85	0.82
4	0.925	0.91	0.905	0.89	0.86
6	0.931	0.925	0.918	0.915	0.89
8	0.942	0.934	0.930	0.925	0.91
10	0.953	0.941	0.938	0.933	0.92
12	0.967	0.95	0.947	0.941	0.931

Table 10.2 The Risk ratios $(\mathcal{R}(\theta, \rho, \delta_{a,S}(X))/\mathcal{R}(\theta, \rho, \delta_0(X)))$ for various values of d and of the parameter θ

$\|\theta\|\backslash d$	$d = 6$	$d = 10$	$d = 15$	$d = 20$	$d = 30$
0	0.92	0.89	0.87	0.85	0.81
2	0.935	0.91	0.903	0.89	0.867
4	0.945	0.925	0.921	0.915	0.891
6	0.952	0.935	0.927	0.921	0.913
8	0.958	0.9444	0.939	0.932	0.926
10	0.965	0.95	0.943	0.939	0.93
12	0.97	0.957	0.951	0.945	0.939

According to Tables 10.1 and 10.2 and Figs. 10.3 and 10.4, the normalized risk $\mathcal{R}(\theta, \ell, \delta_{a,S})/$
$\mathcal{R}(\theta, \ell, \delta_0)$ and $(\mathcal{R}(\theta, \rho, \delta_{a,S}(X))/\mathcal{R}(\theta, \rho, \delta_0(X)))$ are a decreasing functions of d for fixed $\|\theta\|$, and increasing functions of $\|\theta\|$ for d fixed. Thus, a significant improvement of the estimator $\delta_{a,S}$ over δ_0 is obtained when d is large and $\|\theta\|$ is small.

10.4.1 The Influence of the Poisson Parameter on the Risk Values

Table 10.3, contains the values of risk $\mathcal{R}(\theta, \ell, \delta_{a,S}(X)$ under MBLFs **(ii)** by varying the lambda values (parameter of the radial Poisson distribution) using the conditions: $\ell(t) = t^q, q = 1/2, d = 10$ and $\theta = (1, 0, \ldots, 0)$.

Table 10.4 gives the values of the risk $\mathcal{R}(\theta, \rho, \delta_{a,S}(X)$ in terms of to the parameter λ of the Poisson distribution of the variable $R = \|X - \theta\|$. We have conducted the simulation study by the R language with the conditions: $d = 10$, $S(t) = t/()1 + t), a = 1, \omega = 1/3, \theta = (1, 0, \ldots, 0)$ and $\rho(t) = \ln(1 + t)$.

Tables 10.3 and 10.4 and Fig. 10.5 showed that the risks increase when λ increases, the growth is more important for MBLFs (ii).

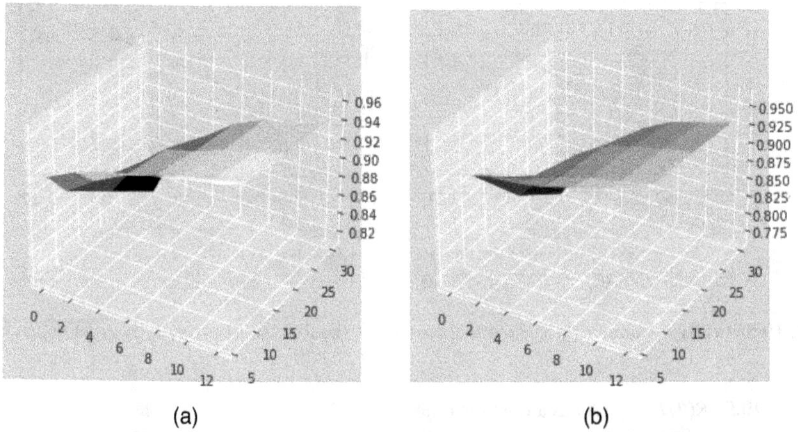

Fig. 10.3 (**a**) Graph of risk ratios $(\mathcal{R}(\theta, \rho, \delta_{a,S}(X))/\mathcal{R}(\theta, \rho, \delta_0(X)))$ as function of the dimension d and $\|\theta\|$ under the MBLFS (i), (**b**) Graph of risk ratios $(\mathcal{R}(\theta, \ell, \delta_{a,S}(X))/\mathcal{R}(\theta, \ell, \delta_0(X)))$ as function of the dimension d and $\|\theta\|$ under MBLFs (ii)

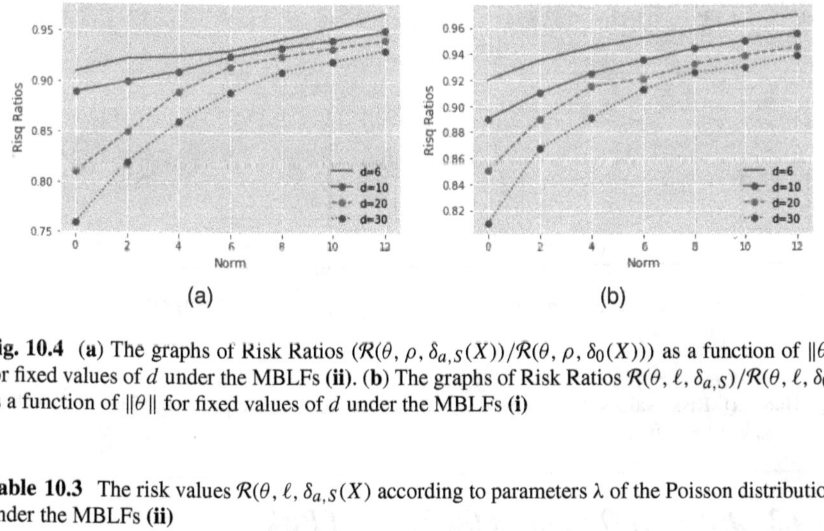

Fig. 10.4 (**a**) The graphs of Risk Ratios $(\mathcal{R}(\theta, \rho, \delta_{a,S}(X))/\mathcal{R}(\theta, \rho, \delta_0(X)))$ as a function of $\|\theta\|$ for fixed values of d under the MBLFs (**ii**). (**b**) The graphs of Risk Ratios $\mathcal{R}(\theta, \ell, \delta_{a,S})/\mathcal{R}(\theta, \ell, \delta_0)$ as a function of $\|\theta\|$ for fixed values of d under the MBLFs (**i**)

Table 10.3 The risk values $\mathcal{R}(\theta, \ell, \delta_{a,S}(X)$ according to parameters λ of the Poisson distribution under the MBLFs (**ii**)

λ	2	4	6	8	10	12	14	16	18	20
Risk	0.913	1.735	2.285	2.7	3	3.25	3.456	3.675	3.81	3.95

Table 10.4 The risk values $\mathcal{R}(\theta, \rho, \delta_{a,S}(X)$ according to parameters λ of the Poisson distribution under MBLFs (i)

λ	2	4	6	8	10	12	14	16	18	20
Risk	1.433	2.692	4.18	5.607	6.933	8.5	9.784	11.21	12.5	14.02

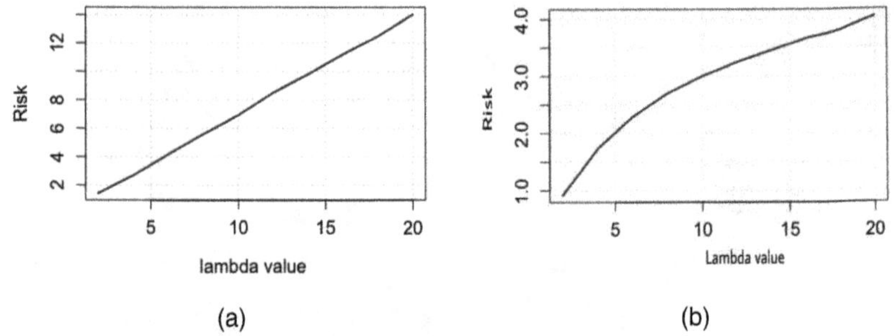

Fig. 10.5 (**a**) Risk values $\mathcal{R}(\theta, \ell, \delta_{a,S}(X)$ based on λ. (**b**) Risk values $\mathcal{R}(\theta, \rho, \delta_{a,S}(X)$ depending on λ

Table 10.5 $\mathcal{R}(\theta, \ell, \delta_{a,S}(X)$ as a function of ω

ω	0	1/8	2/8	3/8	4/8	5/8	6/8	7/8
Risk	9.842	9.225	8.41	7.879	7.06	6.11	5.087	3.545

Table 10.6 $\mathcal{R}(\theta, \rho, \delta_{a,S}(X)$ as a function of ω

ω	0	1/8	2/8	3/8	4/8	5/8	6/8	7/8
Risk	4.455	3.925	3.374	2.825	2.246	1.688	1.317	0.561

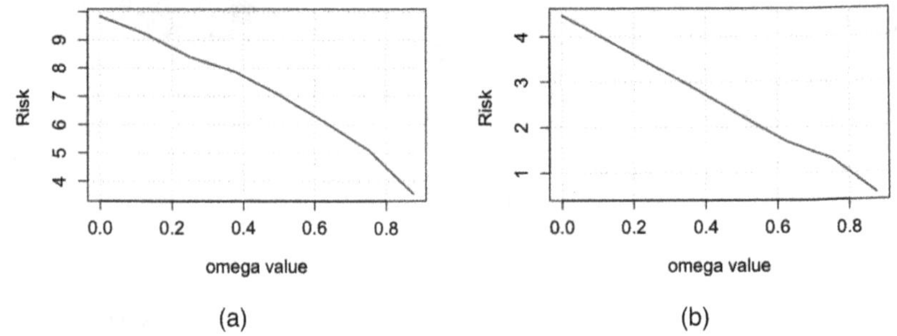

Fig. 10.6 (**a**) Risk values $\mathcal{R}(\theta, \ell, \delta_{a,S}(X)$ as a function of omega. (**b**) The risk values $\mathcal{R}(\theta, \rho, \delta_{a,S}(X)$ as a function of omega

10.4.2 Effect of Omega on the Value of Risk

We will give in Table 10.5, the risk values $\mathcal{R}(\theta, \ell, \delta_{a,S}(X)$ according to the parameter ω of the MBLFs (ii) by implementing the simulations with the parameters $d = 10$, $\lambda = 10$, $\ell(t) = t^q$, $q = 1/2$, $S(t) = t/(1+t)$, $a = 1$ and $\theta = (1, 0 \ldots, 0)$. Table 10.6 gives the values of the risk $\mathcal{R}(\theta, \rho, \delta_{a,S}(X)$ according to the parameter ω of the MBLFs (i) under the conditions $d = 10$, $\lambda = 10$, $a = 1/2$, $S(t) = t/(1+t)$, $\theta = (1, 0, \cdots, 0)$ and $\rho(t) = \ln(1+t)$. In both cases, we find that the risk decreases with ω, which is illustrate in the Fig. 10.6

Appendix

This Appendix presents proofs of the announced results with some supplementary technical lemmas.

Proof of Lemma 10.1 Let h be the radial distribution, we have:

$$
\begin{aligned}
&E_\theta\left[\rho'(\|X-\theta\|^2)(X-\theta)^\top g(X)\right]\\
&=\int_{\mathbb{R}+}\int_{S_{r,\theta}}\rho'(\|x-\theta\|^2)(x-\theta)^\top g(x)dU_{r,\theta}(x)dh(r)\\
&=\int_{\mathbb{R}+}\frac{r\rho'(r^2)}{\sigma_{r,\theta}(S_{r,\theta})}\int_{S_{r,\theta}}\frac{(x-\theta)^\top}{\|x-\theta\|}g(x)d\sigma_{r,\theta}(x)dh(r)\\
&=\int_{\mathbb{R}}\frac{r\rho'(r^2)}{\sigma_{r,\theta}(S_{r,\theta})}\int_{B_{r,\theta}}\nabla g(x)dxdh(r),\ \ (by\ \ Stokes'\,theorem)\\
&=\frac{1}{d}\int_{\mathbb{R}}\rho'(r^2)r^2\int_{B_{r,\theta}}\nabla g(x)dV_{r,\theta}(x)dh(r)
\end{aligned}
$$

since the volume of $B_{r,\theta}$ is $\lambda(B_{r,\theta})=r\sigma_{r,\theta}(S_{r,\theta})/d$. ∎

Proof of Lemma 10.2 Due to the radius conditioning, we get:

$$
\begin{aligned}
&E_\theta\left[(1-\omega)\rho'(\|X-\theta\|^2)[2(1-\omega)g(X).(X-\theta)+(1-\omega)^2\|g(X)\|^2]\right]\\
&=E\left[2(1-\omega)^2\rho'(R^2)E_{R,\theta}\left[g(X).(X-\theta)/\|X-\theta\|=R\right]\right]+\\
&\quad E\left[(1-\omega)^3\rho'(R^2)E_{r,\theta}\left[\|g(X)\|^2/\|X-\theta=R\right]\right]\\
&\le E\left[(1-\omega)^3\rho'(R^2)E_{R,\theta}\left[\frac{S(\|X\|^2)}{\|X\|^2}/\|X-\theta\|=R\right]\right]-\\
&\quad 2a(1-\omega)^2\frac{d-2}{d}\int_{\mathbb{R}+}r^2\rho'(r^2)\int_{B_{r,\theta}}\frac{S(\|X\|^2)}{\|X\|^2}dV_{r,\theta}(x)d\rho(r).
\end{aligned}
$$

Using the subharmonicity of $h: x\longrightarrow\dfrac{S(\|x\|^2)}{\|x\|^2}$, we arrive at:

$$
\int_{B_{r,\theta}}h(x)dV_{r,\theta}(x)d\rho(r)\le\int_{S_{r,\theta}}h(x)dU_{r,\theta}(x)d\rho(r)
$$

Hence,

$$E_\theta\left[(1-\omega)\rho'(\|X-\theta\|^2)[2(1-\omega)g(X).(X-\theta)+(1-\omega)^2\|g(X)\|^2]\right]$$
$$\leq E\left[(1-\omega)^2 a\rho'(R^2)\left(\frac{a(1-\omega)}{R^2}-\frac{2(d-2)}{d}\right)\right.$$
$$\left.\times E_{R,\theta}[R^2\frac{S(\|X\|^2)}{\|X\|^2}/\|X-\theta\|=R]\right]$$

■

Proof of Theorem 10.1 The difference in risk is:

$$\Delta\mathcal{R}(\theta)=\mathcal{R}(\theta,\rho,\delta)-\mathcal{R}(\theta,\rho,\delta_0)$$
$$\leq E_\theta\left[\omega\rho'(0)(1-\omega)^2\|g(X)\|^2\right]+E_\theta\left[(1-\omega)\rho'(\|X-\theta\|^2)\right.$$
$$\left.\times[2(1-\omega)g(X).(X-\theta)+(1-\omega)^2\|g(X)\|^2]\right]$$
$$\leq E\left\{\frac{a^2\omega\rho'(0)(1-\omega)^2}{R^2}E_{R,\theta}\left[R^2\frac{S(\|X\|^2)}{\|X\|^2}/\|X-\theta\|=R\right]\right\}+E\left\{a(1-\omega)^2\right.$$
$$\left.\times\rho'(R^2)\left(a(1-\omega)/R^2-\frac{2(d-2)}{d}\right)E_{R,\theta}\left[R^2\frac{S(\|X\|^2)}{\|X\|^2}/\|X-\theta\|=R\right]\right\}$$
$$=a(1-\omega)^2E\left\{a\left(\omega\rho'(0)+(1-\omega)\rho'(R^2)\right)\right.$$
$$\left.\times\frac{1}{R^2}E_{R,\theta}\left[R^2\frac{S(\|X\|^2)}{\|X\|^2}/\|X-\theta\|=R^2\right]\right\}+$$
$$\times a(1-\omega)^2E\left\{-\rho'(R^2)R^2\frac{2(d-2)}{d}E_{R,\theta}\left[\frac{S(\|X\|^2)}{\|X\|^2}/\|X-\theta\|=R\right]\right\}.$$

Noting that the expectation $E_{R,\theta}\left[R^2S(\|X\|^2)/\|X\|^2\mid\|X-\theta\|=R\right]$ is nondecreasing in R and $\rho'(R^2)/R^2$ is nonincreasing in R, since ρ is concave then ρ' is nonincreasing. Using the covariance inequality, we obtain an upper bound of the first term of the last equality:

$$E\left[a\left(\omega\rho'(0)+(1-\omega)\rho'(R^2)\right)\frac{1}{R^2}\right]\times E\left\{E_{R,\theta}\left[R^2\frac{S(\|X\|^2)}{\|X\|^2}/\|X-\theta\|=R\right]\right\}.$$

For $d\geq 4$, the function $E_{R,\theta}\left[\frac{S(\|X\|^2)}{\|X\|^2}/\|X-\theta\|=R\right]$ is nonincreasing in R and R^2 is increasing in R. Thus, by the covariance inequality we have

$$E\left\{E_{R,\theta}\left[R^2\frac{S(\|X\|^2)}{\|X\|^2}/\|X-\theta\|=R\right]\right\}$$

$$
\begin{aligned}
&= E\left[R^2 E_{R,\theta}\left[\frac{S(\|X\|^2)}{\|X\|^2} / \|X-\theta\| = R\right]\right] \\
&\leq E\left[R^2\right] \times E_\theta\left[\frac{S(\|X\|^2)}{\|X\|^2}\right].
\end{aligned}
$$

In the same way, as above, with the covariance inequality, we obtain an upper bound on the second term of the last equality:

$$
E\left[-\rho'(R^2)R^2\frac{2(d-2)}{d}\right] \times E_\theta\left[\frac{S(\|X\|^2)}{\|X\|^2}\right].
$$

Then the difference in risk is bounded by

$$
a(1-\omega)^2\left\{a\left(\omega\rho'(0)E\left[\tfrac{1}{R^2}\right] + (1-\omega)E\left[\tfrac{\rho'(R^2)}{R^2}\right]\right) - \frac{2(d-2)}{d}E\left[\rho'(R^2)R^2\right]\right\}
$$
$$
\times\, E_\theta\left[\tfrac{S(\|X\|^2)}{\|X\|^2}\right].
$$

From the above, it immediately follows that (10.3) is a sufficient condition for the risk difference to be negative for all θ. ■

Proof of Lemma 10.3 Since $\ell(\cdot)$ is a concave function, for any point α and β, $\ell(\alpha+\beta) - \ell(a) \leq \beta\ell'(\alpha)$. So, for $\alpha = L_{\omega,\delta_0}(\theta,\delta_0)$ and $\alpha+\beta = L_{\omega,\delta_0}(\theta,\delta)$

$$
\begin{aligned}
\Delta L_{\omega,\delta_0,\ell}(\theta,\delta) &= L_{\omega,\ell,\delta_0}(\theta,\delta) - L_{\omega,\ell,\delta_0}(\theta,\delta_0) \\
&\leq (1-\omega)^2\ell'((1-\omega)\|\delta_0 \quad \theta\|^2)\Delta L_{0,\delta_0}(0,\delta).
\end{aligned}
$$

■

Proof of Theorem 10.2 Since $\ell(\cdot)$ is a non-decreasing concave function, then, using Lemma 10.5, we obtain an upper bound for the risk difference as follows:

$$
\begin{aligned}
\Delta\mathcal{R}(\theta) &= E_\theta\left[\Delta_{\omega,\ell,\delta_0}(\theta,\delta_{a,S})\right] \\
&\leq E_\theta\left[(1-\omega)^2\ell'((1-\omega)\|\delta_0-\theta\|^2)\Delta_0(\theta,\delta)\right] \\
&= (1-\omega)^2\left\{E_\theta\left[\ell'((1-\omega)\|X-\theta\|^2)\|g(X)\|^2\right]\right. \\
&\quad \left. + E_\theta\left[2\ell'((1-\omega)\|X-\theta\|^2)g(X)'(X-\theta)\right]\right\} \\
&\leq a(1-\omega)^3 E\left\{\ell'((1-\omega)R^2)\left(a(1-\omega) - \frac{2(d-2)}{d}R^2\right)\right. \\
&\quad \left. \times E_{R,\theta}\left[\frac{S(\|X\|^2)}{\|X\|^2} / \|X-\theta\| = R\right]\right\}
\end{aligned}
$$

$$= a(1-\omega)^3 \Bigg\{ E\Bigg[E_{R,\theta}\left(R^2 \frac{S(\|X\|^2)}{\|X\|^2} / \|X-\theta\| = R \right) \times \ell'((1-\omega)R^2) a(1-\omega)R^{-2} \Bigg] + E\Bigg[E_{R,\theta}\left(\frac{S(\|X\|^2)}{\|X\|^2} / \|X-\theta\|=R \right) \times \frac{-2(d-2)R^2}{d} \ell'((1-\omega)R^2) \Bigg] \Bigg\} \tag{10.7}$$

where the last inequality follows from Lemma 10.5 and the second one from (i). In the same way as for the proof of the last theorem, we use the covariance inequality. Given that $E_{R,\theta}\left[R^2 S(\|X\|^2)/\|X\|^2/\|X-\theta\| = R\right]$ increases in R, $a(1-\omega)R^{-2}\ell'\left((1-\omega)R^2\right)$ is non-increasing in R, according to the first term of Eq. (10.7) and the covariance inequality, we deduce that

$$E\left[E_{R,\theta}\left(R^2 \frac{S(\|X\|^2)}{\|X\|^2} / \|X-\theta\| = R \right) a(1-\omega)R^{-2}\ell'((1-\omega)R^2) \right] \tag{10.8}$$

$$\leq E\left[R^2 E_{R,\theta}\left(\frac{S(\|X\|^2)}{\|X\|^2} / \|X-\theta\| = R \right) \right] E\left[a(1-\omega)R^{-2}\ell'((1-\omega)R^2) \right].$$

According to $E_{R,\theta}\left[\left\{S(\|X\|^2)/\|X\|^2\right\}/\|X-\theta\| = R\right]$ is nonincreasing in R, and R^2 is increasing in R and using the covariance inequality, we obtain an upper bound of the first expectation in the last inequality

$$E_\theta\left[R^2 E_{R,\theta}\left(\frac{S(\|X\|^2)}{\|X\|^2} / \|X-\theta\| = R \right) \right] \leq E_\theta\left[\frac{S(\|X\|^2)}{\|X\|^2} \right] E\left[R^2 \right]. \tag{10.9}$$

Since $\left\{-2((d-2)/d)R^{-2}\ell'((1-\omega)R^2)\right\}$ increases in R according to **H2** and using the covariance inequality, we get also an upper bound of the second expectation in (10.7)

$$E_\theta\left[\frac{S(\|X\|^2)}{\|X\|^2} \right] E\left[\frac{-2(d-2)R^2}{d} \ell'((1-\omega)R^2) \right] \tag{10.10}$$

Now, combining (10.8), (10.9) and (10.10) with (10.7), we obtain

$$\Delta\mathcal{R}(\theta) \leq a(1-\omega)^3 \Big(a(1-\omega) E\left[\ell'((1-\omega)R^2)R^{-2} \right] \times E_\theta\left[R^2 \right] - 2\frac{d-2}{d} E\left[R^2 \ell'((1-\omega)R^2) \right] \Big) \times E_\theta\left[\frac{S(\|X\|^2)}{\|X\|^2} \right].$$

This completes the proof of Theorem 10.1. ■

Lemma 10.5 *If S is a non-negative, differentiable and concave real valued function, then S is non-decreasing on $\mathbb{R}_+$ and the function $t \to S(t)/t$ is non-increasing on $\mathbb{R}_+^*$. Furthermore, if in addition S is twice differentiable, then the function $x \to S(\|x\|^2)/\|x\|^2$ is super-harmonic for $d \geq 4$.*

Proof S is differentiable, its concavity is equivalent to $S(t) - S(t_0) \leq S'(t_0)(t - t_0)$ for any t_0 and any t, the first part of the lemma follows. Hence, if $S'(t_0) < 0$ for some $t_0 \geq 0$, then $\lim_{t\to\infty} S(t) = -\infty$ which contradicts the assumption that $S(t) \geq 0$ for any t. Therefore $S'(t) \geq 0$ for all $t \geq 0$. Similarly, for any $t > 0$, we have $S(0) - S(t) \leq -tS'(t)$. Hence $t^2 \left(S(t)/t\right)' = tS'(t) - S(t) \leq -S(0) \leq 0$ so that $S(t)/t$ is non-increasing.

Finally by straightforward calculation the Laplacian of $S(\|x\|^2)/\|x\|^2$ is given by

$$\Delta\left[S(\|x\|^2)/\|x\|^2\right]$$
$$= \frac{2}{\|x\|^4}\left[2\|x\|^4 S''(\|x\|^2) + (p-4)\left(\|x\|^2 S'(\|x\|^2) - S(\|x\|)^2\right)\right] \leq 0$$

since $S''(.) \leq 0$, $tS'(t) - S(t) \leq 0$ and $d \geq 4$, which is the desired result.

Lemma 10.6 *If g is a superharmonic function on $\mathbb{R}^p$ and Z is a random variable with a uniform distribution on the sphere centered at the origin and of radius τ, then, for any $\theta \in \mathbb{R}^p$, the expectation $E_\theta\left[g(\theta + Z)\right]$ is a non-increasing function of τ.*

Proof See Du Plessis (1970, Theorem 2.30, page 54).

Lemma 10.7 *Let Y be a random variable, $g(y)$ and $h(y)$ any functions for which $E[g(Y)]$, $E[h(Y)]$ and $E[g(Y)h(Y)]$ exist. Then*

(a) If one of the functions $g(.)$ and $h(.)$ is nonincreasing and the other is nondecreasing, then $E[g(Y)h(Y)] \leq E[g(Y)]E[h(Y)]$
(b) If both functions are either nondecreasing or nonincreasing, then $E[g(Y)h(Y)] \geq E[g(Y)]E[h(Y)]$.

Proof See Lehmann and Casella (1998).

Lemma 10.8 *If $g(X)$ is a superharmonic function , $\nabla^2 g(X) = \sum(\partial/\partial X_i^2)g(X) < 0$), and X has a uniform distribution on the sphere centered at θ with radius R denoted $X \sim U(\|X - \theta\| = R^2)$, then $E_\theta\left[g(X)\right] < E_\theta\left[g(Z)\right]$, where $Z \sim U(\|Z - \theta\|^2 \leq R^2)$.*

So

$$\frac{1}{A(S)}\int_S g(X)dA(X) \leq \frac{1}{M(B)}\int_B g(Z)dM(Y),$$

where $A(S)$ and $M(B)$ represent the areas of the sphere S and the ball B respectively.

Proof See Du Du Plessis (1970), page 54.

Lemma 10.9 *Giving X with a spherically symmetric distribution around θ and $\beta : \mathbb{R}_+ \longrightarrow \mathbb{R}_+$ a non-decreasing function. Then, $E\left[R^2 \frac{\beta(\|X\|^2)}{\|X\|^2} / \|X - \theta\| = R\right]$ is a non-decreasing function of $R > 0$.*

Proof See Brandwein and Strawderman (1979) and Brandwein and Strawderman (1980, pages 394–395) within the proof of their Theorem 3.3.1.

Acknowledgments The authors thank the editor and the reviewers for their very helpful and constructive comments.

References

Baranchik, A. (1970). A family of minimax estimators of the mean of a multivariate normal distribution. *Annals of Mathematical Statistics, 41*(2), 642–645.

Brandwein, A., & Strawderman, W. (1979). Minimax estimation of mean of spherically symmetric distributions under general quadratic loss. *Journal of Multivariate Analysis, 9*(2), 579–588.

Brandwein, A., & Strawderman, W. (1980). Minimax estimation of location parameters for spherically symmetric distributions with concave loss. *Annals of Statistics, 8*(2), 279–284.

Du Plessis, N. (1970). *An introduction to potential theory*. Wiley-Interscience.

Hobbad, L. M. E., & Ouassou, I. (2021). On shrinkage estimation of a spherically symmetric distribution for balanced loss functions. *Journal of Multivariate Analysis*, 186.

James, W., & Stein, C. (1961). Estimation with quadratic loss. In *Proceedings of the fourth berkeley symposium on mathematical statistics and probability* (Vol. 1, pp. 361–379).

Kubokawa, M., & Strawderman, W. E. (2017). A on predictive density estimation for location families under integrated absolute error loss. *Bernoulli, 23*(3), 3197–3212.

Lehmann, E. L., & Casella, G. (1998). *Theory of point estimation* (2nd ed.). Springer.

Marchand, E., & Strawderman, W. E. (2020). On shrinkage estimation for balanced loss functions. *Journal of Multivariate Analysis*, 175.

Stein, C. (1956). Inadmissibility of the usual estimator for the mean vector of a multivariate normal distribution. In *Proceedings of the fourth berkeley symposium on mathematical statistics and probability* (Vol. 1, pp. 197–206).

Zellner, A. (1994). Bayesian and non Bayesian estimation using balanced loss functions. In S. S. Gupta, & J. O. Berger (Eds.), *Statistical decision theory and related topics* (pp. 371–390). Springer.

Chapter 11
Modelling the Movement of a South African Cheetah Using a Hidden Markov Model and Circular-Linear Regression

Bianca Gascoyne, Najmeh Nakhaei Rad, and Andriette Bekker

11.1 Introduction

Conservation is a hugely relevant topic in South Africa, Africa, and the rest of the world, and studying animal movement data can aid in the efforts to conserve wildlife. Global Positioning System (GPS) tracking has greatly improved the collection of animal movement data and this, in turn, can be used to model and improve our understanding of animal movement (Rivest et al., 2016; Ramkilawon et al., 2023). In the field of animal ecology, comprehending and modeling how animals move is crucial. For instance, researchers studying animal behavior are interested in determining whether animals display preferred directions of movement or are drawn to various environmental features, like areas abundant in food or locations they have visited before, due to the spatial memory effect. By analysing the GPS data, one can determine the habitats and migratory routes of particular species (Hebblewhite & Haydon, 2010). This is of critical importance when it comes to endangered species. There are many reasons why animal movements take place, e.g. searching for food or water, escaping predators and/or protecting young, and all of these reasons are influenced by environmental factors. It is of ecological importance to statistically

B. Gascoyne · N. Nakhaei Rad (✉)
Department of Statistics, University of Pretoria, Pretoria, South Africa
e-mail: u20538902@tuks.co.za; najmeh.nakhaeirad@up.ac.za

A. Bekker
Department of Statistics, University of Pretoria, Pretoria, South Africa

DSI-NRF Centre of Excellence in Mathematical and Statistical Sciences (CoE-MaSS), Johannesburg, South Africa

Centre for Environmental Studies, Department of Geography, Geoinformatics and Meteorology, University of Pretoria, Pretoria, South Africa
e-mail: andriette.bekker@up.ac.za

C. A. Coelho, D.-G. Chen (eds.), *Statistical Modeling and Applications*, Emerging Topics in Statistics and Biostatistics, https://doi.org/10.1007/978-3-031-69622-0_11

investigate how the movement of an animal is driven by its environment and external factors. Prediction of the direction of movement of animals is a complex task that involves considering both the biological traits of the species and the environmental context. The GPS data can be analyzed to determine patterns such as migration routes, daily movements within a home range, and paths to specific resources. Modeling these movements requires sophisticated statistical tools that can interpret the collected data and predict future locations based on patterns observed. Predictive modeling not only helps in understanding how animals use their habitat but also aids in conservation planning, such as identifying critical habitats that need protection or predicting the impact of environmental changes on animal movements. Furthermore, understanding the directional movement of animals can also shed light on their sensory and cognitive abilities. For instance, how animals perceive their environment and remember valuable locations like food sources or safe havens involves cognitive processes that are reflected in their movement patterns. Overall, the prediction of animal movement direction is a dynamic area of research that combines elements of ecology, behavior, technology, and statistics to address both fundamental scientific questions and practical conservation needs (Rivest et al., 2016).

In this paper, the species of interest is the cheetah (*Acinonyx Jubatus*), an animal whose conservation status is ‘vulnerable’ (one level below ‘endangered’). Cheetah numbers are decreasing drastically, largely due to human-wildlife conflict (HWC), habitat loss, and the illegal wildlife trade (Tricorache et al., 2021). As of 2017, when the last global assessment of cheetah numbers was conducted, it was estimated that as few as 6517 mature individuals were left in the wild, of which 3526 were considered to be in Southern Africa. Not only have their population numbers decreased, but cheetah populations are now known to occur in only 9% of their historical distributional range (Evans et al., 2022). By studying this animal’s movement data specifically, better measures can be put into place to protect them (Tricorache et al., 2021).

From the GPS coordinates, two critical values that reflect an animal’s movement—turning angles (direction) and step lengths (distance)—can be extracted. The turning angle variable makes the study challenging since it is a circular variable. A circular random variable is measured in degrees from 0° to 360° or radians in $(-\pi, \pi]$. The usual linear statistical methods are no longer applicable when looking at circular data due to their periodical nature, and, therefore, different methods need to be developed. Even the usual statistical measures, such as mean and median, are not applicable (Mardia & Jupp, 2000). One of the most useful circular distributions is the von Mises distribution. The probability density function (pdf) of the von Mises with parameters μ and κ is given by

$$f(\theta; \mu, \kappa) = \frac{1}{2\pi I_0(\kappa)} \exp[\kappa \cos(\theta - \mu)], \tag{11.1}$$

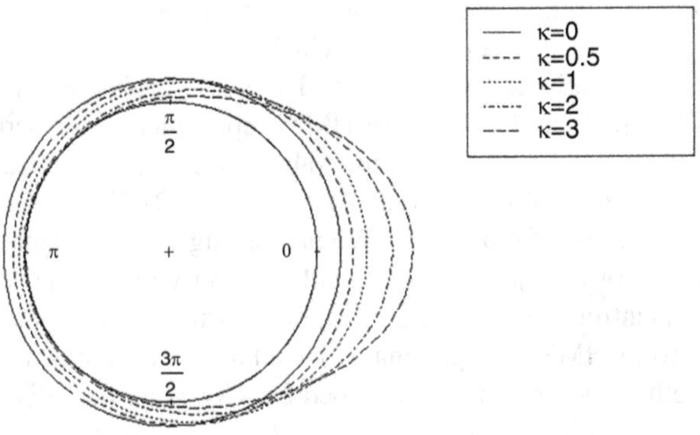

Fig. 11.1 Polar representation of von Mises pdf with mean direction 0 and varying values of the concentration parameter, κ

where $I_0(\cdot)$ is the modified Bessel function of the first kind and order 0 (Mardia & Jupp, 2000). The parameter μ denotes the mean direction, while the parameter κ represents the concentration parameter. See Fig. 11.1 for a graphical display of (11.1) for different values of κ. The von Mises distribution is symmetric about $\theta = \mu$ and unimodal. It is also the unique continuous distribution for which the sample mean direction, $\bar{\theta}$, serves as the maximum likelihood estimate of the population mean direction, μ. The von Mises distribution can thus, be considered as the circular equivalent of the normal distribution on the line (Lee, 2010). The spread of the von Mises distribution is determined by the concentration parameter κ. The larger the value of κ, the greater the concentration and lower the spread of distribution. As $\kappa \to 0$, the von Mises pdf tends towards the pdf of the uniform distribution on the circle (Mardia & Jupp, 2000).

To model animal movement, a combination of robust statistical methods and adaptable models is necessary. The displacement can be described as the distance and direction between consecutive points, and a circular-linear process is used to model movement within a two-dimensional discrete environment (Nicosia et al., 2017). Two different statistical approaches are employed for modelling the movement of Rain, a female cheetah in Pilanesberg National Park, South Africa.

Hidden Markov Models (HMMs) are frequently considered for animal movement modelling to divide movement trajectories into underlying behavioural states, such as "resting", "foraging", "migrating", and "exploratory". For instance, Jonsen et al. (2005) used a hidden correlated random walk to understand how two different species of seal move. Goodall et al. (2019) used an HMM with three states to study the movement of lions. Caribou movement in Canada was studied by Nicosia et al. (2017) and Franke et al. (2004) using HMMs with two and three states, respectively. Langrock et al. (2014) focused on reindeer, using a HMM to separate group-influenced movement from independent movement. The choice of HMM is due to the flexibility of HMMs and the efficient recursive algorithms available for statistical analysis (see Patterson et al. 2009, Langrock et al. 2012, and McClintock et al. 2020). An HMM represents a time series model consisting of two main components: an observable series and a hidden state sequence that cannot be directly observed (Langrock et al., 2012). The observed variables pertain to quantifications of movement. When working with movement data, there are two approaches for modelling: one can either directly model the positions or work with derived quantities. In this paper, the derived quantities are step lengths and turning angles. To fit HMM, the gamma and Weibull distributions are commonly used for step length modelling and the wrapped Cauchy, von Mises distribution in (11.1) and the skew von Mises (Bekker et al., 2022) are applied for turning angle modelling. With the introduction of software packages, such as *moveHMM* (Michelot et al., 2016) and *momentuHMM* (McClintock & Michelot, 2018), modelling HMMs has become far easier.

Secondly, in the presence of turning angle as a circular variable, circular regression is a way to predict the direction of movement. Circular regression is used when at least one variable is circular. This can be the response variable, the predictor variable or both variables which result in circular-linear, linear-circular and circular-circular regressions, respectively. The concept of circular-linear regression was introduced by Gould (1969). Applying this concept, Fisher and Lee (1992) examined the relationship between direction and distance of periwinkle movement. Lund (1999) employed a circular regression model, encompassing both circular and linear predictors, to understand fish spawning timing. In a different context, Marchetti and Scapini (2003) used a circular-linear regression model to explore sandhopper orientation concerning various predictors. Building upon the work of Fisher and Lee (1992) and Rivest et al. (2016) analysed the same periwinkle data using a different circular-linear regression model, and they also extended their study to bison movement data using a model with combined circular and linear predictors.

Subsequently, in this paper, to understand Rain's movement and behavioural pattern, an HMM is used with two hidden states, "encamped" and "exploratory". In addition, to predict the direction of an Rain's movement while taking environmental factors into account, a circular-linear regression model is implemented. The rest of the paper is organized as follows: Sect. 11.2 gives some theoretical background of analysed models, specifically the HMM in Sect. 11.2.1 and circular-linear regression in Sect. 11.2.2. The case study is presented in Sect. 11.3, followed by a discussion and conclusion in Sect. 11.4.

11.2 Methodology

In this section, the HMM and the circular-linear regression model are introduced.

11.2.1 Hidden Markov Model

We provide a brief overview of the key aspects of the HMM approach. To gain a comprehensive understanding of the HMM approach within the context of animal movement modelling, readers are advised to consult the primary sources (e.g., Patterson et al. 2009, Langrock et al. 2012, and Zucchini and MacDonald 2009) for complete details.

The following components specify an HMM:

- A set of N states: $S_1, S_2, \ldots, S_N$.
- A transition probability matrix P : each p_{ij} represents the probability of moving from state i to state j. p_{ij} is given by

$$p_{ij} = P(X_t = j|X_{t-1} = i).$$

  Note that $\sum_{j=1}^{N} p_{ij} = 1\forall i$, i.e. the elements in row i add up to 1.
- A sequence of T observations: $O = \{o_1, o_2, \ldots, o_T\}$.
- A sequence of observation likelihoods (also known as emission probabilities): $B = b_i(o_t)$, where each $b_i(o_t)$ represents the probability of an observation o_t being generated from state i.
- An initial probability vector: $\underline{\pi}$, where $\pi_i = P(X_0 = i)$, $i = 1, \ldots, N$, is the probability that the Markov chain will start in state i. It is possible for a state j to have $\pi_j = 0$, which means that it cannot be an initial state.

  Note that $\sum_{i=1}^{N} \pi_i = 1$, i.e. the sum of all the initial probabilities is equal to 1.

A first-order HMM puts into effect two simplifying assumptions:

1. Markov assumption:

$$P(X_t = j|X_1 - i_1, \ldots, X_{t-1} = i_{t-1}) = P(X_t = j|X_{t-1} = i_{t-1}).$$

2. Output independence:

$$P(o_i|X_1, \ldots, X_T, o_1, \ldots, o_T) = P(o_i|X_i),$$

   i.e. the probability of an output observation o_i depends only on the state that produced the observation, X_i, and not on any other state(s) or observation(s).

The standard approach of HMMs for modelling an individual animal's movement involves analysing bivariate time series that encompass both step length and turning angle at each time point, as depicted in Fig. 11.2. At each time point, the parameters of the step length distribution (such as a gamma distribution) and the turning angle distribution (such as a von Mises distribution in (1)) are determined by an underlying, unobserved state. These states, which are finite in number, serve as broad categorisations of movement characteristics, often reflecting

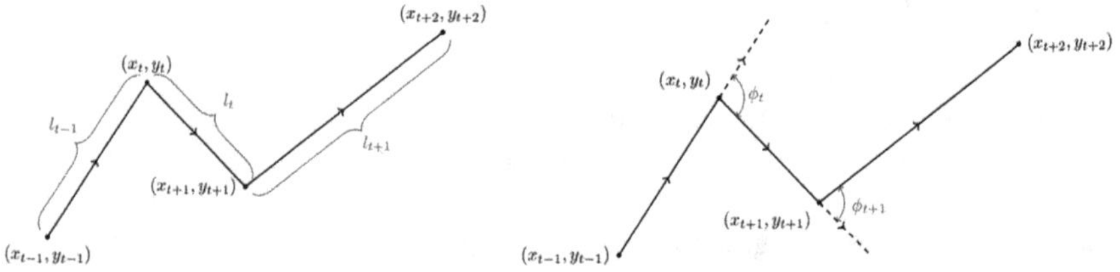

Fig. 11.2 Illustration of step lengths and turning angles

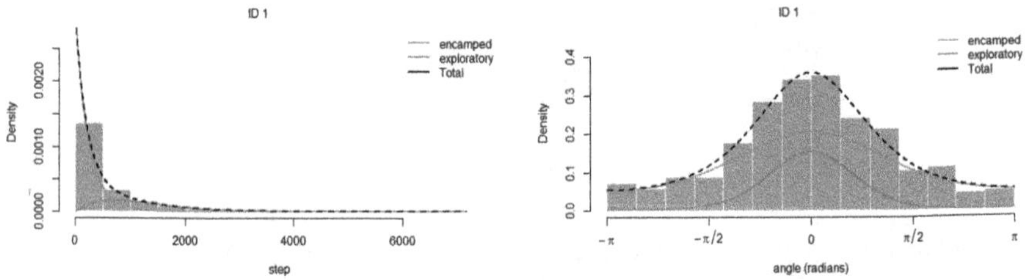

Fig. 11.3 Histograms of step lengths and turning angles for the simulated data, $n = 1000$

proxies for the animal's behavioural states (like encamped versus exploratory). The sequence of these states is assumed to be generated through a Markov chain, often displaying a tendency to remain in a state for an extended period of time before transitioning to another one. The probabilities of transitioning between states, along with the parameters characterizing the state-linked distributions, form the model's parameters that require estimation. The number of states is not known and necessitates user specification. Typically, movement data assume a limited number of states (for instance, ≤ 4). An intriguing biological aspect involves inferring state transition probabilities as functions of environmental factors. After estimating these model parameters, users can compute the most probable sequence of behavioural states.

In short, hidden Markov models should be characterized by three primary problems:

1. Likelihood: For a given HMM $\lambda = (P, B)$ and observation sequence O, the likelihood $P(O|\lambda)$ must be determined using the forward algorithm (Patterson et al., 2009)
2. Decoding: For a given HMM $\lambda = (P, B)$ and observation sequence O, the best hidden state sequence S must be discovered using the Viterbi algorithm (Zucchini & MacDonald, 2009)
3. Learning: Given the observation sequence O and the best hidden states, the HMM parameters P and B must be learned using an Expectation-Maximisation algorithm (Nicosia et al., 2017)

Figure 11.3 shows the histograms and fitted distributions resulting from a 2-state HMM fitted on simulated data. The two states were "encamped" and "exploratory".

Table 11.1 Estimated mean and standard deviation of gamma distributions

Gamma	n	Mean	Standard deviation
Encamped	100	240.8168	270.2295
Exploratory		1164.4892	760.6009
Encamped	500	214.3730	225.2187
Exploratory		1164.8481	845.2315
Encamped	1000	242.0964	245.0002
Exploratory		1235.3700	788.6503

Table 11.2 Estimated parameters of von Mises distributions

von Mises	n	Mean	Concentration
Encamped	100	0.0000	0.7810176
Exploratory		0.0000	1.856003
Encamped	500	0.0000	0.5876482
Exploratory		0.0000	2.213663
Encamped	1000	0.0000	0.6746364
Exploratory		0.0000	2.430559

Step length data for the encamped state was generated from a gamma distribution with mean 229 and standard deviation 238. For the exploratory state, step length data was generated from a gamma distribution with mean 1210 and standard deviation 783. Turning angle data for the encamped state was generated from a von Mises distribution with mean 0 and concentration 0.605 and turning angle data for the exploratory state was generated from another von Mises distribution with mean 0 and concentration 2.397. Datasets of size $n = 100$, 500 and 1000 were generated.

The estimated mean and standard deviation of the gamma distributions for each state of the fitted models can be found in Table 11.1.

The estimated parameters of the von Mises distributions for each state of the fitted models can be found in Table 11.2.

For the remainder of this simulation, the focus is on the case $n = 1000$.

The initial probability vector of the model for $n = 1000$ is given to be:

$$\underline{\pi} = [0.9999, 0.0000]$$

and the transition probability matrix is given as:

$$P = \begin{bmatrix} 0.9292098 & 0.07079019 \\ 0.1850796 & 0.81492044 \end{bmatrix}.$$

Figure 11.4 shows the decoded "track" of the simulated data, for $n = 1000$. It is clear to seen that majority of the movement from the generated dataset is in the "encamped" state.

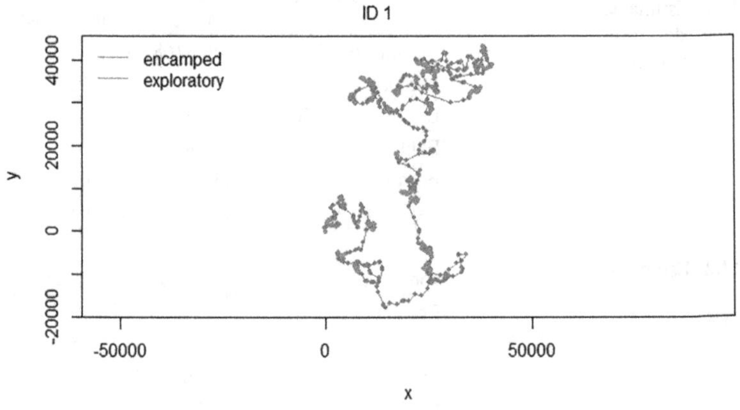

Fig. 11.4 Decoded track of the simulated movement, $n = 1000$

11.2.2 Circular-Linear Regression

The most commonly used circular regression approach is circular-linear regression. In this case, the response variable is of a circular nature and the explanatory variable(s) are linear. The majority of circular-linear regression models assume that the circular variable follows a von Mises distribution (Mohammad et al., 2021).

Gould (1969) proposed a regression model which models the mean direction of Θ, the circular variable, in terms of a vector of linear independent variables, $(x_1, x_2, \ldots, x_k)'$. The mean direction of Θ is given by

$$\mu = \mu_0 + \sum \beta_i x_i. \tag{11.2}$$

This model results in a 'barber's pole' form, a curve spiralling infinitely up the surface of a cylinder (Mohammad et al., 2021).

Fisher and Lee (1992) assumed that the circular observations, $\Theta_1, \Theta_2, \ldots, \Theta_n$, follow von Mises distributions with mean directions $\mu_1, \mu_2, \ldots, \mu_n$ and concentration parameters $\kappa_1, \kappa_2, \ldots, \kappa_n$, with all concentration parameters equal to κ. The mean directions are given by

$$\mu_i = \mu + g(\boldsymbol{\beta}^T \boldsymbol{x}_i), \tag{11.3}$$

where $\boldsymbol{x}_i \in \mathbb{R}^k$ is the vector of explanatory variables for the ith case and $\boldsymbol{\beta}$ is a vector of regression coefficients. The function g is called the link function and its purpose is to map the real line to the circle. It can be assumed known, or to be one of a parametric family of suitable link functions. A possible link function is

$$g(x) = 2\tan^{-1}(\boldsymbol{\beta}^T \boldsymbol{x}_i). \tag{11.4}$$

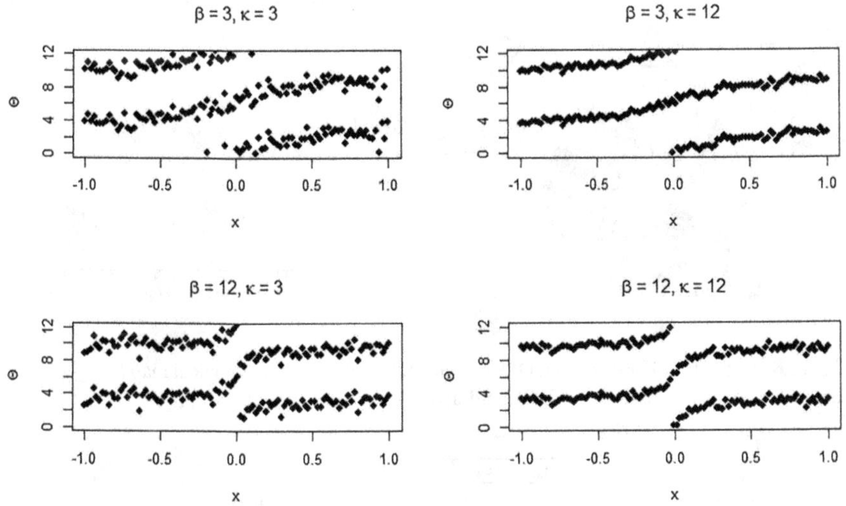

Fig. 11.5 Four simulated datasets showing the behaviour of the regression model proposed by Fisher and Lee (1992), when the distribution of Θ_i is M(μ_i, κ) with $\mu_i = \mu + g(\beta x_i)$, the link function is $g(x) = 2\tan^{-1}(\beta x)$ and x_i's equally spaced on $(-1,1)$

Figure 11.5 graphically shows the behaviour of this model, with link function given in (11.4), for various values of β and κ. Each simulated dataset has size 100. The 100 points (x, Θ), as well as the points $(x, \Theta + 2\pi)$, were plotted on the Cartesian plane.

11.3 Case Study

11.3.1 Data and Study Area

The dataset used in this study consists of the GPS coordinates of one particular female cheetah, named Rain, in Pilanesberg National Park in the North West Province of South Africa. The data was recorded from 10 am on 13 April 2014 to 10 am on 7 February 2015, in 6 hour intervals. This resulted in 1203 observations. (The dataset can be found at https://www.movebank.org/cms/webapp?gwt_fragment=page=studies,path=study1260886163). Figure 11.6 below shows the coordinates displayed on a topographical map, as well as the path followed by Rain. The temperature was also recorded at each interval.

First, step length and direction must be extracted from this data. This was done in R software by using the *momentuHHM* package and following the steps laid out for a similar dataset by McClintock and Michelot (2018).

The histogram in Fig. 11.7 below shows the frequency distribution of the step lengths with an average step length of 672.9495 for a 6 hour period. The mean direction was 0.02577844 and the rose diagram below displays the angle frequencies.

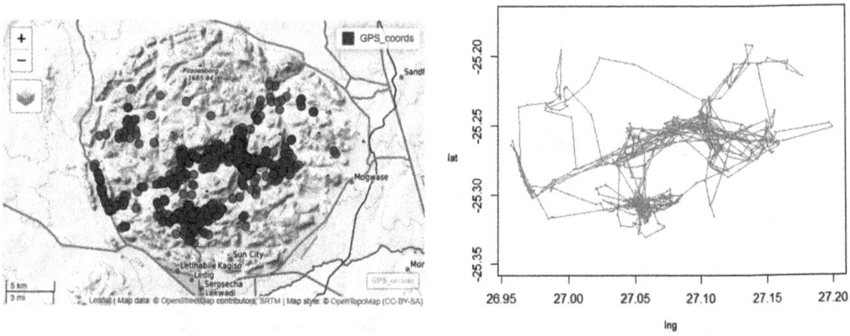

Fig. 11.6 Dataset: Rain movement. GPS coordinates plotted on topographical map along with the path followed according to these GPS coordinates

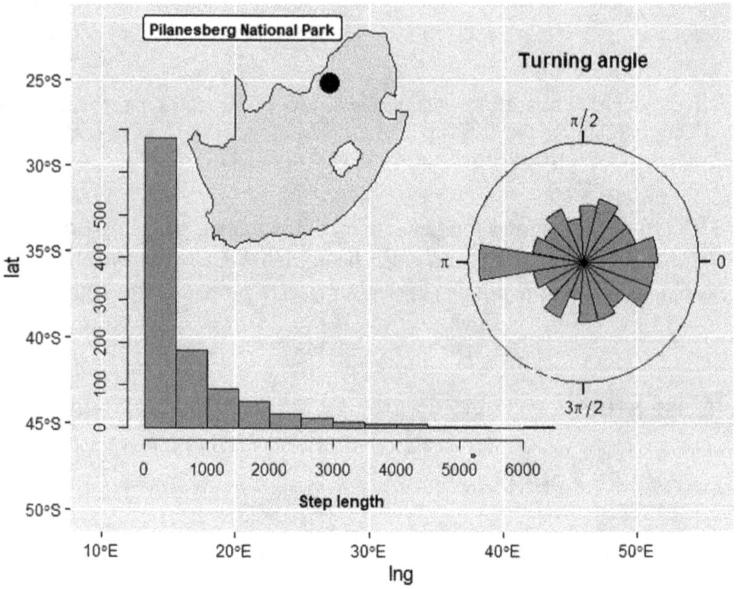

Fig. 11.7 Dataset: Rain movement. Graphical representation of step lengths and angles

11.3.2 Results

11.3.2.1 Covariate-Free 2-State HMM

This model only fits the step lengths and turning angles. We assumed a gamma distribution for the step lengths and a von Mises distribution for the turning angles. Six initial parameters were chosen, namely the mean and standard deviation of the gamma distribution and the concentration of the von Mises distribution (11.1) for

Table 11.3 Estimated parameters of gamma distributions

Gamma	Mean	Standard deviation
Encamped	753.928	1244.291
Exploratory	10217.3879	422.2194

Table 11.4 Estimated parameters of von Mises distributions

von Mises	Mean	Concentration
Encamped	0.0000	0.0332
Exploratory	0.0000	0.4434

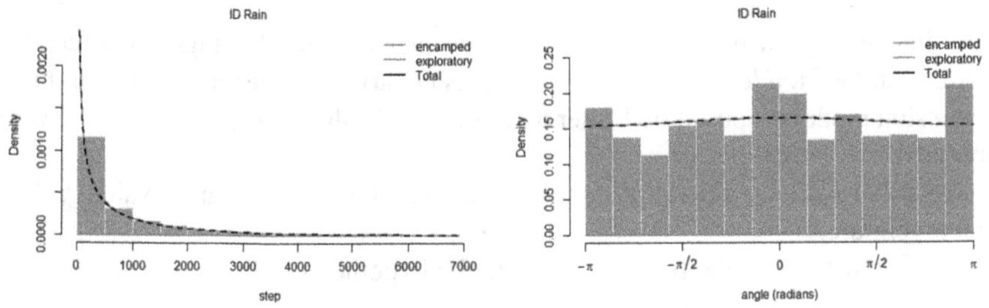

Fig. 11.8 Histograms of step lengths and turning angles for the fitted model

both states. These initial parameters were chosen by looking at the histogram and rose diagram in Fig. 11.7. A mean of 100 and standard deviation of 500 for the gamma distribution for the encamped state and a mean of 100 and standard deviation of 200 for the exploratory were selected. For the von Mises distributions, a mean direction of 0 was selected for both states and concentrations of 0.3 and 0.7 were selected for the encamped and exploratory states, respectively. The *fitHMM* function from the *momentuHMM* package was used to fit the model.

The estimated parameters of the gamma and von Mises distributions for the encamped and exploratory states are shown in Tables 11.3 and 11.4, respectively.

The fitted distributions can be seen in Fig. 11.8.

The initial probability vector of the model is obtained as:

$$\underline{\pi} = [0.9996, 0.0004]\,,$$

and the transition probability matrix is given as:

$$P = \begin{bmatrix} 1.0000 & 0.0000 \\ 0.9549 & 0.0451 \end{bmatrix}.$$

The transition probability matrix shows the probability of moving from one state to another. It can be seen from the transition probability matrix that the probability of Rain moving from the exploratory state to encamped is much higher than probability of transitioning from the encamped state to exploratory. In addition, the probability of remaining in the encamped state is one. Pilanesberg National Park is only 572 km^2, and, thus, is not large enough for any of Rain's movement to be considered exploratory. It must also be mentioned that Rain had cubs during the period of interest. A female cheetah moves her cubs to a new den every few days and, although she will leave them alone for extended periods of time, she will return to her cubs to feed them until they are old enough to travel with her. This could also indicate why Rain remained in the encamped state.

11.3.2.2 Regression Model with Two Covariates

The first regression model was fitted for the mean of Θ, the angle or direction, using both the step length and temperature as covariates. After fitting the circular-regression model proposed by Fisher and Lee (1992), the results in Table 11.5 were obtained.

Note that β_1, t_1 and p-value$_1$ relate to step length and β_2, t_2 and p-value$_2$ relate to temperature.

The t statistics and the p-values relate to the hypothesis

$$H_0 : \beta_i = 0 \quad \text{vs} \quad H_A : \beta_i \neq 0, \quad i = 1, 2,$$

i.e., whether or not each regression coefficient is significantly different from zero.

From p-value$_1$ it can be seen that the null hypothesis for β_1 is rejected at a 5% level of significance, i.e., the regression coefficient for step length differs significantly from zero. However, p-value$_2$ shows that the null hypothesis for β_2 is not rejected, i.e., the regression coefficient for temperature is not significantly different from zero. Thus, temperature is not a significant variable in the model.

Table 11.5 Regression model summary

Regression coefficients:	
$\hat{\beta}_1$	−0.017677219
$\hat{\beta}_2$	0.008420843
Standard error of	
$\hat{\beta}_1$	0.007768487
$\hat{\beta}_2$	0.006025043
Log-likelihood	18.08774
t statistics:	
t_1	2.275503
t_2	1.397640
P-values:	
p-value$_1$	0.01143787
p-value$_2$	0.08111055

Table 11.6 Regression model summary

Regression coefficient, $\hat{\beta}$	−0.01617375
Standard error of $\hat{\beta}$	0.007202231
Log-likelihood	17.33842
t statistic	2.245658
P-value	0.01236295

11.3.2.3 Regression Model with One Covariate

As a follow up study, the second regression model was fitted for the mean of Θ, the angle or direction, using only step length as the covariate. After fitting the circular-linear regression model proposed by Fisher and Lee (1992), the results in Table 11.6 were obtained.

The p-value is less than 0.05, therefore the null hypothesis is rejected at a 5% level of significance and it can be concluded that β is significantly different from zero.

11.4 Discussion and Conclusion

In the field of animal ecology, understanding and accurately modeling animal movement patterns is essential for a range of ecological studies and conservation efforts . Researchers, particularly those focusing on animal behavior, are deeply interested in various aspects of how animals navigate their environments. This includes investigating whether animals exhibit preferred directions of movement and if they are influenced by attractions to multiple environmental stimuli. Such stimuli often include areas rich in food resources or sites that animals repeatedly visit, a behavior that may be driven by what is known as the spatial memory effect (Nicosia et al., 2017; Ramkilawon et al., 2023).

In this case study we used an HMM to determine the behavioural states of Rain, a female cheetah, in Pilanesberg National park. It was found that, during the time period studied, she was considered to only be in the encamped state. This is largely due to the fact that Pilanesberg National Park is a relatively small reserve and, thus, none of the distances travelled by Rain were long enough to be regarded as exploratory. Rain had a cubs during the time period of interest, which could further add reason as to why she was considered to be in the encamped state.

We also used a circular-linear regression model to determine whether a relationship exits between the step lengths and angles of movement of a cheetah, as well as the external temperature. The models investigated show that step length and angle of movement are related, however the external temperature has no effect on the direction travelled by the cheetah.

Studying a cheetah's movement can aid in the conservation efforts needed to protect them. Cheetah numbers are decreasing drastically and, although the species is currently only considered to be vulnerable, they are fast on their way to being classified as endangered. A cheetah is a large predator and the ecosystems in which they live will suffer without them. Cheetah conservation is of economic importance too. Tourism is one of the fastest growing industries in the world, of which ecotourism is the leading segment. Ecotourism plays a major part in conserving nature by creating economic incentives for wildlife conservation (Lindsey et al., 2007). Lindsey et al. (2007) conducted a study to determine the wildlife viewing preferences of visitors to South Africa's protected areas, with the two main areas being Kruger National Park and Pilanesberg National Park. It was found that the second most important feature for tourists was the presence of large predators (which includes the cheetah). Thus, studying and conserving these species is beneficial to our economy.

Furthermore, studying animal movement data can also make a difference in analysing climate change. As climate change directly impacts animals' habitats, studying their movement can give insight into areas that have been greatly affected. Animals move away from areas with scarce resources, particularly water, and may need to travel far in search of better areas (Hebblewhite & Haydon, 2010).

Acknowledgments The research is supported in part by the RDP grant at the University of Pretoria, the National Research Foundation (NRF) of South Africa, Ref.: RA210106581084, grant No. 150170; Ref.: SRUG2204203865, and DSI-NRF Centre of Excellence in Mathematical and Statistical Sciences (CoE-MaSS), South Africa. The opinions expressed and conclusions arrived at are those of the authors and are not necessarily to be attributed to the CoE-MaSS or the NRF.

References

Bekker, A., Nakhaei Rad, N., Arashi, M., & Ley, C. (2022). Generalized skew-symmetric circular and toroidal distributions. In *Directional statistics for innovative applications: a bicentennial tribute to florence nightingale* (pp. 161–186). Springer.

Evans, M. J., Gordon, I. J., Pierson, J. C., Neaves, L. E., Wilson, B. A., Brockett, B., Ross, C. E., Smith, K. J., Rapley, S., Andrewartha, T. A., et al. (2022). Reintroduction biology and the iucn red list: The dominance of species of least concern in the peer-reviewed literature. *Global Ecology and Conservation*, *38*, e02242.

Fisher, N. I., & Lee, A. J. (1992). Regression models for an angular response. *Biometrics*, 665–677.

Franke, A., Caelli, T., & Hudson, R. J. (2004). Analysis of movements and behavior of caribou (rangifer tarandus) using hidden markov models. *Ecological Modelling*, *173*(2–3), 259–270.

Goodall, V. L., Ferreira, S. M., Funston, P. J., & Maruping-Mzileni, N. (2019). Uncovering hidden states in african lion movement data using hidden markov models. *Wildlife Research*, *46*(4), 296–303.

Gould, A. L. (1969). A regression technique for angular variates. *Biometrics*, 683–700.

Hebblewhite, M., & Haydon, D. T. (2010). Distinguishing technology from biology: a critical review of the use of gps telemetry data in ecology. *Philosophical Transactions of the Royal Society B: Biological Sciences*, *365*(1550), 2303–2312.

Jonsen, I. D., Flemming, J. M., & Myers, R. A. (2005). Robust state–space modeling of animal movement data. *Ecology*, *86*(11), 2874–2880.

Langrock, R., King, R., Matthiopoulos, J., Thomas, L., Fortin, D., & Morales, J. M. (2012). Flexible and practical modeling of animal telemetry data: hidden markov models and extensions. *Ecology*, *93*(11), 2336–2342.

Langrock, R., Hopcraft, J. G. C., Blackwell, P. G., Goodall, V., King, R., Niu, M., Patterson, T. A., Pedersen, M. W., Skarin, A., & Schick, R. S. (2014). Modelling group dynamic animal movement. *Methods in Ecology and Evolution*, *5*(2), 190–199.

Lee, A. (2010). Circular data. *Wiley Interdisciplinary Reviews: Computational Statistics*, *2*(4), 477–486.

Lindsey, P. A., Alexander, R., Mills, M. G., Romañach, S., & Woodroffe, R. (2007). Wildlife viewing preferences of visitors to protected areas in south africa: Implications for the role of ecotourism in conservation. *Journal of Ecotourism*, *6*(1), 19–33.

Lund, U. (1999). Least circular distance regression for directional data. *Journal of Applied Statistics*, *26*(6), 723–733.

Marchetti, G. M., & Scapini, F. (2003). Use of multiple regression models in the study of sandhopper orientation under natural conditions. *Estuarine, Coastal and Shelf Science*, *58*, 207–215.

Mardia, K. V., & Jupp, P. E. (2000). *Directional statistics*. John Wiley & Sons.

McClintock, B. T., & Michelot, T. (2018). momentuhmm: R package for generalized hidden markov models of animal movement. *Methods in Ecology and Evolution*, *9*(6), 1518–1530.

McClintock, B. T., Langrock, R., Gimenez, O., Cam, E., Borchers, D. L., Glennie, R., & Patterson, T. A. (2020). Uncovering ecological state dynamics with hidden markov models. *Ecology Letters*, *23*(12), 1878–1903.

Michelot, T., Langrock, R., & Patterson, T. A. (2016). movehmm: an r package for the statistical modelling of animal movement data using hidden markov models. *Methods in Ecology and Evolution*, *7*(11), 1308–1315.

Mohammad, H. H., Satari, S. Z., & Yusoff, W. N. S. W. (2021). Review on circular-linear regression models. In *Journal of Physics: Conference Series*, *1988*, 012108.

Nicosia, A., Duchesne, T., Rivest, L.-P., & Fortin, D. (2017). A general hidden state random walk model for animal movement. *Computational Statistics & Data Analysis*, *105*, 76–95.

Patterson, T. A., Basson, M., Bravington, M. V., & Gunn, J. S. (2009). Classifying movement behaviour in relation to environmental conditions using hidden markov models. *Journal of Animal Ecology*, *78*(6), 1113–1123.

Ramkilawon, G. D., Ferreira, J., & Nakhaeirad, N. (2023). Sine-skewed von mises-and lindley/gumbel models as candidates for direction and distance in modelling animal movement. *Brazilian Journal of Biometrics*, *41*(2), 175–190.

Rivest, L.-P., Duchesne, T., Nicosia, A., & Fortin, D. (2016). A general angular regression model for the analysis of data on animal movement in ecology. *Journal of the Royal Statistical Society Series C: Applied Statistics*, *65*(3), 445–463.

Tricorache, P., Yashphe, S., & Marker, L. (2021). Global dataset for seized and non-intercepted illegal cheetah trade (acinonyx jubatus) 2010–2019. *Data in Brief*, *35*, 106848.

Zucchini, W., & MacDonald, I. L. (2009). *Hidden Markov models for time series: an introduction using R*. Chapman and Hall/CRC.

Index

C. A. Coelho, D. G. Chen (eds.), *Statistical Modeling and Applications*, Emerging Topics in Statistics and Biostatistics, https://doi.org/10.1007/978-3-031-69622-0